INTERNATIONAL CENTRE FOR MECHANICAL SCIENCES

COURSES AND LECTURES - No. 295

CONTINUUM DAMAGE MECHANICS THEORY AND APPLICATIONS

EDITED BY

D. KRAJCINOVIC
UNIVERSITY OF CHICAGO

J. LEMAITRE
UNIVERSITE' DE PARIS VI

SPRINGER-VERLAG WIEN GMBH

Le spese di stampa di questo volume sono in parte coperte da contributi
del Consiglio Nazionale delle Ricerche.

This volume contains 114 illustrations.

Originally published by Springer Verlag Wien-New York in 1987

ISBN 978-3-211-82011-7 ISBN 978-3-7091-2806-0 (eBook)
DOI 10.1007/978-3-7091-2806-0

PREFACE

While no idea is entirely new, it is pretty agreed that the 1958 seminal study of the creep rupture by Lazar M. Kachanov marks the birthday of what we today know as continuum damage mechanics. In the following years the interest in this new branch of continuum mechanics grew beyond the fondest hopes as it became plainly obvious that damage mechanics represents a rational framework for the treatment of a wide range of brittle phenomena.

One of the early enthusiasts and contributors to the development of damage mechanics was also the late professor Anthony Sawczuk, former rector of CISM. It was indeed he who in 1983 broached the idea of organizing a course like this under the auspices of CISM. He planned to take an active participation in the course and only his untimely death prevented him from helping us in this venture. We missed his knowledge and his encouragement.

As organizers we attempted to present a balanced view of this rapidly developing field. We reached far and wide geographically and sought people with different opinions and interests. We were sorry that the limits on the number of lectures prevented us from inviting many other worthy researchers in this field.

In this brief contact with the CISM administration and an extremely knowledgeable and responsive group of bright young people who attended this course, we were once again assured that the late professor Sawczuk was correct in predicting that this course would have been a success. We only hope that the readers of this monograph will share our enthusiasm for this field and join the steadily growing group of contributors to the development of damage mechanics.

The two organizers gratefully acknowledge the financial contributions of CISM and city of Udine, Italy, to the students who attended the course. The financial assistance of the Office of Scientific Research of NATO facilitated our cooperation in organizing this course as well. The help of the administrative staff of CISM during the planning phase of the course and during the course itself was indispensable.

Dusan Krajcinovic

Jean Lemaitre

Contents

INTRODUCTION AND GENERAL OVERVIEW

Jan Hult
Chalmers University of Technology
Gothenburg, Sweden

Continuum damage mechanics (CDM) has evolved as a means to analyse the effect of material deterioration in solids under mechanical or thermal load. Whereas fracture mechanics deals with the influence of macroscopic cracks, CDM deals with the collective effect of distributed cracklike defects. The aim of CDM is to describe the influence of such material damage on stiffness and strength of loaded structures.

A characteristic feature of CDM is the introduction into the constitutive equations of one or more, scalar or tensorial, field quantities as measures of the degradation of the material.

This overview presents and discusses various approaches to this branch of continuum mechanics. Basic concepts are defined, and damage evolution as a time independent or time dependent process is discussed. Experimental identification of damage parameters is also considered.

1 INTRODUCTION

1.1 BACKGROUND AND AIM. KACHANOV CONCEPT

All real materials deform when loaded. The deformation may be elastic or inelastic. It may be time independent or time dependent. Occasionally rupture may occur, being either ductile or brittle.

The deformation characteristics of the material are determined by its composition and atomic structure and also by the temperature and rate of loading. Crystalline and amorphous materials have different properties as have single crystals versus polycrystalline aggregates. Deformation characteristics are different at high and low temperatures and at high and low loading rates. Complete understanding of the deformational behaviour of a particular material would require detailed knowledge of the atomic structure of that material in addition to vast computational facilities. The deformational properties are therefore described by constitutive equations, which are either derived from micromechanical or statistical considerations, or, more often, postulated to fit measurements test specimens.

In general constitutive equations relate to the material modelled as a continuum, i.e. a material without atomic structure. The deformation is described by a field variable, the strain. The distribution of internal forces in the material is described by another field variable, the stress. These concepts have been found extremely useful in analyzing the behaviour of load carrying structures in spite of the fact that they take no account of the discrete structure of real materials.

Under certain kinds of loading the material structure may begin to disintegrate. Small cracks may form between crystal grains or across them. Voids and other forms of small cavities may appear in highly stressed parts. Such deterioration weakens the material, lowers its load carrying capacity. By their very nature these defects are discrete entities. An accurate analysis of their influence would have to consider them as discrete disturbances of the material continuum, a prohibitive task.

In a pioneering paper Kachanov (1958) proposed to describe the collective effect of such deterioration by a field variable, the continuity. Hence an inherently discrete process was modelled by a continuous variable. What was lost in accuracy in modeling the deterioration was then gained in computational simplicity.

A similar idea lies behind the concepts of stress and a material continuum. Only when the discrete nature of the material was disregarded was it possible to develop what became the theory of elasticity. The success of that theory and its later development into general continuum mechanics warrant a further pursuit of Kachanov's basic concept.

The state of the material with regard to deterioration was characterized here by a dimensionless, scalar field variable ψ denoted continuity. To a completely defect free material was ascribed the condition $\psi = 1$, whereas $\psi = 0$ was defined to characterize a completely destroyed material with no remaining load carrying capacity. Kachanov also postulated a law according to which ψ changes with time in a material subject to stress at elevated temperature. The purpose was to describe the phenomenon of brittle creep rupture which occurs in metals subject to stress at elevated temperature during extended times. The details in Kachanov's analysis of brittle creep rupture will be presented in Section 2.1 below.

The continuity ψ may be said to quantify the absence of material deterioration. The complementary quantity $D \equiv 1 - \psi$ is therefore a measure of the state of deterioration or damage. For a completely undamaged material $D = 0$ whereas $D = 1$ corresponds to a state of complete loss of integrity of the material structure. The designation D for a field variable to describe the degree of material damage has lately come into more widespread use and will be used in the sequel. The designation ω is also commonly used in literature.

While Kachanov (1958) assumed ψ to be a scalar field variable later developments have led to the study of tensorial quantities to describe damage, cf. Section 2.3 below.

Even though the Kachanov model was entirely phenomenological, micromechanical studies of deformation in metals at elevated temperatures have lent support to this model, cf. Section 3.2 below. Such results have come to create increasing interest in damage analyses based on mechanics principles. The term <u>continuum damage mechanics</u> (CDM) has been coined for such analyses, cf. the review article by Krajcinovic (1984). The aim of CDM is to develop methods for the prediction of the load carrying capacity of structures subject to material damage evolution. It is a counterpart of <u>fracture mechanics</u> (FM), which deals with structures containing one or several cracks of finite size. In FM the cracks are usually assumed to be embedded in a non-deteriorating material. Studies have also been made where a crack of finite size is embedded in a material undergoing damage growth. The methods of FM and CDM may then be combined to predict the resulting decrease of load carrying capacity, cf. Section 6.1.

1.2 OTHER DAMAGE CONCEPTS

The term damage has also found use in other branches of applied mechanics, notably fatigue failure or creep rupture under varying loading histories.

The Palmgren (1924) damage concept originated in the interpretation of endurance tests with ball bearings. The endurances, i.e. number of revolutions to failure, N_i $(i = 1,2,3,\dots)$ at various constant load levels P_i had first been determined. Tests were then run at varying, but stepwise constant, load levels P_i, each being applied during n_i revolutions. On the basis of these tests Palmgren proposed the failure condition

$$\sum n_i/N_i = 1 \tag{1.1}$$

The ratio

$$n_i = n_i/N_i \tag{1.2}$$

has later come to be known as the <u>fatigue damage</u> produced by n_i load cycles at load level P_i. The failure condition (1.1) can then be stated as

$$\sum D_i = 1 \tag{1.3}$$

The Palmgren paper did not contain any quantitative comparison between actual tests and the proposed failure condition. This may explain why it attracted rather little attention. It was not until Miner (1945) restated the hypothesis that it came to command interest among designers. It has subsequently become known as the Palmgren-Miner law or the linear cumulative damage law of fatigue failure under varying load amplitudes.

A completely analogous definition of creep damage was proposed by Robinson (1952). With T_i denoting the creep rupture lifetime associated with a constant load P_i, the ratio

$$D_i = t_i/T_i \tag{1.4}$$

was taken to define the damage caused by exposure to the load P_i during the time t_i. The condition for creep rupture to occur under stepwise varying load P_i was again stated as (1.3).

From the definitions (1.2) and (1.4) follows that the failure condition (1.3) is exactly fulfilled in cases of constant load and that reasonable agreement is to be expected in cases of only slight load variations. Extensive fatigue tests and creep rupture tests with strongly varying loads have shown the linear damage rule (1.3) to be, at times, rather inaccurate. Particularly strong deviations from the linear hypothesis have been found in certain combined creep-fatigue tests. The deviation may be conservative or non-conservative depending on the load sequence. In spite of this the importance of the linear damage definitions (1.2) and (1.4) prevails, chiefly for their conceptual as well as mathematical simplicity. They are, however, both a posteriori definitions of damage, since they require N_i or T_i to be known. This implies that use of the failure condition (1.3) must be preceded by testing at all the constant load levels P_i $(i = 1,2,3,\ldots)$.

In contrast to this the CDM damage concept is related only to the current mechanical state of the material. Detailed knowledge about the conditions for deterioration of the material structure would make it possible to predict the load carrying capacity or lifetime without first performing extensive testing.

1.3 REFERENCES

L.M. Kachanov, On the Time to Failure under Creep Conditions - Izv. Akad. Nauk. SSR, Otd. Tekhn. N.8 (1958), 26-31.

D. Krajcinovic, Continuum Damage Mechanics - Appl. Mech. Rev. 37 (1984), 1-6.

M.A. Miner, Cumulative Damage in Fatigue - J. Appl. Mech. 12 (1945), A159-A164.

A. Palmgren, Die Lebensdauer von Kugellagern - Zeitschrift VDI 68 (1924), 339-341.

E.L. Robinson, Effect of Temperature Variation on the Long-Time Rupture Strength of Steels - Trans. ASME 74 (1952), 777-780.

2 MACROMECHANICAL MODELS

2.1 KACHANOV MODEL, UNIAXIAL TENSION

The Kachanov damage model was developed to describe brittle creep rupture under uniaxial tension. In a creep rupture test a tensile specimen is subjected to a constant load while kept at elevated temperature. The time to rupture is then recorded.

For metals the behaviour is different at high and low loads. At high load the specimen elongates and becomes thinner, which causes the stress to increase with time leading eventually to a ductile rupture. At low load the elongation may be negligibly small, so that the cross sectional size remains virtually constant in time. Eventually the material structure will deteriorate due to the formation of microcracks and small voids. They increase in size and finally link up to form macrocracks leading to brittle rupture. Kachanov (1958) proposed to model this process as follows.

With $0 \leq \psi \leq 1$ characterizing the state of deterioration ($\psi = 1$ corresponding to intact material and $\psi = 0$ to completely destroyed) the quantity σ/ψ may be interpreted as an 'effective' stress, if σ denotes the nominal stress. Kachanov postulated the following law for the rate of change of the continuity

$$d\psi/dt = -A(\sigma/\psi)^n \tag{2.1}$$

With σ constant in time, as in a brittle creep rupture test, eq. (2.1) may be integrated to give, considering the initial condition $\psi(0) = 1$

$$\psi(t) = [1-(n+1)A\sigma^n t]^{1/(n+1)} \tag{2.2}$$

The rupture condition $\psi(t_R) = 0$ then gives the rupture time

$$t_R = [(n+1)A\sigma^n]^{-1} \tag{2.3}$$

A plot of $\log t_R$ versus $\log \sigma$ gives a straight line of slope $-1/n$ as shown in Fig. 2.1.

The two constants, A and n, appearing in (2.1) may be determined by adapting the expression (2.3) to recorded brittle creep rupture data, which usually show a linear relationship between $\log \sigma$ and $\log t_R$ as in Fig. 2.1.

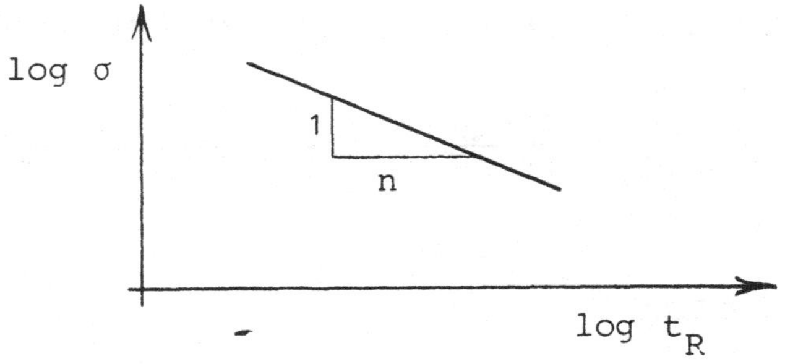

Fig. 2.1. Creep rupture life $t_R(\sigma)$ as derived from Kachanov's model.

2.2 EXTENSIONS OF KACHANOV MODEL, UNIAXIAL TENSION

The deterioration of the material structure caused by microcracks and voids implies an 'internal' decrease in load carrying area. The elongation of the specimen is accompanied by thinning, i.e. an 'external' decrease in load carrying area. Both these effects tend to accelerate the creep deformation, leading to tertiary creep. Rabotnov (1968) proposed to modify the creep law accordingly. With now common notation the governing equations for the creep rupture test may be derived as follows, cf. Fig. 2.2.

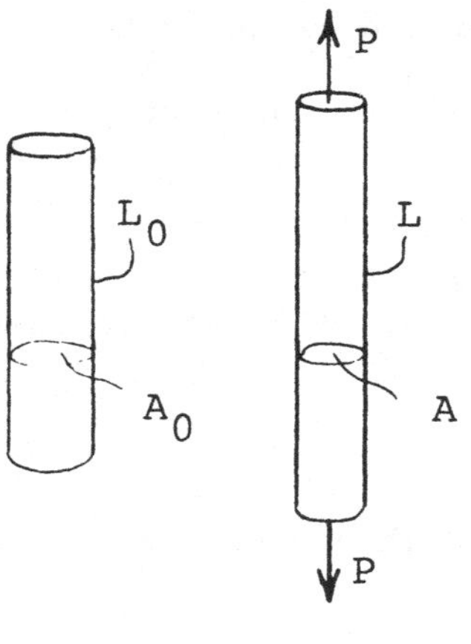

Fig. 2.2. Uniaxial tensile specimen.

Three stress quantities are first defined:

Nominal stress

$$\sigma_0 = P/A_0 \quad (2.4)$$

where A_0 denotes initial cross sectional area before the load is applied.

Real stress

$$\sigma = P/A \quad (2.5)$$

where A denotes cross sectional area as defined by exterior cross sectional size.

Effective stress

$$s = \sigma/(1-D) \quad (2.6)$$

where D is denoted damage. For a completely undamaged material D=0 i.e. $s = \sigma$. Increasing damage implies increasing effective stress. The upper limit D = 1 corresponds to infinite effective stress.

From (2.5) and (2.6) follows

$$s = P/[A(1-D)] \quad (2.7)$$

and hence the quantity

$$A_n \equiv A(1-D) \quad (2.8)$$

may be interpreted as a ficticious net load carrying area, decreasing from A_n=A in the undamaged state (D=0) to A_n=0 in the completely damaged state (D=1). It must be stressed here that A_n, as defined by (2.8) is not, in general, the average cross sectional area between the voids, cf. Section 3.1 below. Disregarding the elastic strain, the creep law is postulated as

$$d\varepsilon/dt = Bs^n \quad (2.9)$$

Here ε denotes the total strain. The damage law is postulated as

$$dD/dt = Cs^{\nu} \quad (2.10)$$

Note the similarity with the expression (2.1) except that A and n have been replaced by C and ν, whereas ψ corresponds to 1-D. Since creep deformation takes place under constant volume the effective stress may be written as follows

$$s = \sigma/(1-D) = [\sigma_0/(1-D)]A_0/A = [\sigma_0/(1-D)]L/L_0$$
$$= [\sigma_0/(1-D)]\ e^{\varepsilon} \tag{2.11}$$

where the logarithmic strain definition has been employed.

In the <u>absence of damage</u> ($D \equiv 0$) the effective stress is

$$s = \sigma_0\ e^{\varepsilon} \tag{2.12}$$

and hence, considering the creep law (2.9)

$$d\varepsilon/dt = B\sigma_0^n\ e^{n\varepsilon} \tag{2.13}$$

After integration, considering the initial condition

$$\varepsilon(0) = 0$$

follows

$$\varepsilon(t) = -(1/n)\ \ln(1 - nB\sigma_0^n\ t) \tag{2.14}$$

Hence the strain increases without limit at time

$$t_{RH} = (nB\sigma_0^n)^{-1} \tag{2.15}$$

This expression for the creep rupture lifetime in the absence of damage was first derived by Hoff (1953), as indicated here by the subscript H.

In the <u>absence of deformation</u> ($\varepsilon \equiv 0$) the effective stress is

$$s = \sigma_0/(1-D) \tag{2.16}$$

and hence, considering the damage law (2.10)

$$dD/dt = C\sigma_0(1-D)^{-\nu} \tag{2.17}$$

After integration, considering the initial condition

$$D(0) = 0 \tag{2.18}$$

follows

$$D(t) = 1 - [1-(\nu+1)C\sigma_0^{\nu}t]^{1/(\nu+1)} \tag{2.19}$$

The damage level $D = 1$ is reached at time

$$t_{RK} = [(\nu+1)C\sigma_0^{\nu}]^{-1} \tag{2.20}$$

which is the previously, (2.3), derived expression for the brittle creep rupture time according to Kachanov (1958), as indicated here by the subscript K.

The mathematical formalism of these creep rupture analyses may be made simpler by a slight redefinition of the effective stress. In analogy to the definition of a logarithmic strain increment

$$d\varepsilon = dL/L = -dA/A \tag{2.21}$$

a damage increment may be defined as

$$dD = -dA_n/A_n \tag{2.22}$$

where A_n is the ficticious area defined by

$$s = P/A_n \tag{2.23}$$

Hence the effective stress takes the form

$$s = \sigma\ e^{D} \tag{2.24}$$

which coincides with (2.6) for small damage. The expression (2.11) will then be replaced by

$$s = \sigma_0\ e^{\varepsilon}\ e^{D} = \sigma_0\ e^{\varepsilon+D} \tag{2.25}$$

From (2.9), (2.10) and (2.25) follows the governing equation for the effective stress

$$(1/s)(ds/dt) - Bs^{n} - Cs^{\nu} = (1/\sigma_0)(d\sigma_0/dt) \tag{2.26}$$

For any given load history $\sigma_0(t)$ this gives the resulting effective stress history $s(t)$.

With the load history shown in Fig. 2.3 follows for the phase 0-1:

$$(1/s)\ ds = (1/\sigma_0)\ d\sigma_0 \tag{2.27}$$

and hence

$$s_0 = \bar{\sigma}_0 \tag{2.28}$$

This reflects the fact that no creep strain or damage develops during this instantaneous load change. For the subsequent phase 1-2 then follows

$$(1/s)(ds/dt) - Bs^{n} - Cs^{\nu} = 0 \tag{2.29}$$

Considering the initial condition $s(0) = \bar{\sigma}_0$ this may be integrated to give

$$t = \int_{\bar{\sigma}_0}^{s} (Bx^{n+1} + Cx^{\nu+1})^{-1}\ dx \tag{2.30}$$

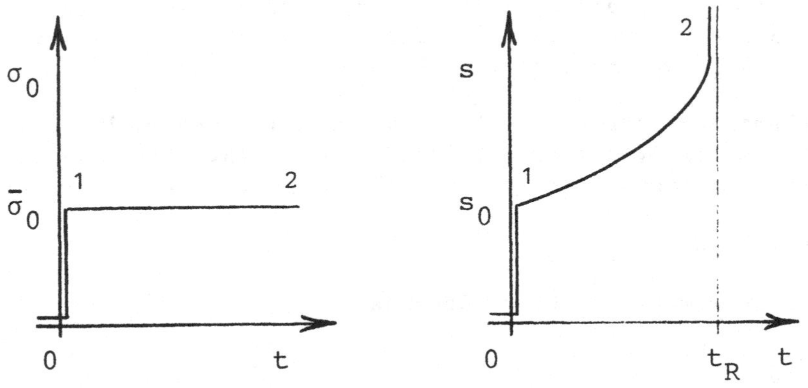

Fig. 2.3. Effective stress history s(t) resulting from a Heaviside type load history $\sigma_0(t)$.

The creep rupture condition $s \to \infty$ gives the creep rupture time

$$t_R = \int_{\bar{\sigma}_0}^{\infty} (Bx^{n+1} + Cx^{\nu+1})^{-1}\, dx \qquad (2.31)$$

With C = 0 (no damage) the result (2.15) is regained. With B = 0 (no creep deformation) there results

$$t_R = (\nu C\bar{\sigma}_0)^{-1} \qquad (2.32)$$

This deviates slightly from the previous result (2.20) due to the altered definition of damage employed here. With B > 0 and C > 0 numerical integration is required. The result is shown in Fig. 2.4.

Refinements of this analysis of creep rupture have been proposed by Broberg (1975) and Hult (1975) who considered the effect of instantaneous strain and damage increments due to instantaneous load increments. In the extended creep law

$$d\varepsilon/dt = (d/dt)G(s) + F(s) \qquad (2.33)$$

and damage evolution law

$$dD/dt = (d/dt)g(s) + f(s) \qquad (2.34)$$

the quantities G(s) and g(s) denote the strain and the damage caused by instantaneous application of the effective stress s. The quantities F(s) and f(s) correspond to the terms Bs^n and Cs^ν in (2.9) and (2.10). From (2.25), (2.33) and (2.34) follows the governing equation

$$(ds/dt)[(1/s)-G'(s)-g'(s)] - F(s) - f(s) = (1/\sigma_0)(d\sigma_0/dt) \qquad (2.35)$$

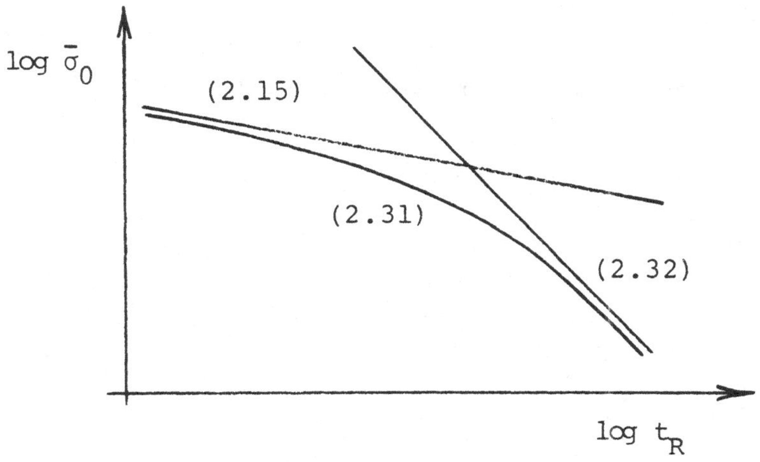

Fig. 2.4. Relation between load $\bar{\sigma}_0$ and creep rupture lifetime according to three models

With the loading history of Fig. 2.3 follows for the load application phase 0-1:

$$[1/s - G'(s) - g'(s)]ds = (1/\sigma_0)\, d\sigma_0 \tag{2.36}$$

Hence $ds/d\sigma_0 \rightarrow \infty$, i.e. instantaneous rupture will occur at the effective stress level s_R, where

$$1/s_R - G'(s_R) - g'(s_R) = 0 \tag{2.37}$$

The corresponding rupture load σ_{0R} is obtained from (2.25):

$$\sigma_{0R} = s_R\, e^{-G(s_R)-g(s_R)} \tag{2.38}$$

If the applied load $\bar{\sigma}_0$ is less than σ_{0R} the subsequent phase 1-2 yields

$$[1/s - G'(s) - g'(s)]\, ds/dt = F(s) + f(s) \tag{2.39}$$

and hence $ds/dt \rightarrow \infty$, i.e. creep rupture will occur at the same critical effective stress s_R as given by (2.37). The time t_R to creep rupture is obtained by integrating (2.39)

$$t_R = \int_{s_0}^{s_R} [1/s - G'(s) - g'(s)][F(s) + f(s)]^{-1} ds \tag{2.40}$$

Here the initial effective stress s_0 is obtained from (2.25)

$$\bar{\sigma}_0 = s_0\, e^{-G(s_0)-g(s_0)} \tag{2.41}$$

These results may be summarized in the diagram of Fig. 2.5

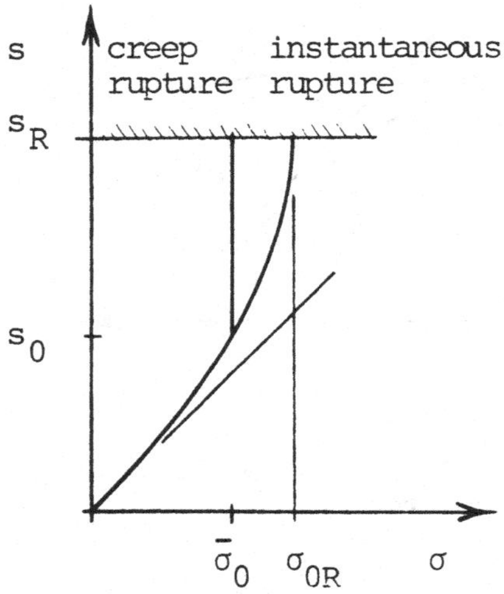

Fig. 2.5. Effective stress and nominal stress in creep rupture tests.

Instantaneous rupture and creep rupture are seen to be of a similar nature, both occurring at a specified critical value of the effective stress. Rupture in both cases is caused by loss of stability due to the simultaneous action of strain and damage.

The presented model, based on the extended creep and damage laws (2.33) and (2.34), covers the purely ductile and purely brittle, time independent or time dependent uniaxial rupture modes.

2.3 EXTENSION TO MULTIAXIAL STRESSES

Multiaxial stress fields dominate in engineering structures. The generalization of the Kachanov and Rabotnov concepts to such cases has therefore commanded wide interest. Two levels of generalization may be discerned for the creep and damage evolution laws depending upon whether damage is assumed to be a scalar or a tensorial quantity. The possibility of ascribing tensorial properties to the damage field variable was mentioned already by Kachanov (1958) but was then not pursued any further.

a) Scalar damage

With $s = \sigma/(1-D)$ the uniaxial creep and damage laws (2.9) and (2.10) take the forms

$$d\varepsilon/dt = B[\sigma^n/(1-D)^n] \tag{2.42}$$

$$dD/dt = C[\sigma^\nu/(1-D)^\nu] \tag{2.43}$$

In the absence of damage the multiaxial creep law is usually stated in the von Mises form. A corresponding creep law in presence of damage may be stated as

$$d\varepsilon_{ij}/dt = 3/2\ B[\sigma_e^{n-1} s_{ij}/(1-D)^n] \tag{2.44}$$

The damage law may be stated as

$$dD/dt = C[\sigma_r^\nu/(1-D)^\nu] \tag{2.45}$$

where σ_r is a 'reference stress', which equals σ in the uniaxial case.

If σ_r is kept constant the damage growth law (2.45) predicts a rupture time, cf. (2.20)

$$t_R = [(\nu+1)C\sigma_r^\nu]^{-1} \tag{2.46}$$

In a state of plane stress, with principal stresses σ_1 and σ_2, the reference stress may be stated as

$$\sigma_r = \varphi(\sigma_1, \sigma_2) \tag{2.47}$$

Hence the expression

$$t_R = [(\nu+1)C\,\varphi^\nu(\sigma_1, \sigma_2)]^{-1} \tag{2.48}$$

defines isochronous rupture curves in the σ_1-σ_2-plane, cf. Leckie and Hayhurst (1974).

Kachanov (1958) proposed the reference stress as

$$\sigma_r = \max(\sigma_1, \sigma_2) \tag{2.49}$$

corresponding to the straight lines in Fig. 2.6.

Hayhurst proposed to state the reference stress as

$$\sigma_r = \alpha\,\sigma_{max} + (1-\alpha)\,\sigma_e, \qquad 0 \leq \alpha \leq 1 \tag{2.50}$$

where σ_{max} is the maximum tensile stress and σ_e is the von Mises effective stress. The corresponding isochronous rupture curve is located between those shown in Fig. 2.6. The parameter α is determined by adaption to experimental data. Biaxial creep rupture tests have shown the following results:

Copper	250 °C	$\alpha \approx 1$
Aluminium	200 °C	$\alpha \approx 0$

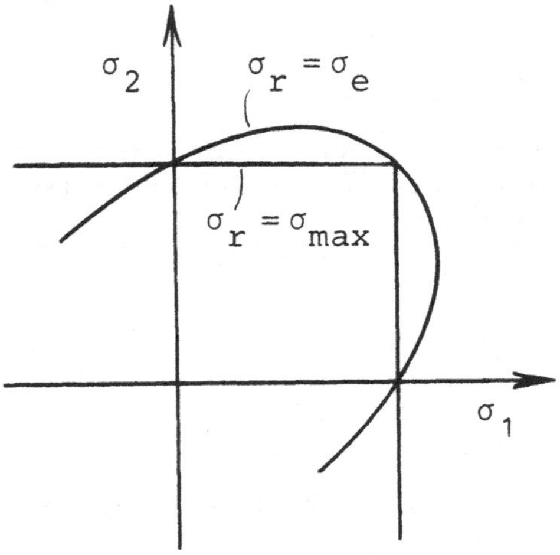

Fig. 2.6. Isochronous rupture curves.

b) Tensorial damage

A tensorial damage model requires knowledge about the geometrical shape and arrangement of the microcracks that cause the damage effect. Detailed treatment of this matter is given in the lectures by Krajcinovic and Murakami.

2.4 INTERNAL VARIABLE APPROACH

The creep and damage laws (2.9) and (2.10) with the effective stress s defined by (2.6) are particular forms of the general evolution laws

$$d\varepsilon/dt = f(\sigma,D) \tag{2.51}$$

$$dD/dt = g(\sigma,D) \tag{2.52}$$

where σ denotes the stress whereas D is an internal state variable yet to be specified. The validity of these expressions and the forms of the functions f and g may be found from experiments with stepwise constant stresses, cf. Leckie (1978).

First a series of creep tests is run with constant stresses σ_1, σ_2, ... and the corresponding strain histories are plotted (full lines in Fig. 2.7).

A new test is then run at stress σ_1 up to time t' (creep curve OP_1'). At time t' the stress is instantaneously raised to σ_2 resulting in the dashed creep curve a. If now the creep law (2.51) is valid the shape of the curve a depends only on the applied stress σ_2 and the magnitude of the internal variable D at time t', i.e. in

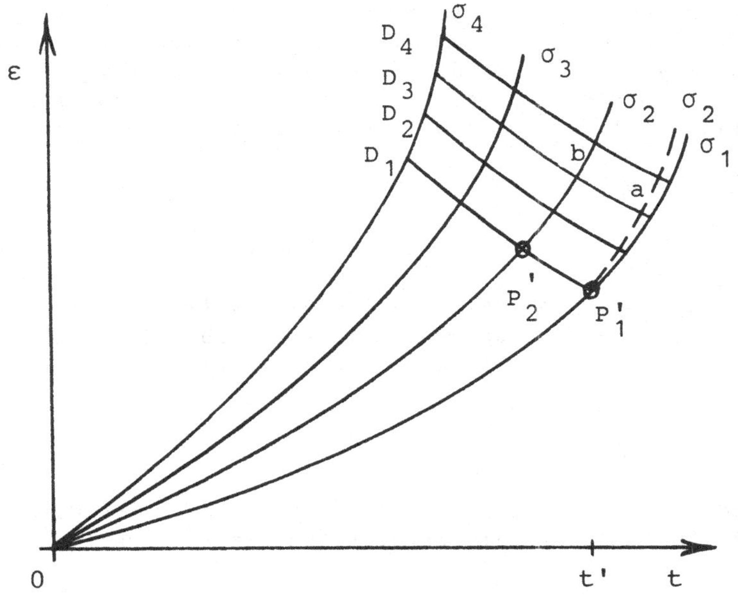

Fig. 2.7. Stepwise constant stress creep tests.

the point P_1'. A point P_2' may then be found on the $O\sigma_2$ curve such that the part b has the same shape as the curve a. This defines a point P_2' where D has the same value as in the point P_1'. In this way loci of constant D-values may be determined as indicated in Fig. 2.7.

If, in particular, the creep and damage laws have the forms

$$d\varepsilon/dt = B[\sigma/(1-D)]^n \tag{2.53}$$

$$dD/dt = C[\sigma/(1-D)]^\nu \tag{2.54}$$

then the D-loci have the forms

$$\varepsilon\ t^{(n-\nu)/\nu} = \text{const} \tag{2.55}$$

If the indicated matching of creep curves cannot be done with desired accuracy, the postulated expressions (2.51) and (2.52) are insufficient to describe the creep and damage histories. A more general set of governing equations would be

$$d\varepsilon/dt = f(\sigma,\ D_1,\ D_2) \tag{2.56}$$

$$dD_1/dt = g_1(\sigma,\ D_1,\ D_2) \tag{2.57}$$

$$dD_2/dt = g_2(\sigma,\ D_1,\ D_2) \tag{2.58}$$

where D_1 and D_2 are a priori undefined internal variables describing the state of the material. The forms of the functions f, g_1 and g_2 may be determined from experiments with step-wise constant stresses analogously to the way described above. Generalization to a larger number of internal variables

$$d\varepsilon/dt = f(\sigma, D_1, D_2, \ldots, D_N) \tag{2.59}$$

$$dD_i/dt = g_i(\sigma, D_1, D_2, \ldots, D_N), \quad (i = 1,2,\ldots,N) \tag{2.60}$$

suggests itself. With increasing N the difficulty in determining g_i (i = 1,2,...,N) soon becomes prohibitive. The single internal variable approach, (2.51) and (2.52), therefore dominates in literature.

2.5 THERMODYNAMICAL ASPECTS

Lemaitre and Chaboche (1985) have proposed to describe damage evolution in thermodynamical terms. A free energy function

$$\Psi = \Psi(\varepsilon, D, T) \tag{2.61}$$

is assumed to exist, with T denoting temperature, such that

$$\sigma = \rho(\partial\Psi/\partial\varepsilon) \tag{2.62}$$

where ρ is the mass density. A variable Y is then defined through the corresponding relation

$$Y = \rho(\partial\Psi/\partial D) \tag{2.63}$$

and the following relations is derived

$$-Y = (1/2)(\partial W_e/\partial D)_{\sigma,T} \tag{2.64}$$

where W_e is the elastic energy density associated with Ψ.

A dissipation potential

$$\varphi^* = \varphi^*(\sigma, Y, T) \tag{2.65}$$

may then be formed such that the damage and plastic strain growth rates may be expressed as

$$dD/dt = -(\partial\varphi^*/\partial Y)_{\sigma,T} \tag{2.66}$$

$$d\varepsilon^p/dt = (\partial\varphi^*/\partial\sigma)_{D,T} \tag{2.67}$$

These relations, which are here stated for a uniaxial stress case, may be directly rewritten in tensorial form for cases of multiaxial stresses. Further development of these concepts is given in the lectures by Lemaitre.

2.6 REFERENCES

H. Broberg, Creep Damage and Rupture - Diss. Chalmers University of Technology, Gothenburg 1975.

N.J. Hoff, Necking and Rupture of Rods Subjected to Constant Tensile Loads - J. Appl. Mech. 20 (1953), 105-108.

J. Hult, Damage-Induced Tensile Instability - Trans. 3rd SMiRT, London 1975, paper L 4/8.

L.M. Kachanov, On the Time to Failure under Creep Conditions - Izv. Akad. Nauk. SSR, Otd. Tekhn. N.8 (1958), 26-31.

F.A. Leckie and D.R. Hayhurst, Creep Rupture of Structures - Proc. Roy. Soc. Lond. A. 340 (1974), 323-347.

F.A. Leckie, The constitutive equations of continuum creep damage mechanics - Phil. Trans. Roy. Soc. Lond. A. 288 (1978), 27-47.

J. Lemaitre and J.-L. Chaboche, Mecanique des Materiaux Solides - Dunod, Paris 1985.

Yu.N. Rabotnov, Creep Rupture - Proc. XII Int. Congr. Appl. Mech. (Stanford 1968). Springer, Berlin 1969.

3 MICROMECHANICAL MODELS

3.1 DUCTILE DEFORMATION OF POROUS MEDIA

Microscopic examination shows that extended creep deformation in polycrystalline metals is accompanied by the formation of pores in the metal structure. The pores may grow in size by material diffusion, which will be considered in Section 3.2, or by ductile creep deformation of the surrounding matrix, which will be considered here.

Continuum mechanics analyses of porous media have been made by Budiansky, Hutchinson and Slutsky (1982), Duva and Hutchinson (1984), Jansson and Stigh (1984, 1985) and others. A central problem in CDM is to calculate the resulting creep rate in a power law material containing a dilute dispersion of pores of given shape, orientation and size distribution. Results are now available for the case of spheroidal pores, all with the same shape and orientation but of varying size, embedded in a uniaxial tensile stress field parallel to one of the rotational symmetry axes, cf. Fig. 3.1.

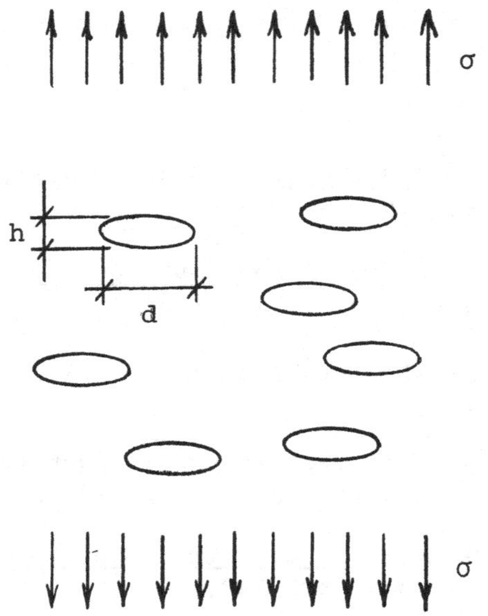

Fig. 3.1. Material with pores.

Analogous results are available for the two-dimensional case of cylindrical pores perpendicular to the loading direction, cf. Jansson and Stigh (1984).

The creep law

$$d\varepsilon/dt = B\sigma^n \tag{3.1}$$

in its multiaxial form is assumed to hold for the matrix material. Writing the creep law for the porous material in the form

$$d\varepsilon/dt = B[\sigma/(1-D)]^n \tag{3.2}$$

the analysis leads to the following results with h and d defined by Fig. 3.1 and N denoting the number of pores per unit volume.

a) <u>Penny shaped cracks perpendicular to load (h=0)</u>

$$D = [(1+1/n)/2\sqrt{1+3/n}]\, Nd^3 \approx Nd^3/2 \tag{3.3}$$

b) <u>Spherical pores (h=d)</u>

$$D \approx (0.86 + 0.14/n)\, Nd^3 \approx Nd^3 \tag{3.4}$$

c) <u>Strongly prolate pores parallel to load</u>

$$D = (\pi/6)(h/d)\, Nd^3 \tag{3.5}$$

These results may be summarized in the diagram of Fig. 3.2

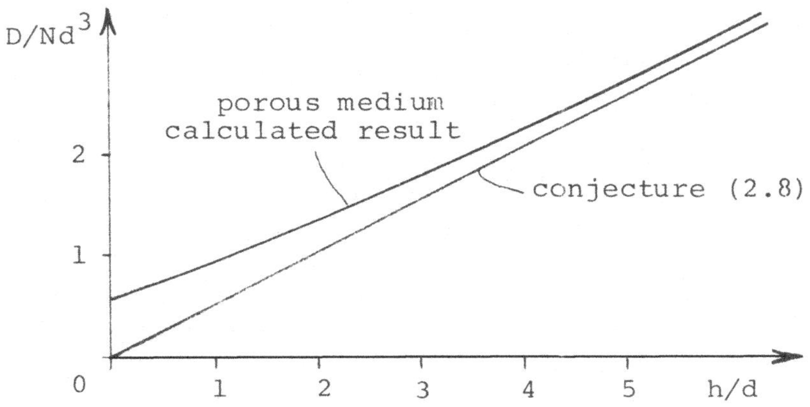

Fig. 3.2. D/Nd^3 versus aspect ratio h/d.

The average relative pore area in case c) is

$$A_p/A = Nh\, \pi d^2/6 \equiv D \tag{3.6}$$

and hence

$$A_n/A = (A_p-A)/A = 1-D \tag{3.7}$$

i.e. the conjecture (2.8) is fulfilled in this case. For other h/d ratios the damage D is greater than obtained from (2.8) with A_n denoting the average relative pore area.

Extensions to cases of interacting pores were developed by Stigh (1985). Based on these results Jansson and Stigh (1985) studied the growth of damage D with time. The pore shape was assumed to be spherical and new pores were assumed to nucleate at a rate proportional to the strain rate. The creep law was taken in the form

$$d\varepsilon/dt = (d/dt)Ks^m + Bs^n \tag{3.8}$$

where the central term corresponds to time independent plastic strain, cf. (2.9). A damage growth law was then derived, assuming the presence of intitial damage D_0 before load application. Loss of equilibrium as defined by $dD/dt \to \infty$ was found to also imply $d\varepsilon/dt \to \infty$ and to define an instability surface in σ-ε-D-space. The pore nucleation rate was found to have a decisive influence on the rupture condition, whereas the initial damage did not. A typical plot of nominal stress σ versus damage D is shown in Fig. 3.3.

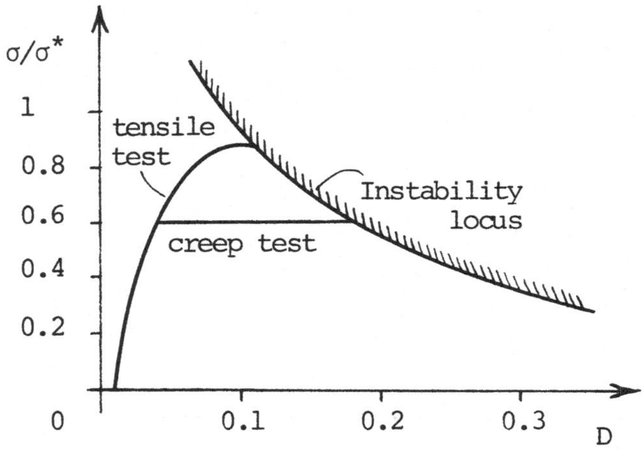

Fig. 3.3. Nominal stress σ and damage D in tensile test and creep test. Hardening exponent m = 2. Initial damage $D_0 = 0.01$.

In the diagram $\sigma^* = (emK)^{-1}$ denotes the critical nominal stress for a non-creeping and non-damaging material. As is to be expected rupture will always occur at a nominal stress less than σ^*. The analysis is limited to cases of relatively large stress where instability is of a ductile nature.

3.2 DAMAGE GROWTH DUE TO MATERIAL DIFFUSION

Detailed analyses of pore growth due to various diffusion mechanisms have been made by Ashby, Cocks, Raj, and others, cf. Cocks and Ashby (1982). Diffusional growth is dominating over ductile creep deformation (considered above) at low applied stresses. Two diffusion mechanisms are given particular attention:

a) Boundary diffusion, where matter diffuses out of the growing pores onto adjacent grain boundaries. The (spherical) pores remain essentially spherical.

b) Surface diffusion, where matter diffuses out of the equatorial region of the pores causing them to become flatter and increasingly crack-like.

In both cases growth equations are derived for the damage, defined here in terms of the relative area ratio of the pores. The resulting expressions are similar in structure to the Kachanov damage growth law (2.1) or (2.10). Ductile pore growth is also analyzed, and the relative importance of the different mechanisms is shown in maps such as the one in Fig. 3.4.

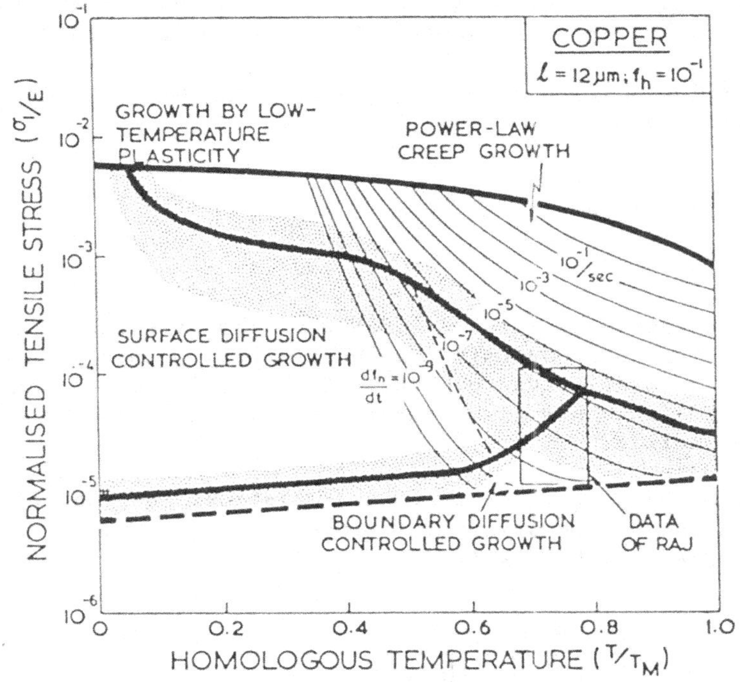

Fig. 3.4. Void growth map for copper, from Cocks and Ashby (1982).

Analyses such as this have served to bridge the gap between phenomenological damage models and theories derived from basic solid state physics.

3.3 REFERENCES

B. Budiansky, J.W. Hutchinson and S. Slutsky, Void Growth and Collapse in Viscous Solids - In 'Mechanics of Solids, the Rodney Hill 60th Anniversary Volume', ed. by H.G. Hopkins and M.J. Sewell. Pergamon Press, Oxford 1982.

A.C.F. Cocks and M.F. Ashby, On Creep Fracture by Void Growth - Progress in Material Science 27 (1982), 189-244.

J.M. Duva and J.W. Hutchinson, Constitutive Potentials for Dilutely Voided Nonlinear Materials - Mechanics of Materials 3 (1984), 41-54.

S. Jansson and U. Stigh, Influence of Cavity Shape on Damage Parameter - J. Appl. Mech. 52 (1985), 609-614.

S. Jansson and U. Stigh, Instability of Dilutely Voided Material in Uniaxial Tension - Proc. IUTAM Symposium on Damage and Fatigue (Haifa 1985). Engineering Fracture Mech., to be published.

U. Stigh, Effects of Interacting Cavities on Damage Parameter - To be published 1986.

4 FIBRE BUNDLE MODEL

4.1 FIBRE CONTINUUM

As already observed by Weibull (1939 a,b) the strength of a solid depends more on local defects than does a bulk property such as stiffness. The statistical analysis of the influence of randomly distributed defects has given important information about the strength of solids.

A standard mechanical model used in statistical analyses of the effect of nonhomogeneous material properties is a set of parallel tensile fibres, cf. Fig. 4.1.

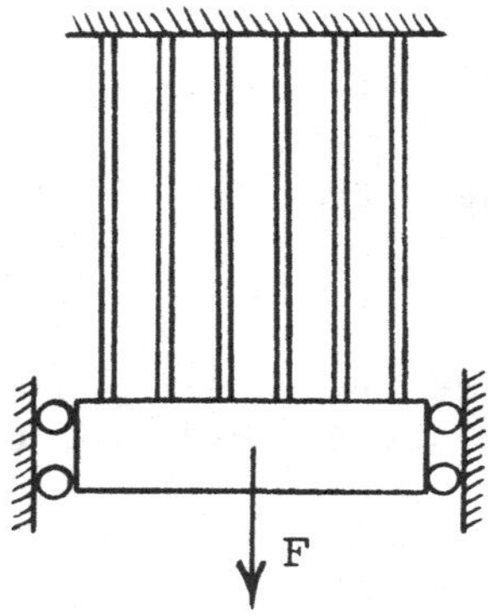

Fig. 4.1. Fibre bundle under tension.

There is no interaction between the fibres, but they are all given the same elongation. With certain stiffness and strength properties ascribed to individual fibres the resulting stiffness and strength of the entire fibre bundle may be determined. Rupture of a certain fibre, which may or may not cause rupture of the entire bundle, corresponds to local damage in form of a microcrack in a continuum.

The analysis of the mechanics of a fibre bundle is simplified by considering a fibre continuum instead of a finite number of fibres, cf. Hult and Travnicek (1983). In the absence of interaction between the fibres they may be rearranged sideways in an arbitrary manner without changing the properties of the bundle. For the present analysis a thin fibre 'sheet' will be analysed where an arbitrary fibre may be identified by its location $0 \leq x \leq 1$, cf. Fig. 4.2.

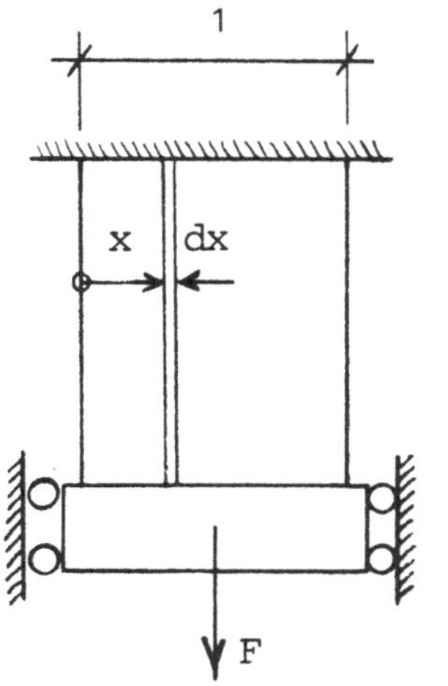

Fig. 4.2. Fibre continuum under tension.

Two kinds of events, which may occur in this system due to a tensile load, will be studied: static rupture and creep rupture.

a) Static rupture

The fibres are assumed to be linearly elastic with Young's modulus $E(x)$ and ultimate strength $\sigma_m(x)$. They are arranged in such a way that $E(x)$ is a monotonically increasing function, i.e. $dE(x)/dx \geq 0$, $0 \leq x \leq 1$. A correlation between $E(x)$ and $\sigma_m(x)$ is assumed to exist, such that $\sigma_m(x)$ is then also a monotonically changing function, which may increasing (positive correlation) or decreasing (negative correlation). To simplify the mathematics of the analysis both $E(x)$ and σ_m are assumed to be linear functions

$$E(x) = \bar{E}[1 + \mu(2x-1)] \tag{4.1}$$

$$\sigma_m(x) = \bar{\sigma}_m[1 + \nu(2x-1)] \tag{4.2}$$

where a bar marks the mean value in $0 \leq x \leq 1$. No restriction is implied in assuming $0 < \mu < 1$, whereas $-1 < \nu < 1$.

All fibres remain intact as long as $\varepsilon E(x) < \sigma_m(x)$ for all x. If $0 < \nu < 1$ and $\mu < \nu$, then a first fibre rupture will occur at $x = 0$. The corresponding load will be denoted F_0. Upon further load increase consecutive fibres will rupture, and a rupture front $x=c$ will move in the positive x-direction. If $0 < \nu < 1$ and $\nu < \mu$, then rupture will begin at $x = 1$ and the rupture front will proceed in the negative x-direction.

The location c of the rupture front is obtained from the condition

$$\varepsilon E(c) = \sigma_m(c) \tag{4.3}$$

and the corresponding load is given by

$$F(c) = \int_c^1 \varepsilon E(x) A dx \tag{4.4}$$

provided $0 < \nu < 1$, $\mu < \nu$. Here A denotes the cross sectional area of the fibre bundle.

With the linear expressions (4.1), (4.2) there results

$$F(c) = A\bar{\sigma}_m[(1-c+\mu c-\mu c^2)(1-\nu+2\nu c)]/(1-\mu+2\mu c) \tag{4.5}$$

The shape of the F(c)-curve depends on the relative magnitudes of μ and ν as shown in Fig. 4.3.

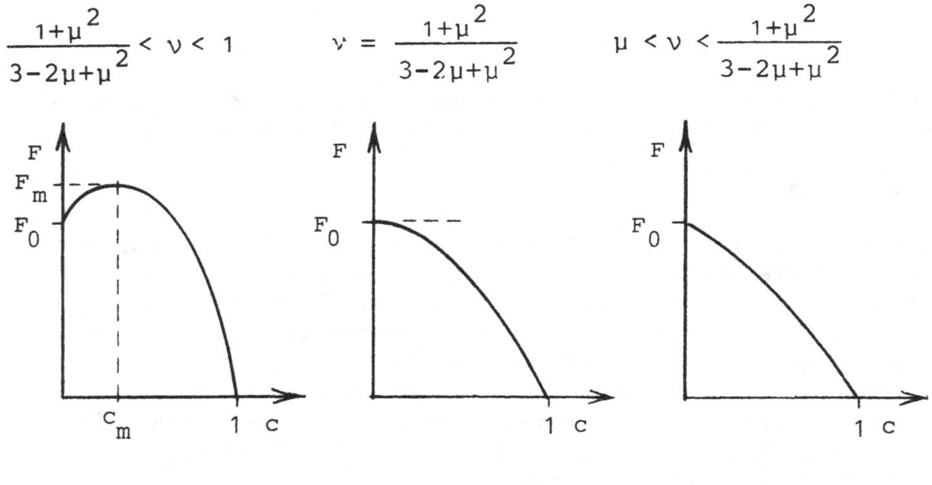

Fig. 4.3. Motion of rupture front c with load F.

In the interval

$$(1+\mu^2)/(3-2\mu+\mu^2) < \nu < 1 \tag{4.6}$$

gradual rupture sets in at $F=F_0$ and continues under increasing load until $\partial F/\partial c = 0$, which occurs at $c = c_m$ being the smallest root of

$$8\mu^2\nu c^3 + (2\mu^2+10\mu\nu-12\mu^2\nu)c^2$$
$$+ (2\mu+4\nu-2\mu^2-10\mu\nu+6\mu^2\nu)c + 1-3\nu+\mu^2+2\mu\nu-\mu^2\nu = 0 \qquad (4.7)$$

The corresponding rupture load is $F_m = F(c_m)$. In the interval

$$\mu < \nu < (1+\mu^2)/(3-2\mu+\mu^2) \qquad (4.8)$$

instantaneous rupture occurs at the load level

$$F_0 = A\bar{\sigma}_m \, [(1-\nu)/(1-\mu)] \qquad (4.9)$$

These results are summarized in Fig. 4.4.

The figures at the various curves show the magnitude of the quantity $F_m/A\bar{\sigma}_m$.

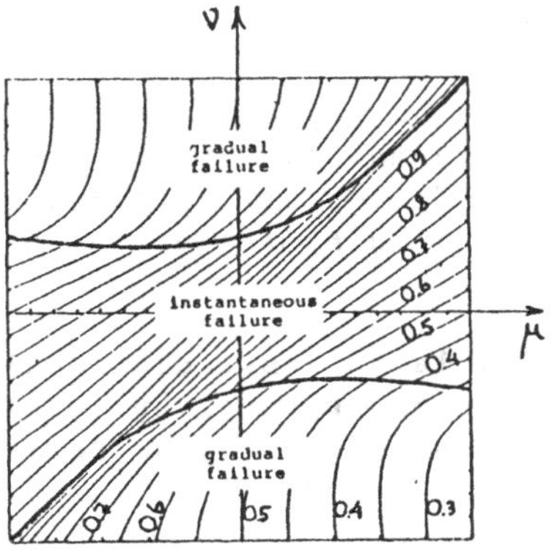

Fig. 4.4. Dependence of failure load on variation μ in Young's modulus and ν in fibre strength.

The quantity 1-c here represents the ratio between remaining fibre area and total bundle area. With the notation in (2.8) then

$$c \equiv D \qquad (4.10)$$

i.e. c is a direct measure of the damage in the fibre bundle. This interpretation of c supports the fibre bundle as a lucid model for analyzing progressive development of damage.

b) Creep rupture

The fibres are assumed to be linearly viscous with the creep law

$$d\varepsilon(t)/dt = \sigma(x,t)/M(x) \tag{4.11}$$

where the creep modulus varies linearly across the fibre bundle, cf. (4.1)

$$M(x) = \bar{M}[1 + \mu(2x-1)] \tag{4.12}$$

The ultimate strenth $\sigma_m(x)$ is again assumed to be given by (4.2). Results obtained by Chrzanowski and Hult (1986) show that the first failure will occur at time

$$t_0 = (A\bar{M}/F)[1 - (F/A\bar{\sigma}_m)\{(1-\mu)/(1-\nu)\}] \tag{4.13}$$

if the load F is applied instantaneously at time t = 0. Hence the first fibre failure will occur immediately upon load application if

$$F \geq A\bar{\sigma}_m \; [(1-\nu)/(1-\mu)] \tag{4.14}$$

Otherwise there will be a certain 'incubation' time $t_0 > 0$ before the first fibre ruptures.

The subsequent failure history may be <u>instantaneous</u> (all remaining fibres rupturing at time t_0) or gradual (a rupture front $x = c$ moving across the bundle until $dc/dt \to \infty$ at c^* implying final rupture). Hence four different failure histories may be identified as shown by Fig. 4.5.

The failure history denoted DI (delayed, instantaneous) is severe since no warning is given of imminent rupture. Up to time t_0'' no fibre failure has occurred, i.e. no damage may be detected prior to rupture.

The parameters μ and ν in (4.12) and (4.2) respectively determine the type of failure history to occur as shown in Fig. 4.6.

The condition for gradual (immediate or delayed) failure to occur is, cf. (4.6)

$$\nu > (1+\mu^2)/(3-2\mu+\mu^2) \tag{4.15}$$

For $\mu=0$ (uniform creep modulus) final rupture occurs at the failure front position

$$c^* = (3\nu-1)/(4\nu) \tag{4.16}$$

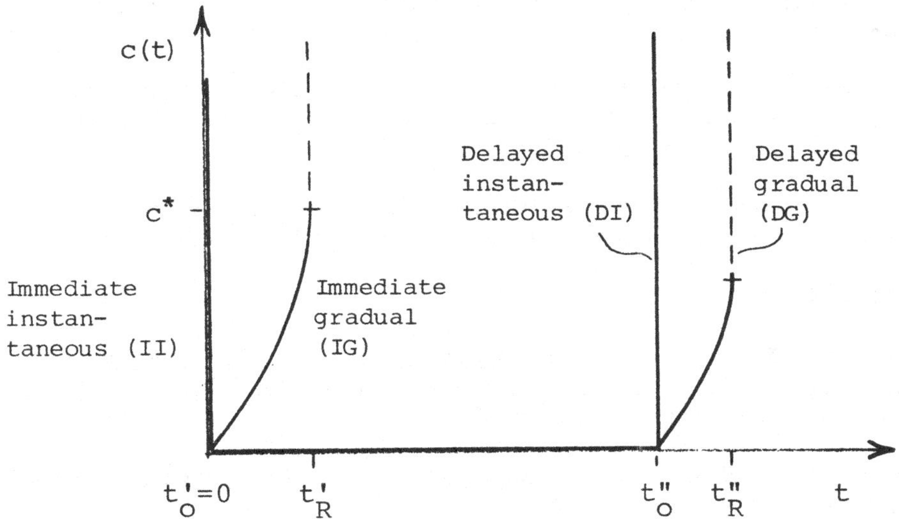

Fig. 4.5. Development of rupture front (damage) in linearly viscous fibre bundle.

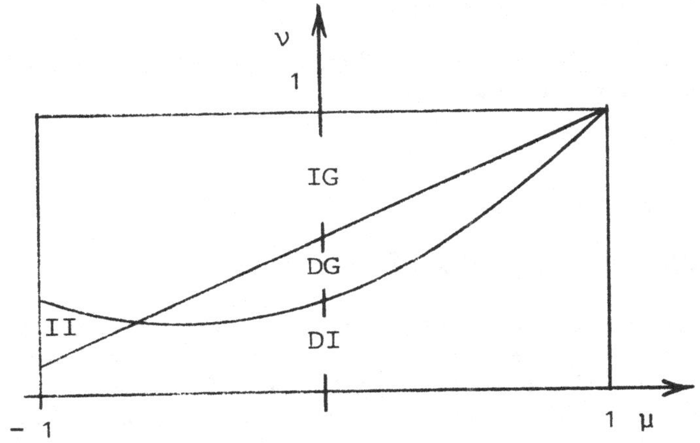

Fig. 4.6. Instantaneous and gradual failure regions.

In the original Kachanov model rupture was assumed to occur at 100 % damage, i.e. D=1, or ψ =0. This appeared to be in contrast to the observed microstructure of ruptured specimens, where damage was of more limited magnitude. As shown here the fibre model indicates that rupture will always occur, due to loss of equilibrium, at some damage level less than 100 %. It, however, also indicates that the rupture load or rupture time is only very slightly overestimated if rupture is assumed to occur at 100 % damage and not less as required from equilibrium conditions as well as from experimental observation.

The largest possible c* is 1/2, which corresponds to $\nu = 1$. For all other cases final rupture will occur att a c* less than 1/2, i.e. at less than 50 % damage.

A related analysis of progressive failure in twodimensional heterogeneous structures was discussed by Burt and Dougill (1977).

4.2 REFERENCES

N.J. Burt and J.W. Dougill, Progressive Failure in a Model Heterogeneous Medium - J. Engineering Mechanics Division, Proc. ASCE, Vol. 103, No. EM3, June 1977, 365-376

M. Chrzanowski and J. Hult, Creep rupture of fibre bundles - To be published 1986.

J. Hult and L. Travnicek, Carrying capacity of fibre bundles with varying strength and stiffness - J. de Mecanique Theorique et Appliquee 2:4 (1983), 643-657.

W. Weibull, A statistical theory of the strength of materials - IVA Proceedings Nr 151, Stockholm 1939.

W. Weibull, The Phenomenon of Rupture in Solids - IVA Proceedings Nr 153, Stockholm 1939.

5 EXPERIMENTAL EVIDENCE

5.1 IDENTIFICATION OF DAMAGE PARAMETERS

Three basic stress concepts were introduced in Section 2.2 viz. nominal stress σ_0, real stress σ and effective stress s. Whereas σ_0 and σ are related to one another through the strain, σ and s are related to one another through the damage as follows.

If a constitutive relation in the fully undamaged state has the form

$$\varepsilon = F(\sigma) \tag{5.1}$$

then s is defined such that the same constitutive relation in the damaged state has the same form expressed in s

$$\varepsilon = F(s) \tag{5.2}$$

An example of this was given in (2.9).

With damage D defined through (2.6), i.e. $s = \sigma/(1-D)$, the constitutive relation (5.2) takes the form

$$\varepsilon = F[\sigma/(1-D)] \tag{5.3}$$

and hence D may be determined by measuring ε vs. σ in a damaged specimen.

Fig. 5.1 indicates a tensile test where Young's modulus is recorded at various stages.

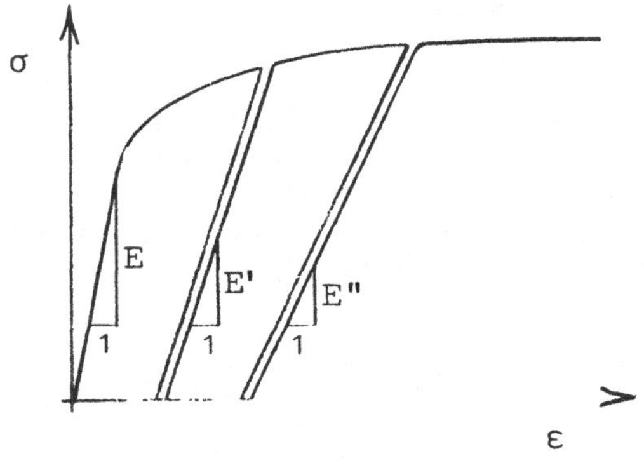

Fig. 5.1. Tensile test with repeated loading and unloading.

From the relation

$$\varepsilon = \sigma/[(1-D)E] \tag{5.4}$$

then follows

$$D' = 1-E'/E$$

$$D'' = 1-E''/E \tag{5.5}$$

- - - - -

- - - - -

and hence the damage D may be related to the strain at the various stages, cf. Fig. 5.2.

Results of such tests have been reported by Lemaitre and Chaboche (1985) and others, cf. lectures by Lemaitre.

Damage evaluation in other kinds of loading may be performed in a similar way.

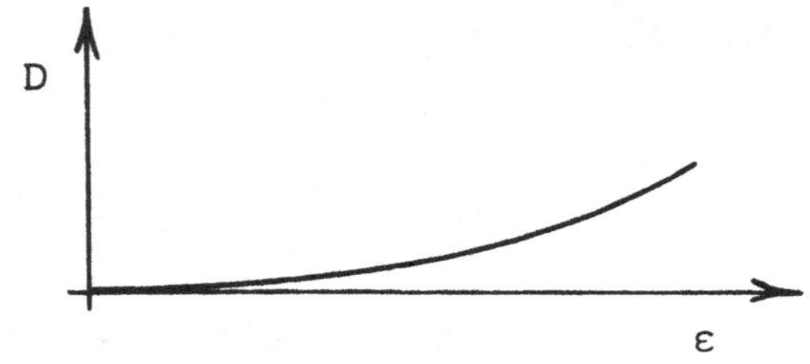

Fig. 5.2. Increase of damage with strain.

The behaviour shown in the diagram of Fig. 5.1 is indicative of simultaneous presence of plastic deformation and damage. Plastic deformation alone (i.e. no damage) would result in a diagram as shown in Fig. 5.3a, whereas damage alone (i.e. no plastic deformation) would result in a diagram as shown in Fig. 5.3b.

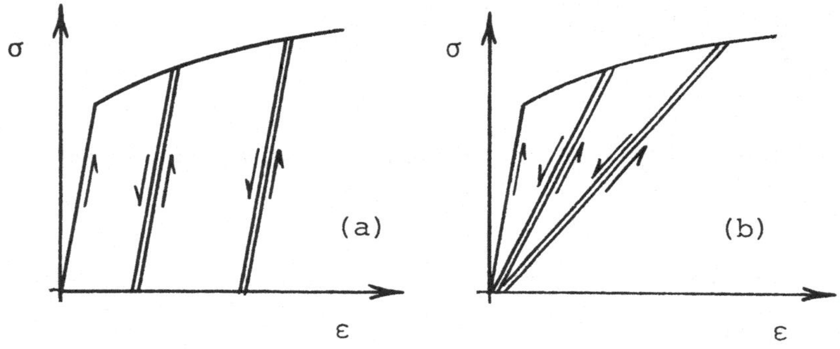

Fig. 5.3. (a) Plastic deformation, no damage
(b) Damage, no plastic deformation

5.2 NON DESTRUCTIVE DAMAGE EVALUATION

A problem of major importance in high temperature power technology is to record the development of damage in severly loaded components during service. Such recording would make it possible to retain the components in service until replacement is indicated by the damage level approaching a critical value.

Sonic inspection methods have a long tradition in engineering practice. The presence of cracks influences resonance frequencies and damping quite strongly and the human ear may often be used to detect such major cracks.

Micro-cracks and pores which constitute material damage may not be detected or measured by audible sonic methods. Ultrasonic techniques may, however, be used in various ways with varying accuracy:

a) Indirect measurement Young's modulus

Young's modulus may be calculated from measured values of longitudinal (v_L) and transversal (v_T) ultrasonic wave speeds according to the expression

$$E' = \rho v_T^2(3v_L^2-4v_T^2)/(v_L^2-v_T^2) \qquad (5.6)$$

and the corresponding damage is then found as

$$D = 1 - E'/E \qquad (5.7)$$

where E is Young's modulus as determined from (5.6) before damage developed, cf. Lemaitre and Chaboche (1985).

b) Recording of damping

Severe micro-cracking will lead to increasing damping of ultrasonic waves. Hence damage can be observed by transmitting ultrasonic waves through components and making comparisons of transmitted amplitudes at various times, cf. Stigh (1985).

5.3 REFERENCES

J. Lemaitre and J.L. Chaboche, Mecanique des Materiaux Solides - Dunod, Paris 1985.

U. Stigh, Damage, Strength and Ultrasonic Velocity, Analysis and Experiments - To be published 1986.

6 DAMAGE MECHANICS AND FRACTURE MECHANICS

6.1 INTERACTION OF MICRO- AND MACROCRACKS

Fracture mechanics deals with the load carrying capacity of structures containing major cracks. The cracks are embedded in a material which is assumed to be a defect free continuum.

Damage mechanics deals with the load carrying capacity of structures without major cracks but where the material itself is damaged due to the presence of microscopic defects such as microcracks or pores.

In many real situations both major cracks and microscopic damage may be present simultaneously. The resulting carrying capacity will then be lower than in cases with only major cracks or microcracks present. To demonstrate this the following related problem may be studied, cf. Jansson and Hult (1977). A tensile bar is considered, which is linearly elastic until it fractures at stress σ_F, cf. Fig. 6.1.

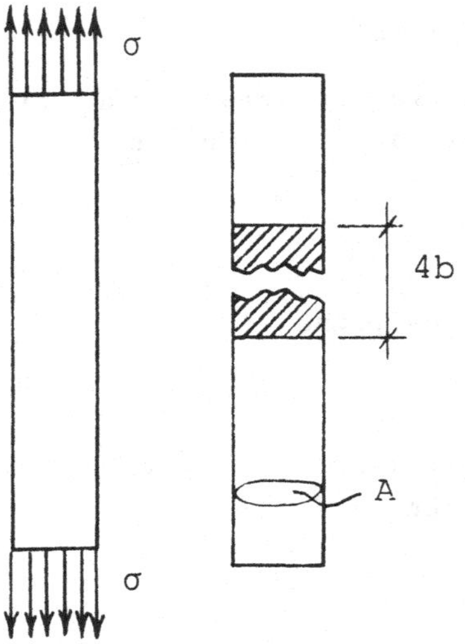

Fig. 6.1. Tensile bar.

The elastic strain energy density just prior to fracture then equals

$$\bar{U} = \sigma_F^2/2E \tag{6.1}$$

In fracturing, the strain energy stored in the immediate vicinity of the two fracture surfaces will be available to supply the work needed to create these surfaces. Since the range of interatomic forces is of the order of the lattice spacing b, the depth of the regions supplying this energy will be but a few times b, say 2b on each side of the separation. The remaining elastic energy, stored outside these regions, will be converted into heat. Hence the strain energy available for the fracturing is

$$U = \bar{U}\ 4bA = \sigma_F^2\ 2bA/E \tag{6.2}$$

where A denotes the cross sectional area of the bar.

The work needed to create the two new material surfaces equals

$$W = \gamma\ 2A \tag{6.3}$$

where γ is the surface energy density or surface tension of the material. From the energy balance $U = W$ then follows the fracture stress of the defect free material

$$\sigma_F = \sqrt{\gamma E/b} \tag{6.4}$$

Then the same problem is studied after the material has suffered damage of magnitude D, which is assumed to increase linearly with strain under load increase but to remain constant under load decrease.

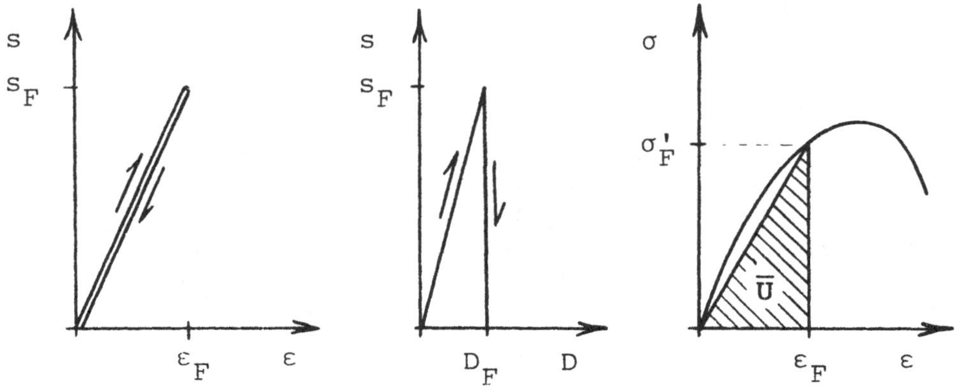

Fig. 6.2. Stress-strain-damage relations for tensile bar.

With the notation of Fig. 6.2 then follows for the unloading stage after fracture

$$\Delta s = \Delta\sigma/(1-D_F), \qquad \Delta\varepsilon = \Delta s/E, \qquad \Delta D = 0 \tag{6.5}$$

from which follows the linear relation

$$\varepsilon = \sigma/(1-D_F)E \tag{6.6}$$

The corresponding strain energy available to create fracture is

$$U = (1/2)\ \sigma_F'\varepsilon_F\ 4bA = (\sigma_F')^2\ 2bA/(1-D_F)E \tag{6.7}$$

With W again given by (6.3) the energy balance U=W yields the fracture stress of the damaged bar

$$\sigma_F' = \sqrt{\gamma E(1-D_F)/b} \tag{6.8}$$

Referring to (6.4) this may be stated as

$$\sigma_F' = \sigma_F \cdot \sqrt{1-D_F} \tag{6.9}$$

and hence the load carrying capacity is lowered by the factor $\sqrt{1-D_F}$ due to the occurrence of damage. The magnitude of D_F

depends on the damage growth law of the material and will not be considered here.

If the bar had contained a major central crack of length 2a the fracture stress σ_F would have been lowered to

$$\sigma_F'' \approx \sqrt{\gamma E/a} \tag{6.10}$$

and it may be conjectured that the presence of damage would have lowered it even further, down to

$$\sigma_F''' \approx \sigma_F''\sqrt{1-D_F} \tag{6.11}$$

An accurate analysis of this problem remains to be performed. The analogous problem of a Dugdale crack embedded in a damaging material has been studied by Janson and Hult (1977) and by Janson (1977, 1978a, 1978b, 1979). These results show the expected effect of the simultaneous presence of a major crack and microscopic damage.

6.2 DAMAGE AND CREEP CRACK GROWTH

The development of ever more creep resistant high temperature alloys has made ductile creep deformation less important in limiting the life of structural elements. Increasing attention is therefore being focussed on creep rupture. A central problem is to find the conditions for creep crack growth. A quantitative theory for the growth of cracks under creep conditions would make it possible to predict the failure time of an element in which a crack has developed and then been found e.g. by ultrasonic echo-pulse testing. Much attention has therefore been given in recent years to the study of creep crack growth. The damage concept is a central ingredient in many such studies, cf. Hayhurst et al (1984 a,b), Smith and Webster (1985), Jansson (1985) and others.

Numerical finite element analyses by Hayhurst et al (1984 a,b) show a strong concentration of damage ahead of sharp cracks loaded in the opening mode (mode I). These results are confirmed by results of microscopic examination. A basic mechanism for creep crack growth is as follows. Voids which nucleate in the plane ahead of the crack grow in size due to diffusion (cf. Section 3.2 above). Eventually the void closest to the crack tip links up with this, which implies that the crack grows, cf. Fig. 6.3.

Models for cavity nucleation and growth can therefore be used to derive expressions for creep crack growth, cf. Riedel (1980, 1983) and Tvergaard (1986). Results are now available for both uniaxial and multiaxial states of stress.

A comprehensive review, including work up to 1985, has been given by Bazant (1986). The maturity of the field is illustrated by the recent appearance of two monographs by Mura (1986) and Kachanov (1986).

Fig. 6.3. Creep crack growth.

6.3 REFERENCES

Z.P. Bazant, Mechanics of distributed cracking - Appl. Mech. Rev. 39:5 (1986), 675-705

D.R. Hayhurst et al, Development of Continuum Damage in the Creep Rupture of Notched Bars - Phil. Trans. Roy. Soc. Lond. A 311 (1984 a), 103-129.

D.R. Hayhurst et al, The Role of Continuum Damage in Creep Crack Growth - Phil. Trans. Roy. Soc. Lond. A 311 (1984 b), 131-158.

J. Janson and J. Hult, Fracture mechanics and damage mechanics, a combined approach - J. de Mecanique Appliquee 1:1 (1977), 69-84.

J. Janson, Dugdale - Crack in a Material with Continuous Damage Formation - Engng. Fracture Mechanics 9 (1977), 891-899.

J. Janson, A Continuous Damage Approach to the Fatigue Process - Engng. Fracture Mechanics 10 (1978 a), 651-657.

J. Janson, Damage Model of Crack Growth and Instability - Engng. Fracture Mechanics 10 (1978 b), 795-806.

J. Janson, Damage Model of Creep - Fatigue Interaction - Engng. Fracture Mechanics 11 (1979), 397-403.

L.M. Kachanov, Introduction to Continuum Damage Mechanics - Kluwer Academic Publishers, Dordrecht 1896.

T. Mura, Micromechanics of Defects in Solids - Kluwer Academic Publishers, Dordrecht 1986.

H. Riedel, The Extension of a Macroscopic Crack at Elevated Temperature by the Growth and Coalescence of Microvoids - Proc. IUTAM Symposium Creep in Structures (Leicester 1980), Springer, Berlin 1981.

H. Riedel, Constrained Grain Boundary Cavitation in a Creeping Body Containing a Macroscopic Crack - Proc. Fourth Int. Conf. Mechanical Behaviour of Materials (Stockholm 1983), Pergamon Press, Oxford 1984.

D.J. Smith and G.A. Webster, Fracture-mechanics interpretation of multiple-creep cracking using damage-mechanics concepts - Materials Science and Technology 1 (1985), 366-372.

V. Tvergaard, Analysis of Creep Crack Growth by Grain Boundary Cavitation - Int. J. Fracture 30 (1986), to be published.

7 CONCLUSION

Continuum damage mechanics (CDM) has come to attract steadily increasing interest. The seminal paper by Kachanov (1958) has led to a vast number of experimental, theoretical and numerical studies of damage in loaded structural elements. The ultimate aim is to develop methods for safe prediction of the carrying capacity or lifetime of such elements, notably components used at high temperature.

It took time for the damage concept to gain foothold in applied mechanics research and application. It describes an inherently discrete process, i.e. the formation of local defects, by means of a continuous field variable. Also it aims to describe several different processes of material deterioration in terms of only one variable. The model was considered by some to be too simple to warrant further development. In spite of this both experimental and theoretical work continued, and results appeared which lent support to the basic ideas in CDM.

The field has been developed as a bridge between microscopic studies of the deterioration in materials under stress and engineering models to be used in design work. Contributions to the development of CDM have come, in equal measure, from both these areas. Recent development of computer methods has made detailed numerical analyses possible, on which a quantitative theory may be based.

Rather few damage parameter data are yet available, but a framework now exists in which such data may be placed to the benefit of high temperature design work.

FORMULATION AND IDENTIFICATION OF DAMAGE KINETIC CONSTITUTIVE EQUATIONS

Jean Lemaitre
Laboratoire de Mécanique et Technologie
Université Paris 6, Cachan, France

ABSTRACT

The damage kinetic constitutive equations are derived from the thermodynamics of irreversible process in which physical considerations and experimental results are introduced in order to choose the proper variables and the analytical forms of the potentials. A general damage model is formulated and then applied to ductile damage, low cycle fatigue, high cycle fatigue and creep damage in order to identified particular kinetic laws to be introduced in structure calculations to predict the initiation and the growth of cracks.

CONTENTS

1. THE PHENOMENOLOGICAL THERMODYNAMICAL APPROACH

In the field of applied mechanics, damage constitutive equations are derived in view of structure calculations in order to predict the state of damage through the remaining life (the time or the number of cycles to initiation of a macro-crack) and the evolution of a crack at the structure scale (the so-called "local approach"). The main tool is then the mechanics of continuous media with stress and strain as the principal variables [J. LEMAITRE-J.L. CHABOCHE, 1985].

The damage constitutive equations must characterize the behavior of the volume element representative in the sense of the mechanics of continuous media. The "size" of the volume has to be large enough in order to represent the local properties by their mean values through continuous variables. Roughly speaking :

- .1 x.1 x.1 mm for metals ;
- 1 x 1 x 1 mm for polymers and composites ;
- 10 x 10 x 10 mm for wood ;
- 100 x 100 x 100 mm for concrete.

Do not speak of damage gradients below these numbers since they have no physical sense, due to the heterogeneities of the materials and the heterogeneities of the nature of damage. This scale is much larger that the one to be considered for elasticity (relative movement of atoms) or for plasticity (movements of dislocations in crystals) because damage is essentially discontinuous in the material at the scale of sets of crystals in metals, molecules in polymers, cells in wood, agregates in concrete.

- micro-cracks, roughly planar, trans or inter-elementary constituants ;
- micro-cavities of volumetric nature growing mainly between crystals, molecules, cells or agregates.

A way to obtain damage constitutive equations within in terms of strains used in elasticity, plasticity or viscoplasticity, is to derive them from the same thermodynamic framework. Working with two potentials, the thermodynamic potential is used to obtain the state laws of the non-dissipative phenomena and the definition of variables, and the potential of dissipation is used to define laws of evolution of the dissipative variables. The result is the set of all constitutive equations for all variables introduced [GERMAIN, 1983]. The coupling between the different phenomena is automatically obtained. For example, the damage, micro-cracks or micro-cavities, reduce the part of the material which effectively supports the loads. The resistance of the material to elastic or plastic deformations also decreases which means that the elastic and plastic properties depend upon the state of damage. This coupling phenomenon appears naturally in the strain constitutive equations if the two potentials previously mentioned are functions of all the variables and especially of a damage variable.

In the classical method of predicting failure this coupling is neglected which give rise to the scheme shown in figure 1

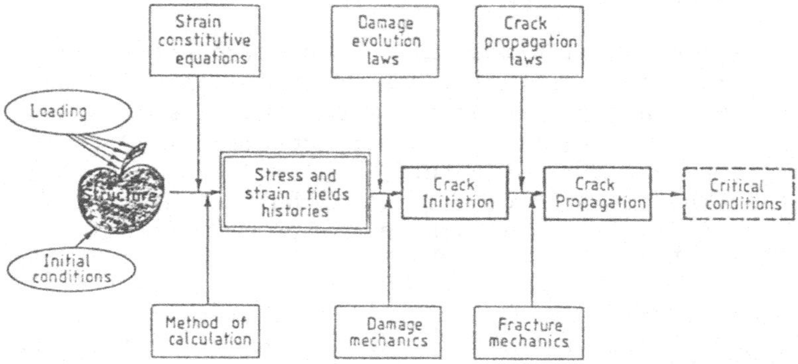

Figure 1 : Scheme of uncoupled structure calculations

In many cases, and especially in the prediction of macro-crack growth, this coupling cannot be neglected because the influence of the damage on the state of stress in the neighbourhood of the crack is of primary importance (zero stress at the crack tip !). Then, the modern way to calculate the rupture conditions is to simultaneously calculated the stress, the strain and the damage field histories with coupled constitutive equations (introduced in a finite element code, for example). Figure 2 [J. LEMAITRE, 1984].

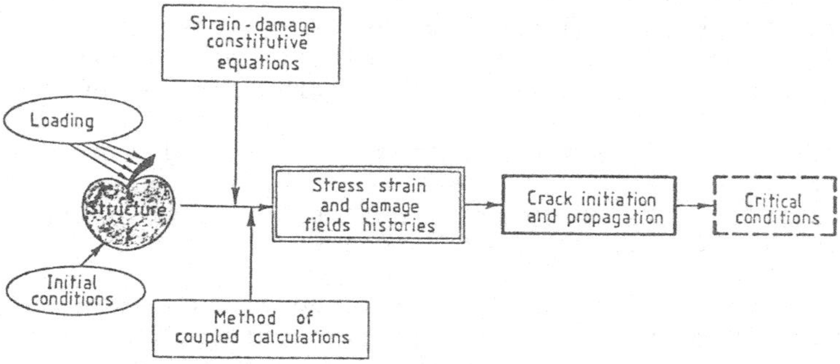

Figure 2 : Scheme of fully coupled calculations

In this phenomenological thermodynamical approach, the difficulties are :

- The choice of "good" variables. For damage, this choice must be in accordance with the physical observation of mechanisms involved in the damage processes and in the micro-mechanics.

- The choice of the analytical forms of the potentials which have to be in accordance with qualitative results of many experiments showing all relations between the variables involved. These choices are mostly subjective which gives the possibility for each researcher to have his own model (no worse, no better than the others !).

- The identification of the coefficients introduced which are characteristic of each material and are most often temperature dependant. Specific tests are used but the range of variation of variables is limited, and it is always difficult to determine the domain of validity of the models : whenever possible, use phenomenological models to interpolate and no to extrapolate !

2. *GENERAL FORMULATION*

2.1. PHENOMENA AND VARIABLES

Damage is a dissipation process always associated with strain and which also involves dissipation. Furthermore, in order to express the coupling between damage and strain, it is necessary to consider the classical variables of continuum mechanics.

2.1.1. Elastoplasticity

- Total strain tensor ε with its partition between
  - elastic strain tensor $\boldsymbol{\varepsilon}^e$
  - plastic strain tensor $\boldsymbol{\varepsilon}^p$

  If we restrict ourselves to small strains

$$\boldsymbol{\varepsilon} = \boldsymbol{\varepsilon}^e + \boldsymbol{\varepsilon}^p$$

- Cauchy stress tensor $\boldsymbol{\sigma}$

- Within the hypothesis of isotropy, the strain hardening may be represented by a scalar variable p associated with its dual R which is related to the "radius" of the loading surface of plasticity in the stress space [D. MARQUIS, 1979]

$$\dot{p} = \left[\frac{2}{3}\,\dot{\boldsymbol{\varepsilon}}^p : \dot{\boldsymbol{\varepsilon}}^p\right]^{1/2} = \left[\frac{2}{3}\,\dot{\varepsilon}^p_{ij}\,\dot{\varepsilon}^p_{ij}\right]^{1/2} \qquad , \quad \dot{p} = \frac{dp}{dt}$$

2.1.2. Damage

The damage variable is defined to be used in conjontion with the concept of stress. Then, it has to consider a surface meaning. Considering only isotropic damage, the scalar damage variable D is the surface density of intersection of micro-cracks and micro-cavities with any plane of $\vec{n}$ orientation in the body (figure 3) corrected from micro-stress concentrations and interaction effects. If δS is the area of the section of a finite volume element and δS_0 the resulting area of defects in the plane considered [KATCHANOV, 1958] [HULT, 1975], then :

$$D = \frac{\delta S_0}{\delta S}$$

D = 0 corresponds to the virgin element

D = 1 corresponds to a ruptured element

In fact, the volume element breaks by atomic decohesion before the value of 1 is obtained. A criterion for macro-crack initiation will be developed later.

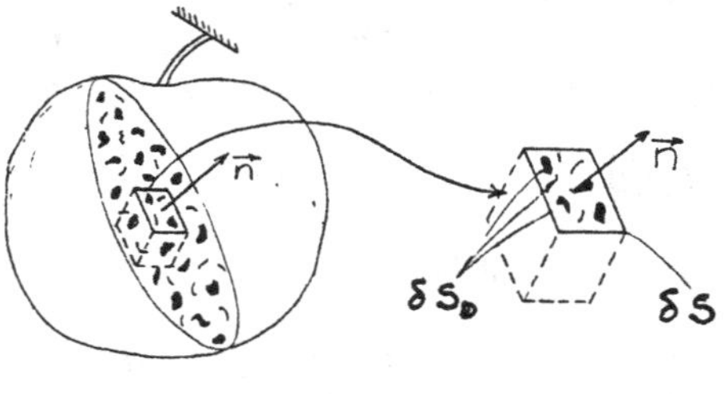

Figure 3 : Micro-macro definition of damage

D is a state variable which can be considered an internal variable in the sense of thermodynamics,as p for strain-hardening. Let Y be its associated variable such that $Y.\dot{D}$ is the power dissipated in the creation of new micro-cracks or micro-cavities [LEMAITRE, CHABOCHE, 1985]. (Notation $\dot{D}$ denotes the time derivative dD/dt).

2.1.3. **Micro-plasticity**

This phenomena, usually neglected, must be considered when damage occurs in the elastic range,even if macroscopic plastic strain does not exist. This is the reversible movement of dislocations which produce a reversible plastic strain $\boldsymbol{\epsilon}^p$ negligible in comparison to the elastic strain $\boldsymbol{\varepsilon}^e$ [FOUGERES, 1985] and the dissipation $\boldsymbol{\sigma}:\dot{\boldsymbol{\epsilon}}^p = \sigma_{ij}\dot{\epsilon}^p_{ij}$ which is responsible for internal damping of all materials. Since very little information exists about this phenomenon, its constitutive equations will be formated with some similitude with macroscopic plasticity.

From the dissipation term $\boldsymbol{\sigma}:\boldsymbol{\epsilon}^p$, the stress $\boldsymbol{\sigma}$ is the associated variable to $\boldsymbol{\epsilon}^p$. Let $\boldsymbol{\pi}$:

$$\dot{\pi} = (\frac{2}{3}\,\dot{\boldsymbol{\epsilon}}^p:\dot{\boldsymbol{\epsilon}}^p)^{1/2}$$

be the accumulated micro-plastic strain and r its associated variable.

Since $\boldsymbol{\epsilon}^p$ strain is rather small, the only measurable quantity related to that phenomenon is the internal damping η defined for one-dimensional loading as

$$\eta = \frac{\Delta W_p}{\Delta W_e} = \int_{1\,Cycle} \frac{\sigma\dot{\varepsilon}_p}{\frac{1}{2E}\,\sigma^2_{MAX}}\,dt$$

η is of the order of 10^{-3} to 10^{-5} for metals [LAZAN, 1960].

2.1.4. Thermal effects

If thermal effects are involved, the temperature T is introduced as an observable state variable together with its associated variable:entropy s.

All these variables are summarized inthe table of figure 4.

State variables		Associated variables
Observable	Internal	
$\boldsymbol{\varepsilon}$		$\boldsymbol{\sigma}$
T		s
	$\boldsymbol{\varepsilon}^e$	$\boldsymbol{\sigma}$
	$\boldsymbol{\varepsilon}^p$	$-\boldsymbol{\sigma}$
	p	R
	D	y
	$\boldsymbol{\alpha}^p$	$-\boldsymbol{\sigma}$
	π	r

Figure 4 : Table of thermodynamic variables

2.2. EFFECTIVE STRESS CONCEPT

2.2.1. Definition

A very useful and interesting concept to express the damage constitutive equations is the one of effective stress $\tilde{\sigma}$ which relates the stress to the area which effectively supports the load [RABOTNOV, 1969] [JANSON, HULT, 1977]. For isotropic damage and a resisting area independant of the sign of the stress (bi-lateral conditions) :

$$\tilde{\sigma}_{ij} = \sigma_{ij}\frac{\delta S}{\delta S-\delta S_0} \quad , \quad \boxed{\tilde{\boldsymbol{\sigma}} = \frac{\boldsymbol{\sigma}}{1-D}}$$

The concept is difficult to generalize to the case of anisotropic damage where the damage variable is of vectorial or tensorial nature [LECKIE, ONAT, 1980] [MURAKAMI, 1981] [KRAJCINOVIC, FONSEKA, 1981] [CHABOCHE, 1982].

2.2.2. Principle of strain equivalence

This principle, associated with the concept of effective stress, gives the key to epxress the potentials from which the constitutive equations are derived [LEMAITRE, 1971].

It states that any strain constitutive equation for a damaged material ($D \neq 0$) is derived from the same potentials as for a virgin material ($D=0$) except that all the stress variables are replaced by effective stresses.

- Example of elasticity :

$$\varepsilon_e = \frac{\tilde{\sigma}}{E} = \frac{\sigma}{E(1-D)}$$

where E is the Young modulus.

2.2.3. Extension to quasi-unilateral conditions

In many cases, the damage may be considered isotropic (equal value of $(\delta S_D/\delta S) = D$ in all directions). However, there exists a large difference in the damage behavior in tension and compression :

- static rupture different in tension and compression (concrete for example) ;
- fatigue failure depending upon the mean stress (Goodman's diagram) ;
- different elasticity modulus in tension and compression in cyclic loadings close to rupture ;
- etc.

These phenomena are associated with the crack closure. When the stress normal to the crack is compressive, even though δS_D being a physical state, always exists, the ability of the cross section to carry loads depends upon the sign of the applied stress. This means that the effective stress must be a different function of the variable D in tension and compression.

In the one-dimensional case, one may write

$$\tilde{\sigma} = \frac{\sigma}{1-D} \quad \text{if} \quad \sigma \geqslant 0$$

$$\tilde{\sigma} = \frac{\sigma}{1-Dh} \quad \text{if} \quad \sigma < 0$$

where h is a closure coefficient which characterizes the closure of the micro-cracks and micro-cavities :

$$0 \leqslant h \leqslant 1$$

$h = 0$ corresponds to the case of the unilateral condition of pure surface micro-cracks ;

$h = 1$ corresponds to the case of bilateral conditions of spherical micro-cavities which do not close ;

$0 < h < 1$ corresponds to intermediary cases.

This simple concept is difficult to expand to the threedimensional case in conformity with the following implications :

- intrincic definition of the effective stress independant of the frame ;
- existence of an elastic potential ;
- coherence with the principle of strain equivalence.

A solution [LADEVEZE, LEMAITRE, 1984] is to divide the Cauchy stress tensor into a "positive" part $\langle\sigma\rangle$ and a "negative" part $\langle-\sigma\rangle$

$$\sigma = \langle\sigma\rangle - \langle-\sigma\rangle$$

defined by the positive and negrative parts of the principal stresses σ_i

$$\langle\sigma\rangle : \begin{bmatrix} \langle\sigma_1\rangle & 0 & 0 \\ 0 & \langle\sigma_2\rangle & 0 \\ 0 & 0 & \langle\sigma_3\rangle \end{bmatrix} \quad , \quad \langle-\sigma\rangle : \begin{bmatrix} \langle-\sigma_1\rangle & 0 & 0 \\ 0 & \langle-\sigma_2\rangle & 0 \\ 0 & 0 & \langle-\sigma_3\rangle \end{bmatrix}$$

where the symbol $\langle\sigma_i\rangle$ denotes

$$\langle\sigma_i\rangle = \sigma_i \quad \text{if} \quad \sigma_i \geqslant 0$$

$$\langle\sigma_i\rangle = 0 \quad \text{if} \quad \sigma_i < 0$$

Then, to be in agreement with the principle of strain equivalence, the damaging term is applied in the elastic potential of complementary energy as :

$$\frac{1}{1-D} \quad \text{for the positive part of the stresses}$$

$$\frac{1}{1-Dh} \quad \text{for the negative part}$$

$$W^* = \left[\begin{array}{l} \dfrac{1}{2(1-D)} \left[\dfrac{1+\nu}{E} \langle \boldsymbol{\sigma} \rangle : \langle \boldsymbol{\sigma} \rangle - \dfrac{\nu}{E} \langle \mathrm{tr}(\boldsymbol{\sigma}) \rangle^2 \right] \\ \\ + \dfrac{1}{2(1-Dh)} \left[\dfrac{1+\nu}{E} \langle -\boldsymbol{\sigma} \rangle : \langle -\boldsymbol{\sigma} \rangle - \dfrac{\nu}{E} \langle -\mathrm{tr}(\boldsymbol{\sigma}) \rangle^2 \right] \end{array} \right]$$

for isotropic elasticity.

Furthermore the law of elasticity derived from :

$$\boldsymbol{\varepsilon}_e = \frac{\partial W^*}{\partial \boldsymbol{\sigma}}$$

is identified with the law of elasticity formely formulated from the principle of strain equivalence :

$$\boldsymbol{\varepsilon}_e = \frac{1+\nu}{E} \tilde{\boldsymbol{\sigma}} - \frac{\nu}{E} \mathrm{tr}(\tilde{\boldsymbol{\sigma}}) \mathbf{1}$$

This gives the definition of the effective stress

$$\boxed{\tilde{\boldsymbol{\sigma}} = \frac{\boldsymbol{\sigma}^{+}}{1-D} - \frac{\boldsymbol{\sigma}^{-}}{1-Dh}}$$

with

$$\boldsymbol{\sigma}^{+} = \langle \boldsymbol{\sigma} \rangle + \frac{3\nu}{1-2\nu} (\sigma_H^{+} - \langle \sigma_H \rangle) \mathbf{1}$$

$$\boldsymbol{\sigma}^{-} = \langle -\boldsymbol{\sigma} \rangle + \frac{3\nu}{1-2\nu} (\sigma_H^{-} - \langle -\sigma_H \rangle) \mathbf{1}$$

and the hydrostatic stress $\sigma_H = (1/3)\mathrm{tr}(\boldsymbol{\sigma})$

$$\sigma_H^{+} = \frac{1}{3} \mathrm{tr}\langle \boldsymbol{\sigma} \rangle \qquad , \qquad \sigma_H^{-} = \frac{1}{3} \mathrm{tr}\langle -\boldsymbol{\sigma} \rangle$$

The terms containing σ_H, σ_H^+ and σ_H^- are terms of coupling in the elastic rigidity due to shear effects. They disappear in the one dimensional case :

$$\tilde{\sigma} = \frac{\langle \sigma \rangle}{1-D} - \frac{\langle -\sigma \rangle}{1-Dh}$$

Identification of the closure coefficient.

The coefficient h characterizes the closure of micro-cracks and micro-cavities and depends upon the density and the shape of the defects. It is material dependant and, as a first approximation for simplicity, we consider h as a constant. The effect of damage itself on closure is neglected.

In the elasticity law, h is directly related to the rigidity in compression :

$$\sigma = (1-D)E\varepsilon_e \quad \text{in tension } (\sigma > 0)$$

$$\sigma = (1-Dh)E\varepsilon_e \quad \text{in compression } (\sigma < 0)$$

if the Young's modulus is known, a measurement of

- the elasticity modulus in tension $\tilde{E}_t = E(1-D)$
- the elasticity modulus in compression $\tilde{E}_c = E(1-Dh)$

on the damaged material allows the coefficient h to be determined

$$\frac{\tilde{E}_c}{\tilde{E}_t} = \frac{1-Dh}{1-D}$$

or, with $D = 1 - \dfrac{\tilde{E}_t}{E}$

$$\boxed{h = \frac{E-\tilde{E}_c}{E-\tilde{E}_t}}$$

An example of the determination of h is given in figure 5, where h, calculated as a function of the number of cycles in a low fatigue process, is almost constant.

The value of h=.2 has been obtained for several other materials

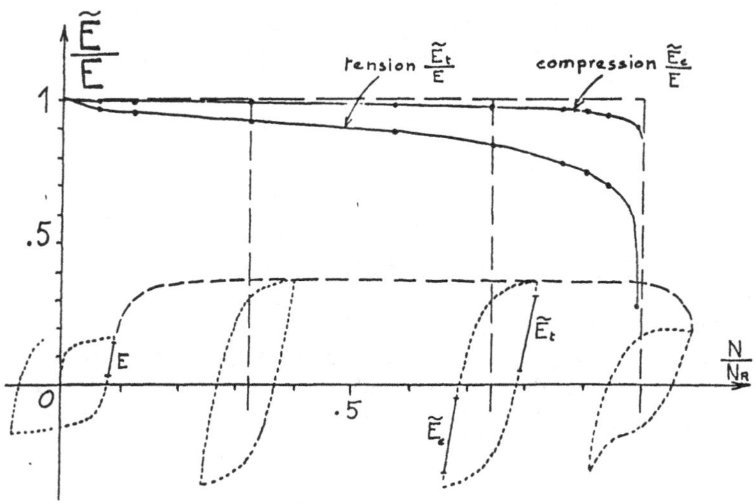

Figure 5 : Example of determination of the closure coefficient : h = .2
Low cycle fatigue A316L stainless steel
T = 550°C , NR = 218 cycles

2.3. THERMODYNAMIC POTENTIAL

2.3.1. Formulation

The existence of a thermodynamic potential is postulated as a convex function of all state variables from which the state laws can be derived as well as the associated variables corresponding to the dissipative phenomena.

One may choose the free energy :

$$\Psi(\varepsilon, T, \varepsilon^e, \varepsilon^p, p, D, \alpha^p, \pi)$$

or its dual Ψ^* by a partial Legendre-Fenchel transform on strain and stress.

For classical elastoplasticity, Ψ depends upon strain by means of ε^e only. Furthermore, there is strong experimental evidence that there is :

- no coupling between plastic strain and elasticity. The elasticity modulus does not depend upon ε^p ;
- a small coupling between micro-plasticity and elasticity, an example is given on figure 6 ;
- no direct coupling between damage and plasticity.

We do not consider thermal effects other than those related to strains and damage behaviors

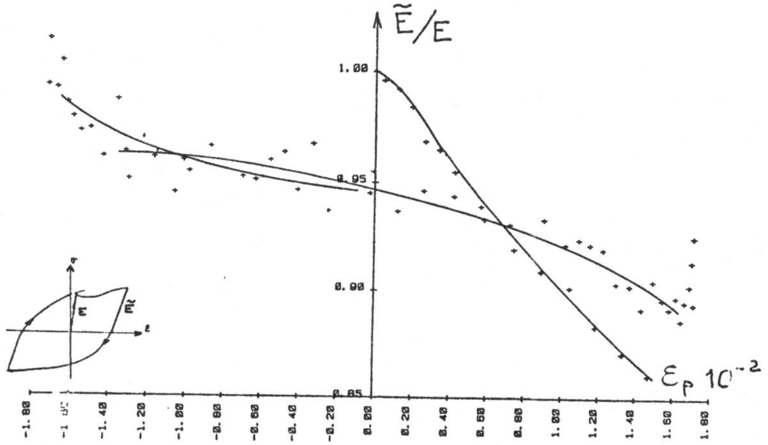

Figure 6 : Influence of micro-plasticity on elasticity modulus XC 38 steel (from M.T. GAUTHERIN)

Then, the thermodynamic potential must be written as :

$$\Psi = \Psi_e(\varepsilon^e, T, D, \pi) + \Psi_p(T, p)$$

or $$\Psi^* = \sup_{\varepsilon^e}(\sigma:\varepsilon^e - \Psi)$$

$$\Psi^* = \Psi_e^*(\sigma, T, D, \pi) - \Psi_p(T, p)$$

The analytic expression is chosen to be in accordance with the principle of strain equivalence for linear elasticity coupled with damage in quasi-unilateral condition and to give the Ramberg-Osgood strain hardening low in tension :

$$\varepsilon_p = \left[\frac{\sigma - \sigma_y}{K_y}\right]^{M_y}$$

where σ_y is the yield stress, and K_y and M_y are material and temperature dependant coefficients.

The function describing the influence of π on elasticity must tend toward a limit very rapidly. We choose :

$$\rho\Psi^* = \left[\begin{array}{l} \dfrac{1}{2(1-D)}\left[\dfrac{1+\nu}{E}\langle\sigma\rangle:\langle\sigma\rangle-\dfrac{\nu}{E}\langle \mathrm{tr}(\sigma)\rangle^2\right] \\ +\dfrac{1}{2(1-Dh)}\left[\dfrac{1+\nu}{E}\langle-\sigma\rangle:\langle-\sigma\rangle-\dfrac{\nu}{E}\langle-\mathrm{tr}(\sigma)\rangle^2\right]\end{array}\right](1-k\pi^{1/m})^{-1} \cdot \frac{M_y}{1+M_y}K_y\, p^{\frac{1+M_y}{M_y}}$$

E and ν are elasticity coefficients varying with temperature.

k and m (m >1) are micro-plasticity coefficients.

ρ is the density.

2.3.2. State laws

The law of elasticity coupled with damage in quasi-unilateral conditions is derived from :

$$\varepsilon^e = \rho\frac{\partial\Psi^*}{\partial\sigma}$$

$$\varepsilon^e = \frac{1}{E(1-k\pi^{1/m})}\left[\frac{1+\nu}{1-D}\langle\sigma\rangle-\frac{\nu}{1-D}\langle \mathrm{tr}(\sigma)\rangle \mathbb{1} + \frac{1+\nu}{1-Dh}\langle-\sigma\rangle-\frac{\nu}{1-Dh}\langle-\mathrm{tr}(\sigma)\rangle 1\right]$$

which is the classical Hooke's law coupled with damage when :

- micro-plasticity is neglected : $\pi = 0$
- unilateral conditions do not occur : $h = 0$

$$\boxed{\varepsilon^e = \frac{1+\nu}{E(1-D)}\sigma - \frac{\nu}{E(1-D)}\mathrm{tr}(\sigma)\mathbb{1}}$$

The variable y associated with D is derived from

$$y = -\rho\frac{\partial\Psi^*}{\partial D}$$

or, with the notation $\overline{y} = -y$,

$$Y = \left[\frac{1}{2(1-D)^2} \left[\frac{1+\nu}{E} \langle\sigma\rangle : \langle\sigma\rangle - \frac{\nu}{E} \langle tr(\sigma) \rangle^2 \right] + \frac{1}{2(1-Dh)^2} \left[\frac{1+\nu}{E} \langle-\sigma\rangle : \langle-\sigma\rangle - \frac{\nu}{E} \langle -tr(\sigma) \rangle^2 \right] \right] (1-k\pi^{1/m})^{-1}$$

which is related to the elastic strain energy of the damaged material.

For $\pi = 0$ and $h = 0$,

$$\bar{Y} = \frac{1}{2(1-D)^2} \left[\frac{1+\nu}{E} \sigma:\sigma - \frac{\nu}{E} [tr(\sigma)]^2 \right]$$

The other associated variables are derived from

$$R = -\varrho \frac{\partial \Psi^*}{\partial p} = K_y \, p^{1/M_y}$$

$$r = \varrho \frac{\partial \Psi^*}{\partial \pi}$$

2.3.3. Damage criterion

The variable Y associated with the damage variable D may be used in order to define a damage equivalent stress σ^* which, for damage, acts as the Von Mises equivalent stress σ_{eq} used in plasticity [LEMAITRE, BAPTISTE, 1982].

A state of one-dimensional tensile stress σ^* equivalent to a three-dimensional state σ is defined as

$$Y(\sigma^*) = Y(\sigma)$$

From the $\overline{Y}$ expression

$$\overline{Y}(\sigma^*) = \frac{\sigma^{*2}}{2E(1-D)^2}$$

and from the above equality :

$$\sigma^* = \left[(1+\nu)\langle\sigma\rangle:\langle\sigma\rangle - \nu\langle tr(\sigma)\rangle^2 + h\frac{(1-D)^2}{(1-Dh)^2}\left[(1+\nu)\langle-\sigma\rangle:\langle-\sigma\rangle - \nu\langle-tr(\sigma)\rangle^2\right] \right]^{1/2} (1-k\pi^{1/m})^{\cdot 1/2}$$

In general, the term $(1-k\pi^{1/m})^{1/2}$ may be considered to be equal to unity in this expression.

For the one-dimensional case :

- in tension σ_t : $\sigma^* = \sigma_t$

- in compression σ_c : $\sigma^* = h^{1/2}\dfrac{(1-D)}{(1-Dh)}|\sigma_c|$

If pure bilateral conditions are considered, h = 1

$$\sigma^* = [(1+\nu)\sigma:\sigma - \nu[tr(\sigma)]^2]^{1/2}$$

or with

- hydrostatic stress defined by

$$\sigma_H = \frac{1}{3} tr(\sigma)$$

- Von Mises equivalent stress defined by

$$\sigma_{eq} = \left[\frac{3}{2}\sigma^D:\sigma^D\right]^{1/2}$$

σ^D being the stress deviator :

$$\sigma^D = \sigma - \sigma_H \mathbb{1}$$

$$\sigma^* = \sigma_{eq}\left[\frac{2}{3}(1+\nu) + 3(1-2\nu)\left[\frac{\sigma_H}{\sigma_{eq}}\right]^2\right]^{1/2}$$

The triaxiality function, R_ν defined by :

$$R_\nu = \left[\frac{\sigma^*}{\sigma_{eq}}\right]^2 = \frac{1}{\sigma_{eq}}\left[\begin{array}{l}(1+\nu)\langle\sigma\rangle:\langle\sigma\rangle - \nu\langle \mathrm{tr}(\sigma)\rangle^2 \\ +h\dfrac{(1-D)^2}{(1-Dh)^2}\left[(1+\nu)\langle-\sigma\rangle:\langle-\sigma\rangle - \nu\langle-\mathrm{tr}(\sigma)\rangle^2\right]\end{array}\right]$$

models the influence of the closure phenomenon and the influence of the triaxiality ratio σ_H/σ_{eq} on damage and rupture.

In pure bilateral conditions, this function reduces to

$$R_\nu = \frac{2}{3}(1+\nu) + 3(1-2\nu)\left[\frac{\sigma_H}{\sigma_{eq}}\right]^2$$

Note that these two very imporant effects lead directly to the thermodynamic potential without any additional hypothesis.

In general, the damage equivalent stress may be written as

$$\sigma^* = \sigma_{eq}\, R_\nu^{-1/2}$$

2.3.4. Rupture criterion

The elastic strain energy is defined as

$$dW_e = \sigma : d\varepsilon^e$$

which, in conjunction with the law of linear elasticity coupled with damage (with π=0 and h=1), gives

$$W_e = \frac{1}{2(1-D)}\left[\frac{1+\nu}{E}\sigma:\sigma - \frac{\nu}{E}[tr(\sigma)]^2\right]$$

This shows that :

$$W_e = \rho\Psi_e$$

$$-Y = \bar{Y} = \frac{W_e}{1-D}$$

and

$$\bar{Y} = \frac{\partial W_e}{\partial D}$$

This last equation is used to define a rupture criterion based on an energetic instability.

It is postulated that the macro-crack initiation which separates the volume element into two parts occurs when the variation of the strain energy due to a variation of damage at constant stress [(∂We)/(∂D)] reaches a critical value

$$\boxed{\bar{Y} = Y_c}$$

with Y_c being characteristic of each material.

The parameter Y_c can be related to the rupture conditions for the one dimensional case :

- in tension σ_t

$\sigma_t = \sigma_t$ the ultimate rupture stress in tension

$D = D_c$ the critical value of damage at rupture in tension

$$Y_c = \frac{1}{2E}\left[\frac{\sigma_u}{1-D_c}\right]^2$$

- in compression $\sigma_c < 0$

$\sigma_c = \sigma_u^-$ the ultimate rupture stress in compression

$D = D_c^-$ the critical value of damage at rupture in compression

$$Y_c = \frac{1}{2E}\left[\frac{\sigma_u^-}{(1-D_c^-)h}\right]^2$$

2.4. POTENTIAL OF DISSIPATION, GENERAL DAMAGE MODEL

2.4.1. Dissipative variables

The dissipation rate is expressed using the Clausius-Duhem inequality which must be positive to satisfy the second principle of thermodynamics :

$$\sigma : \dot{\varepsilon} - \rho(\dot{\Psi}+s\dot{T}) - \vec{q}.\frac{\overrightarrow{\text{grad}T}}{T} \geqslant 0$$

Or, together with the state laws,

$$\sigma:\dot{\varepsilon}^p - R\dot{p} - Y\dot{D} + \sigma\dot{e}_p - r\dot{\pi} - \vec{q}\,\frac{\overrightarrow{\text{grad}T}}{T} \geqslant 0$$

The table 7 lists all fluxes and their associate dual variables (conjugate thermodynamic forces).

Flux variables	conjugate dual variables
$\dot{\varepsilon}^p$	σ
$-\dot{p}$	R
$-\dot{D}$	Y
$\dot{e}^p$	σ
$-\dot{\pi}$	r
$\frac{\vec{q}}{T}$	$\overrightarrow{\text{grad}T}$

Figure 7 : Liste of dissipative variables

2.4.2. Normality rule

The existence of a potential of dissipation is postulated as a convex scalar function of all flux variables and the state variables acting as parameters, from which the complementary kinetic laws of evolution may be derived [LEMAITRE, CHABOCHE, 1985] :

$$\varphi\,(\dot{\varepsilon}^p,\ \dot{p},\ \dot{D},\ \dot{e}^p,\ \dot{\pi},\ \frac{\vec{q}}{T}\ ;\ \underbrace{T,\ \varepsilon^e,\ \varepsilon^p,\ p,\ D,\ e^p,\ \pi}_{\text{parameters}})$$

or any function obtained by partial or total Legendre-fenchel transform

$$\varphi^*(\sigma, \dot{p}, Y, \dot{\epsilon}^p, \dot{\pi}, \overrightarrow{\text{grad}T} \; ; \underbrace{T, \varepsilon^e, \varepsilon^p, p, D, \epsilon^p, \pi})$$

This yields the generalized normality rule

$$\dot{\varepsilon}^p = \frac{\partial\varphi^*}{\partial\sigma} \qquad R = -\frac{\partial\varphi^*}{\partial\dot{p}}$$

$$\dot{D} = -\frac{\partial\varphi^*}{\partial Y}$$

$$\sigma = \frac{\partial\varphi^*}{\partial\dot{\epsilon}^p} \qquad r = -\frac{\partial\varphi^*}{\partial\dot{\pi}}$$

$$\frac{\vec{q}}{T} = -\frac{\partial\varphi^*}{\partial\overrightarrow{\text{grad}T}}$$

For phenomena which do not depend explicitly upon time (such as plasticity), the function φ is homogeneous positive of degree 1. The partial derivatives of φ^* do not exist, and it is only possible to write (for the case of plasticity) :

$$\dot{\varepsilon}^p \in \partial\varphi^* \longrightarrow \text{loading}$$

$$\dot{\varepsilon}^p = 0 \longrightarrow \text{unloading}$$

with $\partial\varphi^*$ being the subdifferential of φ^*.

If the loading criterion is defined by a convex function (Von Mises criterion for example)

$$\text{loading if} \quad \begin{cases} f = 0 \\ \dot{f} = 0 \end{cases}$$

$$\text{unloading if} \quad f < 0 \quad \text{or} \quad \dot{f} < 0$$

φ^* is the indicative function of f, and from a derivation available elswhere, it is equivalent to state :

$$\dot{\varepsilon}^p = \frac{\partial f}{\partial\sigma}\dot{\lambda}$$

$$\dot{p} = -\frac{\partial f}{\partial R}\dot{\lambda}$$

where $\dot{\lambda}$ is a scalar multiplier with the other equations remaining the same.

2.4.3. Properties of the potential of dissipation

After the choice of variables is made, the most important step in the determination of strain and damage constitutive equations is the selection of analytical expressions for the potential of dissipation used to write the kinetic law (relative fluxes and their associated dual variables).

Concentrating only on damage, these properties are :

- irreversible nature of damage $\dot{D} \geq 0$;
- non linearities of damage with regard to stress, time or number of cycles in fatigue ;
- effect of triaxiality on rupture ;
- effect of mean stress or difference in the behavior of damaged material in tension and compression ;
- existence of damage thresholds ;
- non linear accumulation of damage due to different loadings.

By empirical, phenomenological, or micor-mechanical considerations, more than one hundred laws of evolution of rupture parameters, some more complex than others, have been proposed in the past ten or fifteen years ! Using thermodynamics, it is now possible to reassemble most of them in a simple potential function with different identification for the different kinds of damage.

This may be considered as shocking by physicists or metallurgists but it is nothing more than considering only one constitutive equation for macroscopic plasticity describing several kinds of mechanisms of dislocation movement !

It must be considered that the various kinds of damage are all governed by some dissipation energy.

This "outstanding" potential function is as follows :

- Firstly we assume that the coupling in the dissipation potential exists only among the state variables which act as parameters, and not through the fluxes themselves; accordingly, the potential may be considered to be the sum of three functions

$$\varphi^* = \underbrace{\varphi^*_p}_{\text{plasticity or viscoplasticity}} + \underbrace{\varphi^*_D(Y,\dot{p},\dot{\pi}\ ;\ T,\epsilon_e,D)}_{\text{damage}} + \underbrace{\varphi^*_\pi}_{\text{micro-plasticity}}$$

- For plasticity, φ^*_p (equal to zero or infinity) is the indicative function of a convex function $f(\sigma,R,D) = 0$ which, for the Von Mises criterion for plasticity coupled with damage, is given in the form of :

$$f = \frac{\sigma_{eq}}{1-D} - R - \sigma_y \leqslant 0$$

where $\sigma_y(T)$ is the yield stress affected by damage in accordance with the principle of strain equivalence :

$$\sigma_y(D\neq 0) = (1-D)\sigma_y(D=0)$$

and where $R = K_y \cdot p^{1/M_y}$

- For visco-plasticity, the classical law of isotropic strain hardening is obtained from a function which contains only σ as a variable, T and p as parameters and D as the parameter in accordance with the principle of strain equivalence :

$$\varphi_p^*(\sigma\ ;\ T,p,D) = \frac{K}{N+1}\left[\frac{\sigma_{eq}}{K(1-D)}\right]^{N+1} p^{-N/M}$$

where K(T), M(T), N(T) are material creep characteristic functions of the temperature.

- Modelling φ_π^* for micro-plasticity is out of the scope of the present work. Since very little information exists on that subject, a simple expression will be proposed when needed in section 5.2.1..

- It will be shown that a damage potential function of the three variables Y, $\dot{p}$ or $\dot{\pi}$ and the two parameters T and D are sufficient to model all the main properties, within the hypothesis of isotropy.

Furthermore, it is sufficient to consider the following very simple analytical form function of 2 coefficients characteristic of the material : S_0 and α_0

$$\varphi_D^*(Y,\dot{p},\dot{\pi}\ ;\ T,\varepsilon^e,D) = \frac{1}{2}\frac{Y^2}{S_0}\frac{\dot{p}+\dot{\pi}}{(1-D)^{\alpha_0}}$$

from which

$$\boxed{\dot{D} = -\frac{\partial\varphi^*}{\partial Y} = \frac{\partial\varphi_D^*}{\partial\overline{Y}} = \frac{\overline{Y}}{S_0}\frac{\dot{p}+\dot{\pi}}{(1-D)^{\alpha_0}}}$$

Since the variables $\overline{Y}$, $\dot{p}$ and $\dot{\pi}$ are positive, the damage rate is always positive.

The variable $\bar{Y}$ containing the triaxiality ratio (σ_H / σ_{eq}) models the effect of triaxiality on rupture.

If $\bar{Y}$ is calculated from the quasi-unilateral expression, it models the difference in the damage behavior in tension and compression and also mean stress effects in fatigue.

The non linearity of damage with regard to stress is given by its dependance on $\dot{p}$ and $\dot{\pi}$ which are non linear functions of stress (in fact $\dot{p}$ or $\dot{\pi}$ since π may be neglected when p exists).

The non linearity of damage with regard to time or number of cycles is given by the term $(1-D)^{\alpha_0}$ which corresponds to a non linear differential equation in D. If the coefficient α_0 is a function of strain ε_e ,then from the law of elasticity :

$$\alpha_0 = \alpha_0(\sigma^*)$$

The differential equation in D is no longer an equation with separated variables leading to the proprety of non linear accumulation.

All these properties are fully defined by 3 coefficients which are material and temperature dependant :

- the exponent $\alpha_0(\sigma^*)$;
- the closure coefficient h ;
- the coefficient S_0 ;
+ the Poisson's ratio ν and a damage threshold ;

which a good quality/price ratio !

Then, the general damage model which may be used for theoretical demonstrations or qualitative results is

$$\dot{D} = \frac{\bar{Y}}{S_0} \frac{\dot{p}}{(1-D)^{\alpha_0}}$$

$\dot{p} \equiv \dot{\pi}$ (micro-plasticity) for damage in the elastic range

3. *DUCTILE DAMAGE*

3.1. PHENOMENOLOGICAL ASPECTS

3.1.1. Physical mechanisms

Ductile damage in metals is essentially related to large deformations. It has to be considered in metal forming process for which it can limit the maximum possible strains without appreciable damage or without crack initiation. It also has an influence on constitutive equations of materials after forming.

From a physical point of view, ductile damage develops as a result of initiation, growth and the coalescence of cavities in the scale of microns. The initiation of these micro-cavities occurs through a decohesion process at the interface between the defects (such as inclusions) and the matrix. The growth of cavities which are generally well distributed in a relatively large volume, are governed by plastic deformation in those regions where the micro-stress concentrations are the highest.

Finally, a macro-crack initiation occurs by the coalescence of several micro-cavities through a process of local instability [ENGEL, KLINGELE, 1981].

During these processes, the local density decreases but by a slightly different amount : from several per thousands to several percents.

3.1.2. Micro-mechanics of void growth

A volume element damaged by a ductile process may be geometrically modelled by a cube with holes inside it. The constitutive equations at the macro-scale may be derived from the constitutive equations of the material matrix and the geometry of defects using a structural calculation at the micro-scale and a homogenization technique using some mean values of the variables at the micro-scales.

The first approach to ductile damage is the early works of [MAC CLINTOCK, 1968] and [RICE AND TRACEY, 1969]. Considering a cylindrical or a spherical hole of radius R in a perfectly plastic matrix, the growth rate (dR /R) of the radius is derived from analytical calculations :

$$\frac{dR}{R} = B \exp\left[C \frac{\sigma_H}{\sigma_{eq}} \right] dp$$

where B and C = 1,5 are determined from the theory.

This formula shows the influence of the triaxiality ratio (σ_H / σ_{eq}) on ductile failure and considers the radius evolution as an exponential function of the accumulated plastic strain

$$p = \left[\frac{2}{3} \varepsilon_p : \varepsilon_p \right]^{1/2}$$

The coalescence of cavities may be considered to occur for a critical value of the cavities growth rate depending upon material properties only [MUDRY, 1982]: $(R/R_0)c$

For the cleavage phenomenon, the rupture criterion may be related to a critical value of stress at a characteristic distance from the crack tip [DEVAUT, ROUSSELIER, MUDRY, PINEAU, 1985].

The homogenization technique may be used to find the strain properties of the volume element [SUQUET, 1980] but not (or not yet !) to derive the constitutive equations for damage evolution.

3.1.3. Experimental measurements

Considering several possible techniques to evaluate damage (static elasticity modulus ultrasonic waves, cyclic softening, electrical resistance, ...) [LEMAITRE, CORDEBOIS, DUFAILLY, 1979], the most appropriate one for ductile damage is the measurement of the elasticity modulus.

This technique is derived directly from the coupling between elasticity and damage and is deduced from the principle of strain equivalence :

$$\tilde{\sigma} = \frac{\sigma}{1-D} = E\varepsilon_e$$

or $\sigma = E(1-D)\varepsilon_e$

$E(1-D) = \tilde{E}$ may be interpreted as the elastic modulus of the damaged material :

$$\tilde{E} = \frac{\sigma}{\varepsilon_e}$$

and, if E, Young's modulus is known, and $\tilde{E}$ can be measured, the damage is found to be :

$$\boxed{D = 1 - \frac{\tilde{E}}{E}}$$

The measurement of $\tilde{E}$ is often difficult due to :

- the localisation of damage which requires the use of very small strain guages ;
- the small non linearities which always exist in the elastic range even during unloading. Figure 8 schematically describe this method

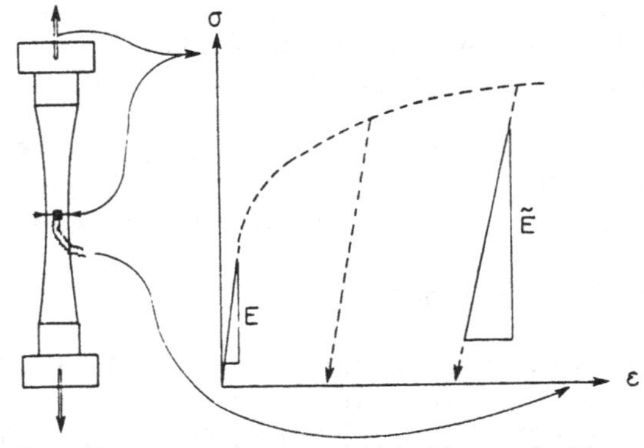

Figure 8 : Damage measurement through the variation of elasticity modulus

- many experiments on various metallic materials have shown that the one-dimensional damage scalar D has a linear evolution with respect to strains. Some results are shown in figure 9

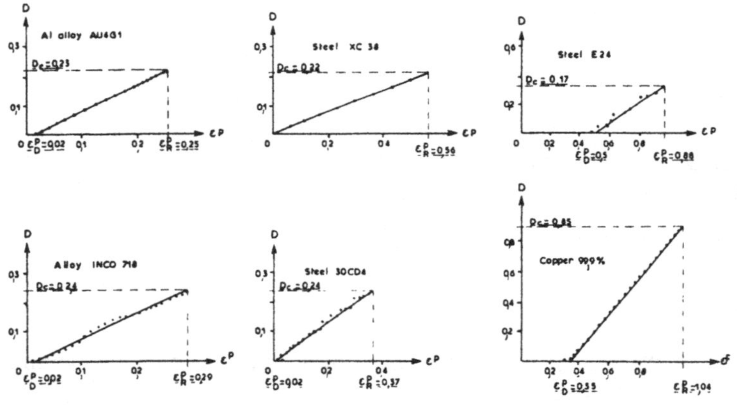

Figure 9 : Ductile damage evolutions on various materials

3.2. MODELS

3.2.1. Formulation

Starting with the general damage model derived in section 2.4.4. in which micro-plasticity is neglected (π=0)

$$\dot{D} = \frac{\overline{Y}}{S_0} \frac{\dot{p}}{(1-D)^{\alpha_0}}$$

- Since ductile damage is always associated with large deformations, the material may be considered perfectly plastic. From the plasticity criterion of section 2.4.3. :

$$\frac{\sigma_{eq}}{1-D} - \sigma_y = 0$$

Then, for perfect bilateral conditions (h=1), or not (h≠1) :

$$\bar{Y} = \frac{\sigma^{*2}}{2E(1-D)^2} = \frac{\sigma_{eq}^2 R_v}{2E(1-D)^2} = \frac{\sigma_y^2 R_v}{2E}$$

which means that $\bar{Y}$ is constant when R_v is constant or when the triaxiality ratio (σ_H / σ_{eq}) is constant if perfect bilateral conditions are considered.

- With the variable responsible for damage being p, the thresold is introduced as an initial condition on p in the differential equation :

$$\dot{D} = \frac{\sigma_y^2 R_v}{2ES_0} \frac{\dot{p}}{(1-D)^{\alpha_0}} \qquad \text{if} \qquad p \geqslant p_0$$

$$\dot{D} = 0 \qquad \text{if} \qquad p < p_0$$

where p_0 is the strain threshold depending upon the material.

- The linearity of the damage evolution with respect to strain observed in experimental results imposes the condition

$$\alpha_0 = 0$$

then,

$$\dot{D} = \frac{\sigma_y^2 R_v}{2ES_0} \dot{p} \qquad \text{if} \qquad p \geqslant p_0$$

3.2.2. Identification

A convenient method to identify this model is to use one-dimensional experiments in tension such as those of figure 8.

In this case

$$\frac{\sigma_H}{\sigma_{eq}} = \frac{1}{3} \qquad \text{and} \qquad R_v = 1$$

Furthermore, in the range of large deformations, the elastic strain may be neglected relative to ε_p :

$$\varepsilon_p \equiv \varepsilon$$

then

$$\dot{D} = \frac{\sigma_y^2}{2ES_0}\dot{\varepsilon}$$

or, by integration, with $D = 0$ for $\varepsilon < \varepsilon_0$ (ε_0 being the one dimensional strain threshold) :

$$D = \frac{\sigma_y^2}{2ES_0}(\varepsilon-\varepsilon_0)$$

Writing this equation for the critical condition of rupture in one dimension :

$$\varepsilon = \varepsilon_R \quad , \quad D = D_c$$

$$D_c = \frac{\sigma_y^2}{2ES_0}(\varepsilon_R-\varepsilon_0) \quad \text{or} \quad \frac{\sigma_y^2}{2ES_0} = \frac{D_c}{\varepsilon_R-\varepsilon_0}$$

and

$$D = D_c \frac{\varepsilon-\varepsilon_0}{\varepsilon_R-\varepsilon_0}$$

The three constants D_c, ε_R, ε_0 are easily determined from a ductile damage test ; some examples of these measurements are shown in figure 9.

Some results are given on the table of the figure 10

Material	Temperature °C	ε_0	ε_R	D_c
Copper	20	.35	1.04	.85
2024 Alloy	20	.03	.25	.23
E24 Steel	20	.50	.88	.17
INCO 718	20	.02	.29	.24

Figure 10 : Ductile damage coefficients

The influence of the triaxiality ratio on the ductile rupture given by the parameter R_v is numerically in agreement with the Mac Clintok-Rice and Tracey model and with experimental results [MUDRY, 1982].

To summarise the formulas used in the model, we need to relate the threedimensional threshold p_0 to ε_0.

Assuming that the influence of the triaxiality function is the same on the threshold as on the strain to rupture p_R

$$\frac{p_D}{\varepsilon_D} = \frac{p_R}{\varepsilon_R}$$

p_R being calculated in the next section as:

$$p_R = \frac{\varepsilon_R}{R_v} \quad , \text{ then} \quad p_D = \frac{\varepsilon_D}{R_v}$$

The general model is [LEMAITRE, 1985] :

$$\dot{D} = \begin{cases} \dfrac{D_c}{\varepsilon_R - \varepsilon_D} R_v \, \dot{p} \ldots \text{ if} & p \geq \dfrac{\varepsilon_D}{R_v} \\ 0 \ldots\ldots\ldots\ldots \text{ if} & p < \dfrac{\varepsilon_D}{R_v} \end{cases}$$

$$\text{with } \dot{p} = \left[\frac{2}{3}\, \dot{\boldsymbol{\varepsilon}} : \dot{\boldsymbol{\varepsilon}}\right]^{1/2} \qquad R_v = \left[\frac{\sigma^*}{\sigma_{eq}}\right]^2 \qquad \sigma_{eq} = \left[\frac{3}{2}\, \boldsymbol{\sigma}^D : \boldsymbol{\sigma}^D\right]^{1/2}$$

$$\sigma^* = \left[(1+\nu)\langle\boldsymbol{\sigma}\rangle : \langle\boldsymbol{\sigma}\rangle - \nu\langle \mathrm{tr}(\boldsymbol{\sigma})\rangle^2 + h \frac{(1-D)^2}{(1-Dh)^2} \left[(1+\nu)\langle-\boldsymbol{\sigma}\rangle : \langle-\boldsymbol{\sigma}\rangle - \nu\langle-\mathrm{tr}(\boldsymbol{\sigma})\rangle^2 \right] \right]^{1/2}$$

case of pure bilateral conditions :

$$R_v = \frac{2}{3}(1+\nu) + 3(1-2\nu)\left[\frac{\sigma_H}{\sigma_{eq}}\right]^2$$

3.3. APPLICATIONS TO DUCTILE FRACTURE

3.3.1. Strain condition to macro-crack initiation

The rupture criterion of the volume element developed in section 2.3.4. is related to a critical value of the strain energy density release rate :

$$Y = Yc \longrightarrow \text{macro-crack initiation}$$

For ductile fracture, it is more convenient to relate macro-crack initiation to strain.

In general, the differential model may be integrated as

$$D = \frac{D_c}{\varepsilon_R - \varepsilon_D} \int_{p_D}^{p} R_v \, dp$$

or for the critical condition of rupture $D = D_c$, $p = p_R$:

$$\boxed{\int_{p_D}^{p_R} R_v \, dp = \varepsilon_R - \varepsilon_D}$$

which allows p_R to be calculated.

If we now are restricted to the case of proportional loading for which R_v = cte

$$D = \frac{D_c R_v}{\varepsilon_R - \varepsilon_D} (p - p_D)$$

and the strain to rupture is given by

$$p_R = p_D + \frac{\varepsilon_R - \varepsilon_D}{R_v}$$

or together with the hypothesis $p_D = \dfrac{\varepsilon_D}{R_v}$

$$\boxed{p_R = \frac{\varepsilon_R}{R_v}}$$

3.3.2. **Master curve of ductile fracture**

The condition $p_R = \varepsilon_R / R_v$ is valid for proportional loading only. Furthermore, if we consider loadings for which the principal stresses are all positive or negative, p_R is a unique function of (σ_H / σ_{eq})

$$p_R = \varepsilon_R \left[\frac{2}{3} (1+v) + 3(1-2v) \left[\frac{\sigma_H}{\sigma_{eq}} \right]^2 \right]^{-1} \qquad \text{if} \quad \sigma_1 \geqslant 0$$

$$p_R = \varepsilon_R \frac{(1-Dh)^2}{h(1-D)^2}\left[\frac{2}{3}(1+v) + 3(1-2v)\left[\frac{\sigma_H}{\sigma_{eq}}\right]^2\right]^{-1} \quad \text{if} \quad \sigma_i < 0$$

with $D = D_c$

As Poisson's ratio remains fairly consistent (.25<v<.33), a master curve can be drawn to give, as a function of (σ_H / σ_{eq}) :

1) $\frac{p_R}{\varepsilon_R}$ in the case of positive principal stresses

2) $\frac{p_R}{\varepsilon_R} h \frac{(1-D_c)^2}{(1-D_c h)^2}$ in the case of negative principal stresses.

See figure 11.

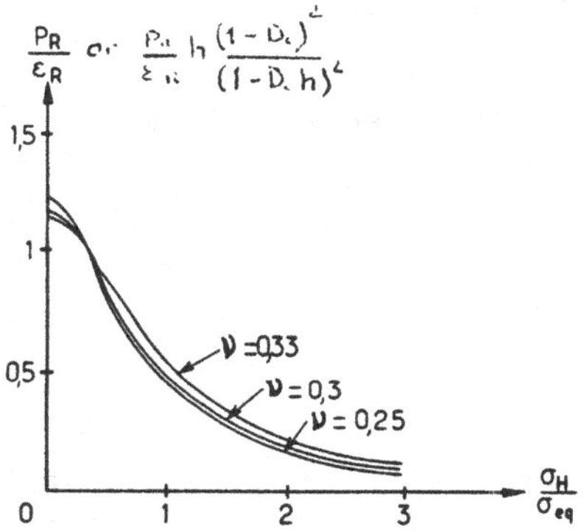

Figure 11 : Ductile fracture limits

This approach is used to determine fracture limits in metal forming processes, for deep drawing forming limits [CORDEBOIS, 1983] and for extrusion and wire drawing.

4. *LOW CYCLE FATIGUE DAMAGE*

4.1. PHENOMENOLOGICAL ASPECTS

4.1.1. Physical mechanisms

Fatigue is a damaging process which occurs when stresses vary between maximum and minimum values due to periodic cycles (or not). Low cycle fatigue is considered when the plastic strain involved is high enough to be measured. This corresponds to stresses roughly higher than the yield stress and to the number of cycles to failures :

NR $\leqslant$ 10 000 cycles

In metals, the damage is mainly in the form of micro-cracks which are mostly trans-granular with a greater localization than for ductile damage [ENGEL, KLINGELE, 1981].

Fatigue is essentially related to cycles. Although a cycle is easy to define for one dimensional loading, it is usually quite complex to define a cycle for dimensional loadings where a "maximum" or a "minimum" of a tensor may only be defined by a scalar norm. It is proposed to define cycle with the variable Y or σ^* and to use the corresponding amplitude of accumulated plastic strain Δp.

4.1.2. Experimental measurements

Measurements of low cycle fatigue damage may be performed by means of the variation of the elastic modulus, but a most convenient method here is to evaluate the value of damage from its coupling effect with cyclic plasticity.

One may use the one dimensional rlation between the stress range $\Delta\sigma$ and the plastic strain range $\Delta\varepsilon_p$ at stabilization.

This relation may be derived from the Ramber-Osgood strain hardening law already mentioned and from Masing's rule (symmetry with a ratio of 2 between tension and compression curves)

$$\Delta\varepsilon_p = \left[\frac{\Delta\sigma}{K_c}\right]^{M_c}$$

where K_c and M_c are material parameters identified from the cyclic hardening curve.

The coupling with damage is introduced through the principle of strain equivalentce with the effective stress :

$$\Delta\varepsilon_p = \left[\frac{\Delta\sigma}{K_c\,(1-D)}\right]^{M_c}$$

If a cyclic test at constant amplitude of plastic strain $\Delta\varepsilon_p$, is considered up to stabilization $\Delta\sigma^*$, the damage may be assumed to be zero ; hence,

$$\Delta\sigma^* = K_c\,\Delta\varepsilon^{1/M_c},$$

Subsequently, the damage has a softening effect :

$$\Delta\sigma = (1-D)K_c \Delta\varepsilon^{1/M_c}$$

Then, from the two relations :

$$\boxed{D = 1 - \frac{\Delta\sigma}{\Delta\sigma^*}}$$

An example is given in figure 12.

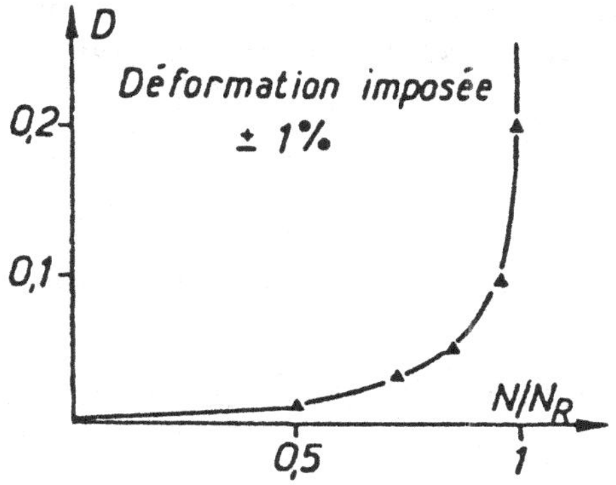

Figure 12 : Example of law cycle damage evolution
316 stainless steel. Temperature 20°C

4.2. MODELS

4.2.1. Formulation

From the general damage model, it is possible to derive several classes of constitutive equations. The most general one is of course :

$$\dot{D} = \frac{\tilde{Y}}{S_0} \frac{\dot{p}}{(1-D)^{\alpha_0}}$$

where micro-plasticity is neglected ($\pi = 0$).

To obtain models expressed in terms of strain, the idea is to express the stress σ in the $\tilde{Y}$ expression as a function of p using the cyclic law of plasticity already mentioned in section 4.1.2 :

$$\Delta\varepsilon_p = \left[\frac{\Delta\sigma}{K_c(1-D)}\right]^{M_c}$$

In the case of proportional loading, the corresponding threedimensional relation is :

$$\frac{\Delta p}{2} = \left[\frac{2\Delta\sigma_{eq}}{K_c(1-D)}\right]^{M_c} \quad \text{or} \quad \frac{\Delta\sigma_{eq}}{1-D} = K'_c\,\Delta p^{1/M_c} \quad \text{with} \quad K'_c = K_c\,2^{-(1+1/M_c)}$$

which may be considered as an integration over one stabilized cycle, with $D \simeq$ cte, of an instantaneous law :

$$\frac{\sigma_{eq}}{1-D} = K_c\,(p-p_i)^{1/M_c}$$

with p_i being the value of p at the begining of a cycle i of amplitude Δp.

Introducing that relation in the $\overline{Y}$ expression :

$$\overline{Y} = \frac{\sigma_{eq}^2 R_v}{2E(1-D)^2} = \frac{R_v}{2E} K_c'^{\,2} (p-p_i)^{2/M_c}$$

and then

$$\dot{D} = \frac{R_v}{2ES_0} K_c'^{\,2} (p-p_i)^{2/M_c} \frac{\dot{p}}{(1-D)^{\alpha_0}}$$

or with a simple change of variables in the coefficients ($2ES_0/K_c'^2 = \Gamma$, $2/M_c = \gamma$, $\alpha_0 = \alpha_1(\Delta p)$)

$$\boxed{\dot{D} = \frac{R_v(p-p_i)^{\gamma}}{\Gamma} \frac{\dot{p}}{(1-D)^{\alpha_1}}}$$

This model is used to calculate (by integrating with respect to time) the low cycle fatigue of a volume element about which any history of strain p(t) and stress σ(t) (for R_v) is known without any restriction on quasi-unilateral conditions or proportionality of loading.

The threshold is given by the condition D = 0 if p = 0.

Integration over one cycle

In fatigue life calculations, it is often convenient to integrate the kinematic variable over the number of cycles ; by doing so we may obtain a cyclic constitutive equation for damage.
Two hypotheses are needed :

- the loading is considered proportional over one cycle and no quasi-unilateral condition is involved : R_v = cte ;
- the damage is very small, so that $(1-D)^{\alpha_1}$ is considered constant over one cycle.

If $\frac{\delta D}{\delta N}$ is then the variation of damage per cycle :

$$\frac{\delta D}{\delta N} = \frac{R_v}{\Gamma(1-D)^{\alpha_1}} \int_{p_i}^{p_i+\Delta p} (p-pi)^{\gamma}\, dp$$

$$\boxed{\frac{\delta D}{\delta N} = \frac{R_v}{\Gamma(\gamma+1)} \frac{\Delta p^{\gamma+1}}{(1-D)^{\alpha_1}}}$$

Integration in the case of a periodic loading

Consider a proportional periodic loading of constant amplitude Δp and R_v = cte. The model may be integrated to obtain :

- the number of cycles to failure defined here by :

  $$D = 1 \longrightarrow N = N_R$$

  Due to the non linearity there is not a large difference between these calculations of N_R corresponding to $Y = Y_c$ (or $D = D_c$) and $D = 1$

  and with the initial condition $N = 0 \longrightarrow D = 0$

  $$\int_0^1 (1-D)^{\alpha_1} \delta D = \frac{R_v}{\Gamma(\gamma+1)} \Delta p^{\gamma+1} \int_0^{N_R} \delta N$$

  $$N_R = \frac{\Gamma(\gamma+1)}{\alpha_1+1} R_v^{-1} \Delta p^{-(\gamma+1)}$$

- the evolution of damage as a function ofthe number of cycles :

  $$\int_0^D (1-D)^{\alpha_1} \delta D = \frac{R_v}{\Gamma(\gamma+1)} \Delta p^{\gamma+1} \int_0^N \delta N$$

$$D = 1 - \left[1 - \frac{N}{N_R}\right]^{1/(\alpha_1+1)}$$

If α_1 is a constant, then the model has the property of linear cumulation. In the case of non-periodic loading $\Delta p(N)$:

$$\int_0^{N_R} \frac{\delta N}{N_F(\Delta p)} = 1$$

where N_F is the function $N_F = \frac{\Gamma(\gamma+1)}{\alpha_1+1} R_v^{-1} \Delta p^{-(\gamma+1)}$.

If α_1 is a function of Δp, the cumulation in non linear.

4.2.2. Law of Manson-Coffin

Consider now the case of one-dimensional loading symmetric in tension and compression $\Delta p = 2\Delta\varepsilon_p$

$$N_R = \frac{\Gamma(\gamma+1)}{\alpha_1+1} (2\Delta\varepsilon_p)^{-(\gamma+1)} = \left[\frac{\Delta\varepsilon_p}{C}\right]^{-(\gamma+1)}$$

This is the law proposed by [MANSON, 1954] and [COFFIN, 1954] which was later improved by considering the "strain range partionning".

4.2.3. Identification

Three coefficients need to be identified for each material : γ, Γ and α_1 eventually function of Δp. They can be obtained from one-dimensional tests.

- Some tests performed at constant amplitude of plastic strain up to rupture give the Manson-Coffin curve $N_R(\Delta\varepsilon_p)$ from which one may evaluate C and $\gamma+1$

$$N_R = \left[\frac{\Delta\varepsilon_p}{C}\right]^{-(\gamma+1)}$$

- To obtain α_1 it is necessary to measure damage, for example, by the evolution of the stress amplitude during the tests used for the Manson-Coffin curve :

$$D = 1 - \left[1 - \frac{N}{N_R}\right]^{1/(\alpha_1+1)}$$

from which values of α_1 can be derived.
α_1 = cte if only one curve is founded for $D(N/N_R)$ corresponding to different values of $\Delta\varepsilon_p$.
$\alpha_1(\Delta p)$ if not (with $\Delta p = 2\Delta\varepsilon_p$).

- Then $\Gamma = \frac{\alpha_1+1}{\gamma+1}(2C)^{\gamma+1}$

Some examples of coefficients are given in the table of fig. 13 (α_1 = cte)

Material	Temperature °C	$\gamma+1$	C	α_1
A 316	20	2.07	.34	5
INCO 718	550	1.615	.407	10.3
Maraging	20	1.22	1.63	4
IN 100	700	2	.244	2.6
IN 100	900	2	.037	3
IN 100	1110	2	.021	4.7

Figure 13 : Low cycle fatigue material coefficients

to summarize, the different models are :

general model	$\dot{D} = \frac{R_v(p-p_i)^\gamma}{\Gamma}\frac{\dot{p}}{(1-D)^{\alpha_1}}$
quasi proportional loading over one cycle	$\frac{\delta D}{\delta N} = \frac{R_v}{\Gamma(\gamma+1)}\frac{\Delta p^{\gamma+1}}{(1-D)^{\alpha_1}}$
one dimensional Manson-Coffin's law	$N_R = \left[\frac{\Delta\varepsilon_p}{C}\right]^{\gamma+1}$
with	$\Gamma = \frac{\alpha+1}{\gamma+1}(2C)^{\gamma+1}$

with $\dot{p} = [\ 2/3\ \dot{\varepsilon}^p : \dot{\varepsilon}^p]^{1/2}$ $\quad p_i = p$ at the begining of a cycle

$$R_v = \left[\frac{\sigma^*}{\sigma_{eq}}\right]^2 \qquad \sigma_{eq} = (2/3\ \sigma^D : \sigma^D)^{1/2} \qquad \Delta p = p_{MAX} - p_i$$

$$\sigma^* = \left[(1+\nu)\langle\sigma\rangle : \langle\sigma\rangle - \nu\langle tr(\sigma)\rangle^2 + h\frac{(1-D)^2}{(1-Dh)^2}\left[(1+\nu)\langle-\sigma\rangle : \langle-\sigma\rangle - \nu\langle-tr(\sigma)\rangle^2\right] \right]^{1/2}$$

4.3. APPLICATIONS

4.3.1. Crack initiation

To calculate the time or the number of cycles to crack initiation in mechanical components, a history of loading at the most loaded point M* is needed. This point is the point at which Y or σ* has the maximum value

$$\underset{M}{\mathrm{Sup}}(\sigma(M)) \longrightarrow M^*$$

The strain and the stress histories ($p(t)$ and $\sigma(t)$ in R_v) are introduced into the model for integration, leading to $D(t)$. The macro-crack initiation is reached when

$$\bar{Y} = \frac{\sigma^{*2}}{2E(1-D)^2} = Y_c$$

For details see [CHABOCHE, 1982] [CAILLETAUD, 1982].

4.3.2. Fatigue crack growth

The behavior of macro-cracks at the structure scale is generally studied by means of the global concepts of fracture mechanics (stress intensity factors, Rice integral, strain energy release rate) ; however, if plasticity occurs to a large extent, the global concepts are no longer valid. A relatively new method called the "local approach of fracture" based on damage mechanics can be used [LEMAITRE, 1985].

This approach considers the mechanics of continuous media in which the crack at the structure scale is that set of points for which the critical damage conditions of macro-crack initiation has already been reached. It is usually applied in the finite element method either with or without coupling between strain and damage.

For details see [BILLARDON, 1983].

5. *HIGH CYCLE FATIGUE DAMAGE*

5.1. PHENOMENOLOGICAL ASPECTS

5.1.1. Physical mechanisms

High cycle fatigue in metals must be considered separately from low cycle fatigue bacause the irreversible strain involved is only micro-plastic which is not measurable and difficult to calculate. Therefore, the damage evolution models must be written in term of stress. The stress level is at worst the order of magnitude of the yield stress which corresponds to the number of cycles to failure

$$NR \geqslant 10^5 \text{ cycles}$$

The damage is very localised which is not always compatible with continuum mechanics for which the damage is "uniformly" distributed in a volume element of a "finite" size !

Even in a uniform field of stress, the high cycle fatigue micro-cracks alway start from the surface boundary of the body (stage 1) along the length of one or two crystals in a direction of about $\pi/4$ with respect to the direction of the maximum principal stress. Later, the micro-crack growth perpendicular to this direction (stage 2). These two stages define the domain of the so called "short cracks". The third stage is the development of one crack by the coalescence of several micro-cracks [ENGEL, KLINGELE, 1981].

5.1.2. Experimental measurements

Due to the high level of localization of the micro-cracks mainly at the surface, it is difficult to evaluate the high cycle fatigue damage through techniques of continuum mechanics. Nevertheless, the two techniques already described may be used.

- variation of elasticity modulus during fatigue tests in tension and compession

$$\boxed{D = 1 - \frac{\tilde{E}}{E}}$$

- variation of the stress range in a periodic fatigue test and a constant strain range $\Delta\varepsilon$.
  Micro-plastic strain π is neglected with respect to elastic strain $\varepsilon_e = \varepsilon$

$$\frac{\Delta\sigma}{1-D} = E\,\Delta\varepsilon$$

If $\Delta\sigma^*$ is the stress amplitude after some cycles of stabilization when D remains zero, then

$$\Delta\sigma^* = E\,\Delta\varepsilon$$

From which

$$\boxed{D = 1 - \frac{\Delta\sigma}{\Delta\sigma^*}} \qquad \text{at} \qquad \Delta\varepsilon = \text{cte}$$

If tests of constant amplitude of stress $\Delta\sigma$ are considered ($\Delta\varepsilon^*$ being the stress amplitude after stabilization)

$$\boxed{D = 1 - \frac{\Delta\varepsilon^*}{\Delta\varepsilon}} \qquad \text{at} \qquad \Delta\sigma = \text{cte}$$

An example is given in figure 14

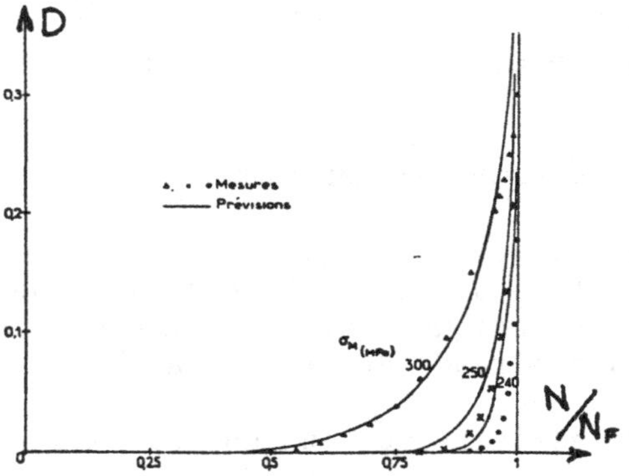

Figure 14 : Example of high-cycle fatigue damage evolution 316 stainless steel. Temperature 20°C

5.2. MODELS

5.2.1. Formulation

Reconsidering the general model, it is possible to derive several classes of models as it has been performed for low cycle fatigue. Here, macro-plasticity does not exist (p = 0), the irreversible strain is micro-plastic

$$\dot{D} = \frac{\overline{Y}}{S_0} \frac{\dot{\pi}}{(1-D)^{\alpha_0}}$$

To obtain a model written in terms of strain, $\dot{\pi}$ must be expressed as a function of stress.

On close examination of the loops (σ, ε) of internal damping of metals, one may see that the reversibility of the micro-plastic strain e_p arises for the mean value of the stress (figure 1). Then a possible one-dimensional constitutive equation of micro-plasticity coupled with damage is

$$\dot{e}_p = \left[\frac{|\sigma-\bar{\sigma}|}{k(1-D)}\right]^{\beta} \frac{|\dot{\sigma}|}{(1-D)} \operatorname{Sgn}(\sigma-\bar{\sigma})$$

where k and β are material constants.

This equation can be verified only by its global effect as internal damping (section 2.1.3.). A possible three-dimensional expansion for the accumulated micro-plastic strain is :

$$\dot{\pi} = \left[\frac{|\sigma_{eq}-\bar{\sigma}_{eq}|}{k(1-D)}\right]^{\beta} \frac{|\dot{\sigma}_{eq}|}{1-D}$$

The corresponding schematic loops are seen in figure 15.

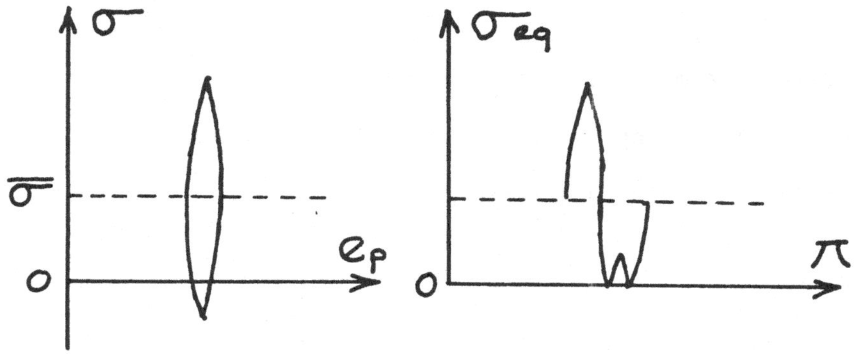

Figure 15 : Loops of internal damping

The damage model is now written as

$$\dot{D} = \frac{\sigma_{eq}^2 R_v}{2ES_0(1-D)^2} \left[\frac{|\sigma_{eq}-\bar{\sigma}_{eq}|}{k(1-D)}\right]^{\beta} \frac{|\dot{\sigma}_{eq}|}{(1-D)(1-D)^{\alpha_0}}$$

or with some change of notations in the coefficients,

$$\dot{D} = \frac{\sigma_{eq}^2 \left|\sigma_{eq} - \bar{\sigma}_{eq}\right|^{\beta}}{B} \frac{R_v |\dot{\sigma}_{eq}|}{(1-D)^{\alpha_2}}$$

with $B = 2ES_0 k$, $\alpha_2 = \alpha_2(\Delta\sigma_{eq}) = \alpha_0 + \beta + 3$

This is a general constitutive equation for high cycle fatigue which is valid for any kind of loadings and which has to be integrated numericaly over time for each cycle if the cycles are different.

The threshold "fatigue limit" is introduced as an initial condition on effective stress. $\sigma^* = R_v \sigma_{eq}$ in the differential equation

$$D = 0 \quad \text{if} \quad \sigma^* < \sigma_l$$

where σ_l is the fatigue limit in tension

Integration over one cycle

Analytical integration is possible only if some assumptions are made. Let us consider

- mean stress $\bar{\sigma}_{eq} = \frac{1}{2}\sigma_{eqM}$

  σ_{eqM} being the maximum of σ_{eq} over a cycle ;
- proportional loading over the all cycle : $R_v = cte$;
- variation of $(1-D)^{\alpha_2}$ neglected

$$\int_D^{D+\frac{\delta D}{\delta N}} dD = \frac{R_v}{B(1-D)^{\alpha_2}} \int_0^{\sigma_{eqM}} \sigma_{eq}^2 \left[2\,\frac{\sigma_{eq}}{2}\right]^{\beta} d\sigma_{eq}$$

$$\frac{\delta D}{\delta N} = \frac{R_v \sigma_{eqM}^{(\beta+3)}}{B(\beta+3)(1-D)^{\alpha_2}} \quad \text{for} \quad \sigma_{eq} = \frac{1}{2}\sigma_{eqM}$$

Integration in the case of periodic loading

Within the same hypothesis as previously mentioned, if the loading is periodic, it is possible to integrate for the conditions

$N = 0 \longrightarrow D = 0$

$D = 1 \longrightarrow N = N_R$ Number of cycles to rupture

The result is :

$$N_F = \frac{B(\beta+3)}{\alpha_2+1} R_v^{-1} \sigma_{eqM}^{-(\beta+3)}$$

And the evolution of damage is given by

$$D = 1 \left[1 - \frac{N}{N_R}\right]^{\frac{1}{\alpha_2+1}}$$

5.2.2. **Woehler Miner's law. Goodman's rule**

If the model is written for the one-dimensional case of a "0,+" loading : $\sigma_{min} = 0$, $\bar{\sigma} = (\sigma_M)/2$

$$\frac{\delta D}{\delta N} = \frac{\sigma_M^{(\beta+3)}}{B(\beta+3)(1-D)^{\alpha_2}}$$

If α_2 is constant, this differential equation has the property of linear cumulation which is known as Palmgreen-Miner's rule. In the case of a loading by blocks of N_i cycles of different amplitudes $\Delta\sigma_i$:

$$\Sigma \frac{N_i}{N_{Ri}(\Delta\sigma_i)} = 1$$

Furthermore,

$$N_R = \frac{B(\beta+3)}{\alpha_2+1} \sigma^{-(\beta+3)}$$

is the equation of the Woehler curve.

An important phenomenon in fatigue is the influence of the mean stress known as Goodman's rule or Goodman's diagram. In the one-dimensional case :

$$\dot{D} = \frac{\sigma^2 |\sigma-\bar{\sigma}|^\beta}{B} \frac{|\dot{\sigma}|}{(1-D)^{\alpha_2}}$$

with

$$R_v = 1 \qquad \text{if} \quad \sigma \geqslant 0$$

$$R_v = h \frac{(1-D)^2}{(1-Dh)^2} \qquad \text{if} \quad \sigma < 0$$

By numerical integration, it is possible to show that the integration of these equations for cyclic loading gives, at least qualitatvely, the Goodman's diagram. Figure 16.

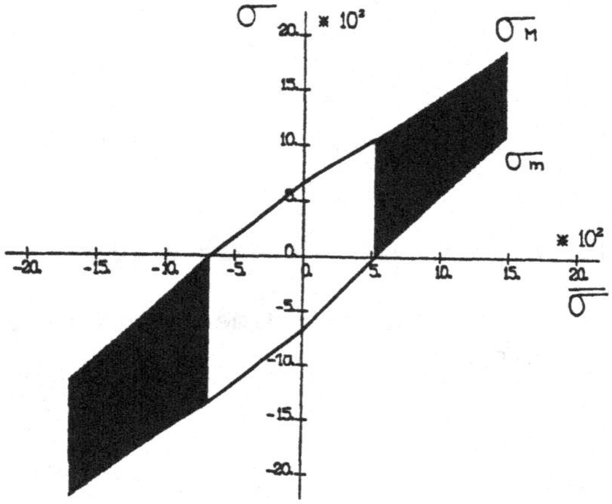

Figure 16 : Goodman's diagram

5.2.3. Identification

The identification of the coefficient α_2, as for α_1 for low cycle fatigue, needs the measurement of damage derived by means of fatigue test.

for a "0,+" periodic test α_2 is deduced from

$$D = 1 - \left[1 - \frac{N}{N_R}\right]^{\frac{1}{\alpha_2+1}}$$

α_2 is found to be constant or a function of σ_M.

The Woehler curve $N_R(\sigma_M)$ in the case of "0,+" tests represented by

$$N_R = \frac{B(\beta+3)}{\alpha_2+1} \sigma_M^{-(\beta+3)}$$

allows the parameters $(\beta+3)$, $[B(\beta+3)/(\alpha_1+1)]$ and the fatigue limit σ_l to be identified.

Then, together with the value of α_2, it is possible to obtain β and B.

Finally, the models may be summarized as follows

general model

$$\dot{D} = \frac{\sigma_{eq}^2 \left|\sigma_{eq} - \overline{\sigma}_{eq}\right|^{\beta}}{B} \frac{R_v|\dot{\sigma}_{eq}|}{(1-D)^{\alpha_2}} \quad \text{if} \quad \sigma^* \geqslant \sigma_l$$

$$\dot{D} = 0 \quad \text{if} \quad \sigma^* < \sigma_l$$

proportional loading
$\overline{\sigma}_{eq} = (1/2)\sigma_{eqM}$

$$\frac{\delta D}{\delta N} = \frac{R_v \sigma_{eqM}^{(\beta+3)}}{B(\beta+3)(1-D)^{\alpha_2}}$$

One dimensional Woehler-Miner law
$\sigma_{min} = 0$

$$N_R = \frac{B(\beta+3)}{\alpha_2+1} \sigma_M^{-(\beta+3)}$$

with

$$\sigma_{eq} = \left[\frac{3}{2}\sigma^D:\sigma^D\right]^{1/2} \qquad \overline{\sigma}_{eq} = \left[\frac{3}{2}\overline{\sigma}_{ij}^D:\overline{\sigma}_{ij}^D\right]^{1/2} \qquad R_v = \left[\frac{\sigma^*}{\sigma_{eq}}\right]^2$$

$$\sigma^* = \left[\begin{array}{l} (1+v)\langle\sigma\rangle:\langle\sigma\rangle - v\langle tr(\sigma)\rangle^2 \\[2mm] + h\frac{(1-D)^2}{(1-Dh)^2}[(1+v)\langle-\sigma\rangle:\langle-\sigma\rangle - v\langle-tr(\sigma)\rangle^2] \end{array} \right]^{1/2}$$

5.3. APPLICATIONS

The same kinds of applications can be proposed as those of low cycle fatigue :

- calculation of macro-crack initiation ;
- local approach of structure-cracks growth [BILLARDON, 1983].

6. CREEP DAMAGE

6.1. PHENOMENOLOGICAL ASPECTS

6.1.1. Physical mechanisms

Creep damage occurs in metals when they are loaded at temperatures mainly above about 1/3 of the melting temperature, when visco-plasticity induces time dependant phenomena.

Monotonic or cyclic loadings first induce the accumulation of dislocations at the joints between crystals leading to micro-cavities and then intergranular micro-cracks and micro-voids on grain boundaries or at the junction of several crystals. These micro-cavities may grow as a function of time even if the stress remains constant [ENGEL, KLINGELE, 1981].

6.1.2. Experimental measurements

The most convenient method is again to consider the coupling between creep damage and visco-plasticity using the principle of strain equivalence in the Norton's law for secondary creep in tension :

$$\dot{\varepsilon}^*_p = \left[\frac{\sigma}{\lambda^*}\right]^{N^*}$$

where λ^* and N^* are material coefficients.

If the damage is considered as negligible up to the beginning of tertiary creep (but not after), the constitutive equation for tertiary creep is

$$\dot{\varepsilon}_p = \left[\frac{\sigma}{\lambda^*(1-D)}\right]^{N^*}$$

For a monotonic creep test under a constant stress σ :

$$\frac{\dot{\varepsilon}^*_p}{\dot{\varepsilon}_p} = (1-D)^{N^*}$$

or

$$\boxed{D = 1 - \left[\frac{\dot{\varepsilon}^*_p}{\dot{\varepsilon}_p}\right]^{\frac{1}{N^*}}}$$

An example of the measurement of D is given in figure 17

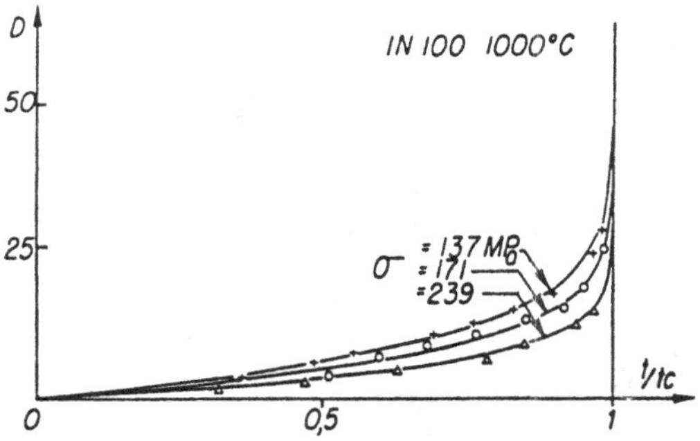

Figure 17 : Example of damage evolution in a creep test IN 100 refractory alloy, Temperature 1000°C

6.2. MODELS

6.2.1. Formulation

As for other kinds of damage, let us consider the general model :

$$\dot{D} = \frac{\overline{Y}}{S_0} \frac{\dot{p}}{(1-D)_{\alpha_0}}$$

Creep damage occurs mainly during tertiary creep where it is usualy difficult to calculate the accumulated plastic strain rate $\dot{p}$. Thus, for practical purposes, it is more convenient to develop a model in which the variables are stresses only.

It is trivial to replace $\dot{p}$ by the Odqvist's law of perfect visco-plasticity valid for secondary creep and for tertiary creep if it derives from the coupled potential of section 2.4.3. (saturated strain hardening : $Kp^{-N/M} \simeq cte = K^*$) :

$$\varphi_p^* = \frac{K^*}{N+1} \left[\frac{\sigma_{eq}}{K^*(1-D)} \right]^{N^*+1}$$

$$\dot{\varepsilon}^p = \frac{\partial \varphi_p^*}{\partial \sigma} \qquad \text{with} \qquad \dot{p} = \left[\frac{2}{3} \dot{\varepsilon}^p : \dot{\varepsilon}^p \right]^{1/2}$$

$$\dot{p} = \frac{1}{1-D} \left[\frac{\sigma_{eq}}{K^*(1-D)} \right]^{N^*}$$

then

$$\dot{D} = \frac{\sigma_{eq}^2 R_v}{2ES_0(1-D)^2} \frac{1}{(1-D)^{\alpha_0+1}} \left[\frac{\sigma_{eq}}{K^*(1-D)}\right]^{N^*}$$

or with an obvious change in notations for the materials coefficients ($r = N^*+2$, $\alpha_3 = \alpha_0+N^*+3$, $A^r = 2ES_0K^{*N^*}$)

$$\boxed{\dot{D} = \left[\frac{\sigma_{eq}}{A}\right]^r \frac{R_v}{(1-D)^{\alpha_3}}}$$

6.2.2. Kachanov's creep damage law

In the three-dimensional model, the influence of the triaxiality ratio together with the effect of quasi-unilateral conditions is given by the factor R_v. The different non-linearities in stress and damage are given by the different exponents r and α_3 while the non-linear cumulation is given by the eventual dependence of α_3 on the damage equivalent stress : $\alpha_3(\sigma^*)$.

If the model is restricted to one-dimensional tension : $R_v = 1$ and to the material condition :

$$r = \alpha_3 = cte$$

This leads to the KACHANOV's law [KACHANOV, 1958]

$$\dot{D} = \left[\frac{\sigma}{A(1-D)}\right]^r$$

6.2.3. Identification

The three coefficients A, r, α_3 can be identified from creep tests in tension (σ = cte) in which the damage is derived from the tertiary creep by :

$$D = 1 - \left[\frac{\dot{\varepsilon}_p^*}{\dot{\varepsilon}_p}\right]^{N^*}$$

From the curve $D(t)$, the coefficient α_3, eventually a function of σ, is obtained by identification with the analytical expression

D(t) integrated from the model

$$\dot{D} = \left[\frac{\sigma}{A}\right]^{r} \frac{1}{(1-D)^{\alpha_3}}$$

together with $t = 0 \longrightarrow D = 0$

$$D = 1 - \left[1 - (\alpha_3+1)\left[\frac{\sigma}{A}\right]^{r} t\right]^{\frac{1}{\alpha_3+1}}$$

The coefficients r and A are identified from the time to rupture tR as a function of stress in creep tests :

$$D = 1 \longrightarrow t = t_R$$

$$t_R = \frac{1}{\alpha+1}\left[\frac{\sigma}{A}\right]^{-r}$$

6.3. APPLICATIONS

6.3.1. Creep fatigue interaction

In the field of high temperature applications such as turbine blades of airplane engines or nuclear pressure vessels, the interaction between fatigue damage and creep damage mechanisms may significantly accelerate the fracture process.

The simple linear interaction rule of TAIRA is in general non-conservative due to the non-linearity (one more !) of the interaction

$$\Sigma \frac{t_i}{t_{ci}} + \Sigma \frac{N_j}{N_{Fj}} = 1$$

where t_i and N_i are the times and the numbers of cycles corresponding to certain stress levels while t_{ci} and N_{Fi} are the times and numbers of cycles to failure for the same stress levels considered as constants.

The non-linear interaction may be obtained by the used of the differential models. The definition of damage D based on the effective area concept permits the addition of intergranular micro-cracking associated with creep damage and transgranular micro-cracking associated with the low cycles fatigue damage :

$$D = D_{creep} + D_{fatigue}$$

$$\dot{D} = \left[\frac{\sigma_{eq}}{A}\right]^r \frac{R_v}{(1-D)^{\alpha_3}} + \frac{R_v (p-p_i)^{\gamma}}{\Gamma} \frac{\dot{p}}{(1-D)^{\alpha_1}}$$

In practical applications, σ_{eq}, R_v, p are known to be functions of time by structural calculations, and the time to rupture is obtained through an integration of that differential equation.

It may be interesting to represent the high non-linearity of the phenonenon by an example of an interaction diagram showing in one dimension of normalized number of cycles to failure $(N_R)/(N_F)$ as a function of the normalized time to rupture $(t_R)/(t_c)$ in the case of a periodic loading with a holding Δt for each cycle (figure 18) [LEMAITRE, CHABOCHE, 1974].

- N_R and t_R are the real number of cycles and time to rupture ;
- N_F and t_c are the number of cycles and time to rupture that would exist repectively for pure fatigue and pure creep

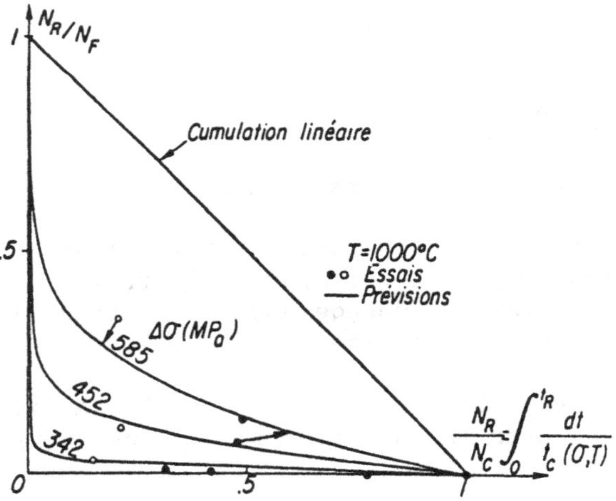

Figure 18 : Example of interaction diagram IN 100 refractory alloy

6.3.2. Creep crack initiation and growth

As mentioned several times previously, classical fracture mechanics has some difficulties in predicting the cracks behavior with plasticity, time, and damage effects. The local approach, here again, is more appropriate.

The creep damage model may be incorporated in a finite element code with or without coupling with elasto-visco-plasticity strains. Using a step by step procedure in time, it is possible to evaluate the growth of the damage zone [HAYHURST, 1975] [BENALLAL, 1984] [SAANOUNI, 1985].

REFERENCES

BENALLAL A., 1984 - Calculs couplés élasto-viscoplasticité-endommagement - Conf. GAMNI, Méthodes Numériques en Fissuration et Endommagement.

BILLARDON R., LEMAITRE J., 1983 - Numerical prediction of crack growth under mixed-mode loading by continuum damage theory - Congrès AUM, Lyon.

CAILLETAUD G., CHABOCHE J.L., 1982 - Life prediction in 304 SS by damage approach - Conf. ASME-PVP, Paper n°82-PVP-72, Orlando, 198 2, T.P. ONERA, 1982.

CHABOCHE J.L., 1982 - Lifetime predictions and cumulative damage under high-temperature conditions - Symp. on Low-Cycle Fatigue and Life Prediction, Firminy (France), 1980, ASTM-STP 770.

CHABOCHE J.L., 1982 - The concept of effective stress applied to elasticity and visco-plasticity in the presence of anisotropic damage - Mech. Behav. of Anis. Solids, Martinus Nijhoff, The Hague.

COFFIN L.F., 1954 - Study of the effects of cyclic thermal stresses in a ductile metal - Trans. ASME 76931.

CORDEBOIS J.P., 1983 - Critères d'instabilité plastique et endommagement ductile en grandes déformations - Thèse d'Etat, Université PARIS 6.

DEVAUX J.C., ROUSSELIER G., MUDRY F., PINEAU A., 1985 - An experiment program for the validation of global ductile fracture criteria using axisymmetrically cracked bars and compact tension specimens - Engng. Fract. Mech. J., Vol.21, n°2.

ENGEL-KLINGELE, 1981 - An atlas of metal damage - Wolf Science Books.

FOUGERES R., HAMEL A., SIDOROFF F., VINCENT A., 1985 - Rôle de l'anélasticité dans les variations de raideur en sollicitations cycliques - Rapport GRECO n°165.

GERMAIN P., 1973 - Cours de mécanique des milieux continus - Masson.

HAYHURST D.R., 1975 - Estimates of the creep rupture life time of structures using the finite element method - J. Mech. Phys. Solids, Vol.23.

HULT J., 1975 - Damage induced tensile instability - Trans. 3rd SMIRT, London.

JANSON J., HULT J., 1977 - Fracture mechanics and damage mechanics, a combined approach - J. Mécanique Appliquée, n°1, pp.69-84.

KACHANOV L.M., 1958 - Time of the rupture process under creep conditions - TVZ Akad. Nauk. S.S.R. Otd. Tech. Nauk., Vol.8,.

KRAJCINOVIC K., FONSEKA G.U., 1981 - Continuous damage theory of brittle materials (parts I & II) - J. Appl. Mech., Vol.48, p.809.

LAZAN J., 1960 - Energy dissipation mechanisms in structures with particular reference to material damping. Structural damping - RUZICKAJE, Pergamon Press.

LECKIE F.A., ONAT E.T., 1980 - Tensorial nature of damage measuring internal variables - Proceedings IUTAM Symp. on Physical Non-Linearties in Structural Analysis, Springer-Verlag, J. HULT, J. LEMAITRE.

LADEVEZE P., LEMAITRE J., 1984 - Damage effective stress in quasi-unilateral condition - IUTAM Congress Lyngby (Denmark).

LEMAITRE J., 1971 - Evaluation of dissipation and damage in metals, submitted to dynamic loading - Proc. I.C.M.1, Kyoto (Japan).

LEMAITRE J., 1984 - How to use damage mechanics! - Nuclear Engineering and Design Journal, Vol.80.

LEMAITRE J., 1985 - A continuous damage mechanics model for ductile fracture - Art. Trans. ASME, Journ. of Engng. Mat. and Techn., Vol.107/1.

LEMAITRE J., 1985 - Local approach of fracture - Proceedings of IUTAM Symposium on Mechanics of Damage and Fatigue, Haifa (Israël).

LEMAITRE J., BAPTISTE D., 1982 - On damage criteria - Proc. and Worshop N.S.F. on Mechanics of Damage and Fracturem Atlanta (USA,

LEMAITRE J., CHABOCHE J.L., 1974 - A nonlinear model of creep-fatigue damage cumulation and interaction - Proc. IUTAM Symp. of Mechanics of visco-elastic media and bodies, Springer-Verlag, Gothenburg.

LEMAITRE J., CHABOCHE J.L., 1985 - Mécanique des matériaux solides - Ed. Dunod, pp.532, (to be translated in english and in chineese).

LEMAITRE J., CORDEBOIS J.P., DUFAILLY J., 1979 - Sur le couplage endommagement-élasticité - Compte Rendu de l'Académie des Sciences, Paris, B.391.

MANSON S.S., 1954 - Behavior of materials under conditions of thermal stress - NACA Tech., Note 2933.

MARQUIS D., 1979 - Identification et modélisation de l'écrouissage anisotrope des métaux - Thèse de 3ème Cycle, Université PARIS 6.

Mc CLINTOCK F., 1968 - A criterion for ductile fracture by the growth of hole - ASME Journal of Applied Mechanics.

MUDRY F., 1982 - Etude de la rupture ductile et de la rupture par clivage d'aciers faiblement alliés - Thèse de l'Ecole des Mines, Université de Technologie de Compiègne.

MURAKAMI S., OHNO N., 1981 - A continuum theory of creep and creep damage - Creep in Structures, 1980, Proc.3rd IUTAM Symp. Creep in Structures, PONTER A.R.S. and HAYHURST D.R., Eds. Springer, pp.422-444, Berlin.

SAANOUNI K., CHABOCHE J.L., 1985 - Numerical aspects of a local approach for predicting creep crack growth - Colloque "Tendances actuelles en calcul des structures", Bastia (Corse).

SUQUET P., 1980 - In comp. rheologiques et structure des matériaux - C.R. 15ème Colloque GFR, Paris.

RABOTNOV Y.N., 1969 - Creep rupture - Proc. XII, Inter. Congres Appl. Mech., Stanford, Springer Berlin.

RICE J., TRACEY D., 1969 - On ductile enlargement of voids in triaxial stress fields - Journal of Mechanics Physics and Solids, Vol.17.

ANISOTROPIC ASPECTS OF MATERIAL DAMAGE AND APPLICATION OF CONTINUUM DAMAGE MECHANICS

Sumio Murakami
Department of Mechanical Engineering
Nagoya University, Chikusa-Ku Nagoya, Japan

ABSTRACT: The application of continuum mechanics to the anisotropic aspect of material damage is discussed. The microstructual change due to material damage usually depends significantly on the direction of the local stress and local strain, and is intrinsically anisotropic. Thus, the oriented nature observed in various kinds of damage and its effect on mechanical behaviour of the materials are first reviewed. Then, the modeling of the anisotropic damage states of materials in terms of mechanical variables is discussed. Definition of damage variables in terms of effective area reduction, change of elastic constants and microscopic character of cavity configuration are reviewed. Damage models based on scalar, vector, and tensor variables are presented. Finally, application of anisotropic damage theories developed by use of these variables will be discussed with special emphasis on elastic damage, elastic-plastic damage, spall damage, creep damage and the coupling of these kinds of damage.

1 INTRODUCTION

As we have learned so far, continuum damage mechanics provides a rational approach to elucidate the comprehensive phenomena of damage and fracture which are beyond the scope of fracture mechanics. These phenomena include the process of nucleation and growth of distributed microscopic cavities in materials under various loading, the passage of their coalescence to macroscopic cracks, the nucleation of microscopic voids in front of crack tips or in shear bands, the effects of material damage on the mechanical and physical properties of material, etc. The shape, orientation and configuration of these microscopic cavities, or their evolution usually depend significantly on the direction of stress and strain; the process of damage and fracture is generally characterized by their marked anisotropy. Thus, the anisotropic aspects of material damage often plays an important role in elaborating the theories and discussions made so far, especially for non-proportional loading, and in reinforcing the foundation of continuum damage mechanics as a new engineering discipline for damage and fracture.

This lecture will be focused on the application of continuum damage mechanics to the anisotropic aspects of damage observed in various engineering materials. Chapter 2 will start with a brief review of oriented nature of cavity formation in material damage, as well as the resulting anisotropic features of macroscopic properties of the materials. In Chapter 3, we will discuss the modeling of the anisotropic damage states of materials in terms of mechanical variables (internal state variables). Finally, Chapter 4 will be concerned with the development of continuum damage mechanics theories to describe the evolution of material damage under typical loading conditions and the mechanical behaviour of the damaged materials; elastic damage, elastic-plastic damage, spall damage, creep damage and the coupling of these kinds of damage will be discussed.

2 ANISOTROPIC ASPECTS OF MATERIAL DAMAGE

2.1 Elastic Damage and Elastic-Plastic Damage

The material damage due to quasi-static increase of loading is brought about by the progressive nucleation and growth of microscopic cavities in the materials as a result of its elastic or elastic-plastic deformation, and is called elastic damage (elastic-brittle damage) or elastic-plastic damage. The microscopic features of the nucleation and growth of such cavities vary significantly depending on the type of materials, their mesostructures, loading conditions, environment and other factors. In the case of metals, the dependence of the fracture mechanisms on temperature and stress has been classified in terms of maps[Ashby and Brown 1983] as shown in Fig.3.4 in Part I (Introduction and General Overview, by J.Hult)

Damage in geological materials, such as rocks and concretes, is characterized by the formation of microcracks on planes generally perpendicular to the direction of the maximum principal stress. Fig.2.1

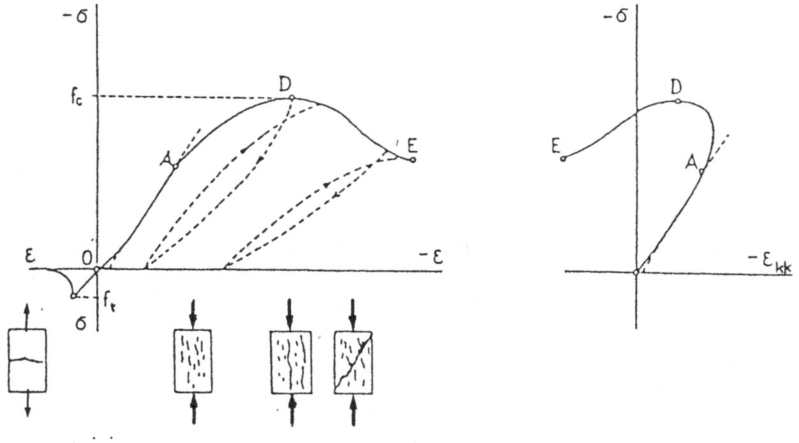

(a) stress-strain relation and damage pattern

(b) macroscopic dilatancy due to microcrack growth

Fig.2.1 Damage process of rocks and concretes under uniaxial compression and tension [From Dragon and Mroz (1979)] illustrates the damage process of uniaxial tests of rock-like materials [Dragon and Mroz 1979]. In the case of compression tests, as the stress increases beyond the discontinuity point A, microcracks appear progressively at randomly distributed points in the specimen. In the region of stable deformation AD, the microcracks develop mainly on planes parallel to the direction of maximum compressive principal stress. In the region of unstable deformation DE, on the other hand, the cracks nucleate further and coalesce into macroscopic cracks. The development of such microcracks is accompanied by a gradual decrease in the elastic stiffness as well as by the macroscopic dilatancy of the material. The final fracture in this case results in macroscopic cracks parallel to the compressions loads, or in the formation of concentrated shear zone in the material [Lemaitre and Chaboche 1985, Jaeger and Cook 1979]. In uniaxial tension tests, on the other hand, materials fracture in brittle way due to the formation of cleavage cracks on a plane perpendicular to the applied stress.

Fig.2.2 shows the schematic crack patterns observed in rocks and concretes under various states of stress under plane stress [Dragon 1976]. Though the density, configuration and the pattern of crack growth depend largely on the stress-ratio, every crack develops mainly on planes perpendicular to the direction of the maximum principal stress; the damage shows apparent anisotropy.

Microscopic mechanisms of the elastic-plastic damage in metals and alloys depend significantly on temperatures, and hence are more complicated [Murakami 1986]. The cavity formation below the cleavage fracture-fibrous fracture transition temperature is characterized either by cleavage crack formation within grains due to the intersection of slips or twins, or by the cleavage crack formation or the cracking of brittle carbide particles on grain boundaries due to the slip intersection with the boundary [Knott 1973, Tetelman and McEvily 1967].

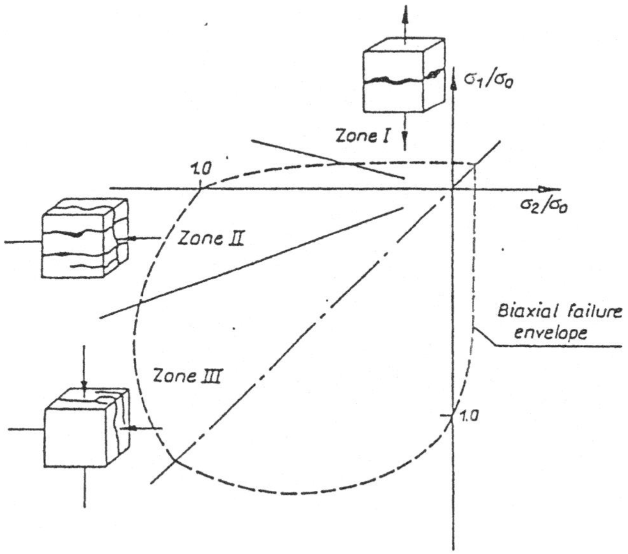

Fig.2.2 Typical patterns of brittle fracture of rocks and concretes in plane stress [From Dragon (1976)]

The damage in metals above the cleavage-fibrous transition temperature, on the other hand, is induced by the void formation within grains or on grain boundaries. The cavities within grains are attributable mainly to the decohesion of the particle-matrix interface, or to the particle cracking. Fig.2.3 illustrates the particle in carbon steel caused by tension and torsion [Gurland 1972]. It will be observed from these figures that the crack surfaces caused by particle cracking within grains are almost perpendicular to the direction of the maximum tensile stress.

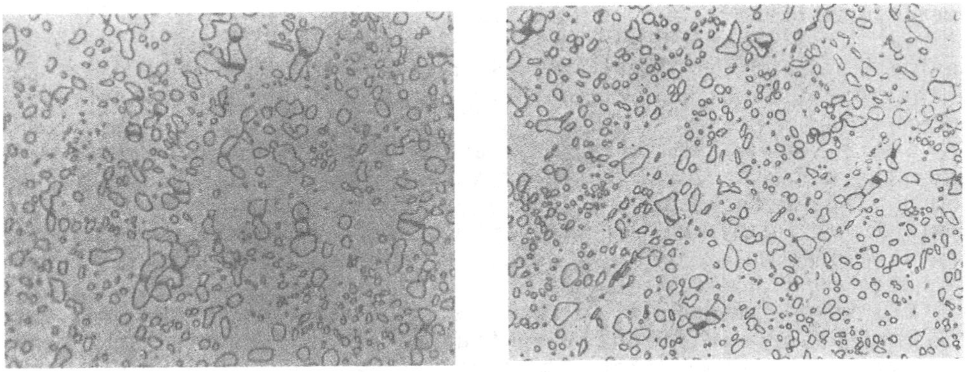

(a) Tension
(Stress direction is vertical)

(b) Torsion
(Torsional axis is horizontal)

Fig.2.3 Cracking of cementite particles in spheroidized 1.05% C steel caused by tension and torsion [From Gurland (1972)]

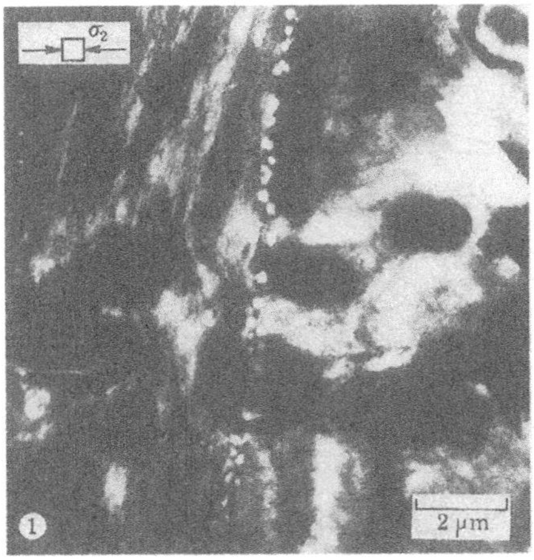

Fig.2.4 Submicrometer grain boundary cavities in Nimonic 80A subjected to compressive plastic strain of 13.7% at room temperature [From Dyson, Loveday and Rodgers (1976)]

The grain boundary cavitation brought about by the intersection of slip bands with a grain boundary or with precipitates on a grain boundary occurs also in this temperature range [Goods and Brown 1979, Knott 1973]. Fig.2.4 illustrates such grain boundary cavities in Nimonic 80A. The grain boundary cavities in this case are observed mainly on grain boundary parallel to the direction of the maximum principal stress.

The development of these cavities in metals has marked influence on various mechanical properties of metals, such as macroscopic elastic constant [Lemaitre and Chaboche 1978], plastic properties [Cordebois and Sidoroff 1982], cyclic plastic properties [Lemaitre and Chaboche 1978], fracture toughness [Janson 1977, Janson and Hult 1977], creep rates and creep rupture times [Dyson, Loveday and Rodgers 1976, Hayhurst, Trampczynski and Leckie 1980]. Fig.2.5 is an example of the effects of elastic-plastic damage on the mechanical properties of the material,and shows the torsional creep curves of Nimonic 80A subjected to prior plastic straining [Dyson, Loveday and Rodgers 1976]. As observed from this figure, the reverse plastic pretorsion has more deleterious effects on creep rates and creep ductility than the forward pretorsion. This is attributable to the anisotropic feature of the elastic-plastic damage observed in Fig.2.4.

Finally,the elastic-plastic damage in composite materials consists of various faults such as matrix cracking, delamination, fiber breakage, fiber-matrix debonding, and constitutes much more complicated mechanisms. This is in contrast to the damage of the geological materials and the metals mentioned above, where the damage under a specific loading condition develops according to a single or a few mechanisms. Fig.2.6 shows the typical crack patterns and the relation between crack density, stiffness reduction and applied stress, observed in the elastic-plastic damage in a glass/epoxy laminate[Highsmith and

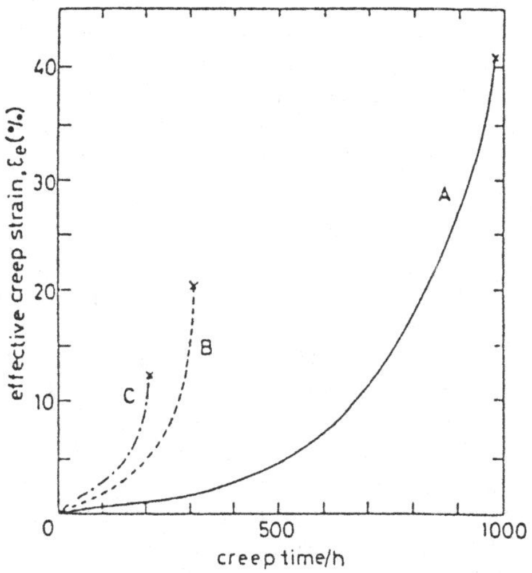

Fig.2.5 Torsional creep curves of Nimonic 80A subjected to plastic pretorsion (750°C, σ_e=234 MPa)
Curve A: Zero plastic prestrain
Curve B: 15% plastic prestrain (forward creep)
Curve C: 15% plastic prestrain (reverse creep)
[From Dyson, Loveday and Rodgers (1976)]

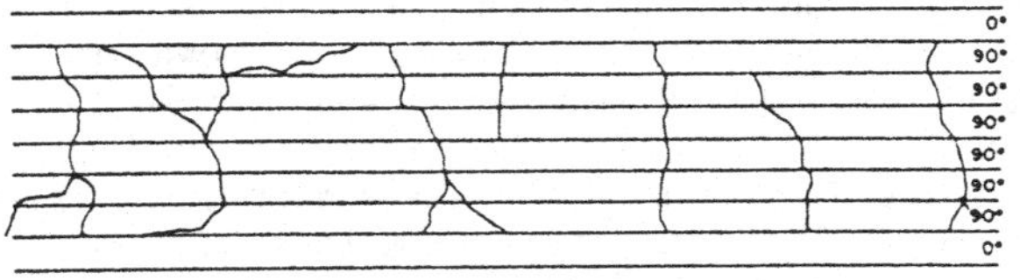

(a) Typical saturation crack patterns

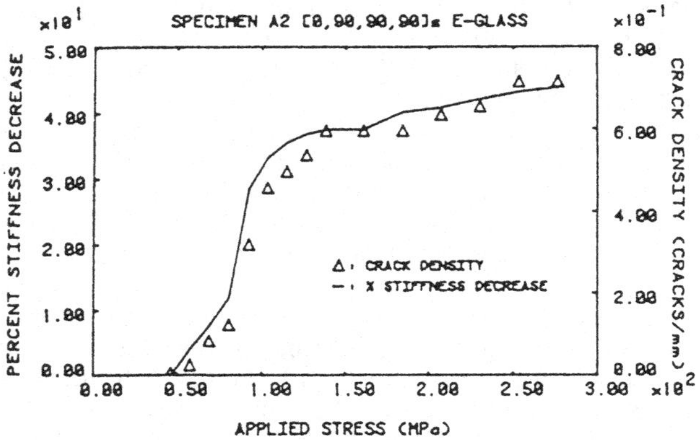

(b) Stiffness decrease and crack density versus stress level

Fig.2.6 Elastic-plastic damage in a glass/epoxy $[0,90_3]_s$ laminate [From Highsmith and Reifsnider (1982)]

Reifsnider 1982]. The stiffness decreased about 45 per cent of its initial value, and is in a close correlation with increase of the crack density. The cracks are distributed almost uniformly in six 90-deg plies. Finally, it should be mentioned that the damage patterns in laminates generally depend not only on the orientation of each lamina, but also on the stacking sequence.

2.2 Spall Damage

Elastic-plastic damage under impulsive loads show considerably different features than those due to quasi-static loading mentioned above, and brings about numerous distributed microvoids or microcracks in materials [Barbee, Seaman, Crewdson and Curran 1972, Butcher, Barker, Munson and Lundergan 1964, Davison, Stevens and Kipp 1977, Seaman, Curran and Shockey 1976]. This difference arises because, while the intensive impulsive load usually induces a large number of cavities in materials, the duration of its application is too short to permit the growth of these cavities. These impulsive loads include the projectile impact, air shock loading, explosion, or intense thermal radiation from x-ray, electron-beam and laser sources. This type of damage is often called spall damage. Fig.2.7 shows an example of spall damage which was induced in 1145 aluminium plates by impact of projectile plates at several impact velocities. The extent of the damage depends on the loading condition, and ranges from the nucleation of a few small cavities to the complete separation of the material.

(a) v = 132 m/sec

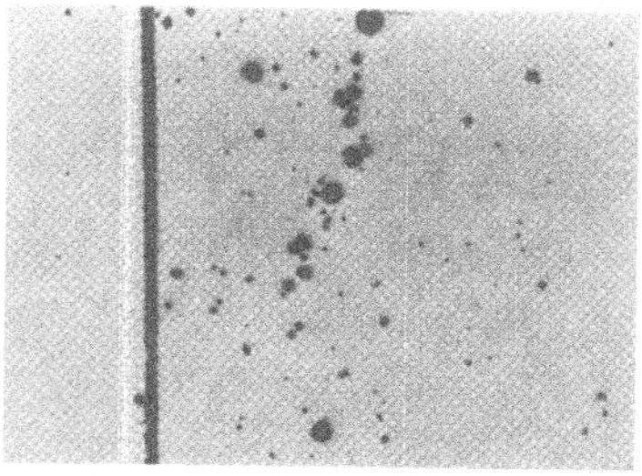

(b) v = 143 m/sec

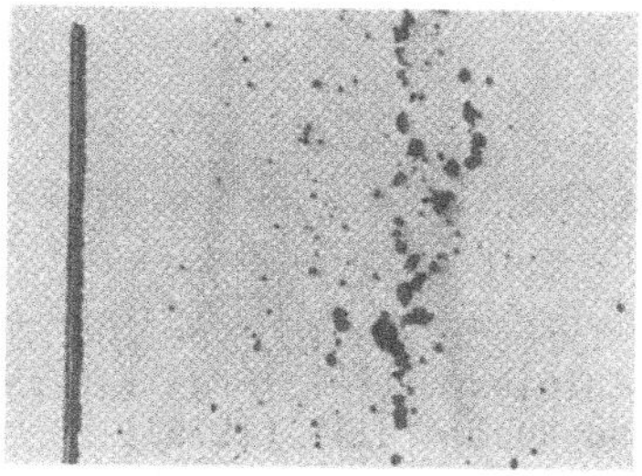

(c) v = 154 m/sec

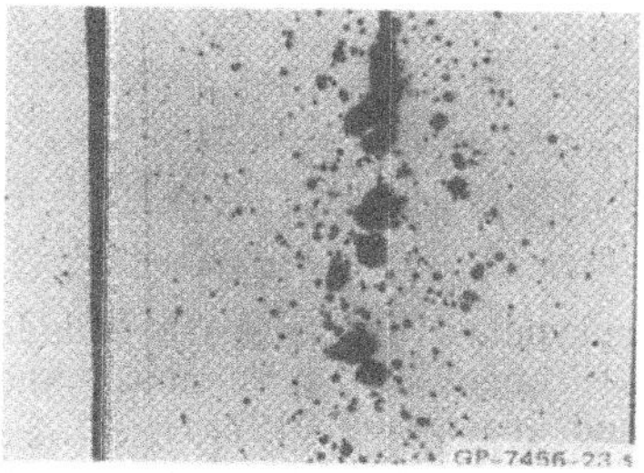

(d) v = 204 m/sec

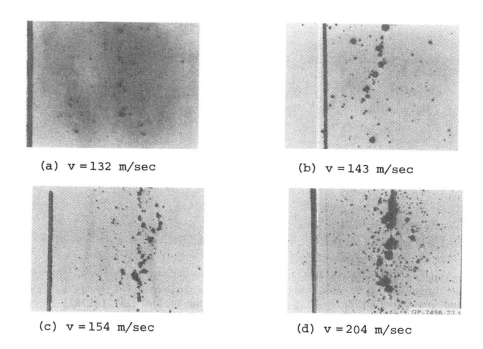

Fig.2.7 Spall damage in 1145 aluminium subjected to impact of projectile plates at various impact velocities v [From Barbee, Seaman, Crewdson and Curran (1972)]

The principal mechanisms of this damage are identical to those of the quasi-static damage, and consist of nucleation, growth and coalescence of voids or microcracks. The damage due to void formation as shown in Fig.2.7 is observed in copper, aluminium and tantalum, while iron and beryllium are damaged by the formation of microcracks. The characteristic feature of this damage is rather uniform distribution of cavities over the materials.

2.3 Creep Damage

Creep damage in polycrystalline metals is induced by the nucleation and growth of microscopic round voids (r-type voids) or wedge-shaped cracks (w-type cracks) on grain boundary [Ashby and Raj 1975, Evans,H.E. 1984, Garofalo 1965, Lagneborg 1981, Tetelman and McEvily 1967]. The coalescence of these cavities into macroscopic cracks leads to the final fracture of the material. Fig.2.8 (a),(b) shows micrographs of grain boundary voids in copper and grain boundary cracks in type 304 stainless steel.

Though w-type cracks tend to appear at lower temperature and higher stress than r-type voids, they both originate mainly on grain boundaries perpendicular to the maximum tensile stress, and the grain boundary sliding is a prerequisite for their nucleation [Evans, H.E. 1984, Lagneborg 1981]. Namely,the nucleation of r-type voids occurs as a result of stress concentration caused by grain boundary sliding at certain irregularities along grain boundaries, such as grain corners, ledges and steps due to transgranular slips, and grain boundary

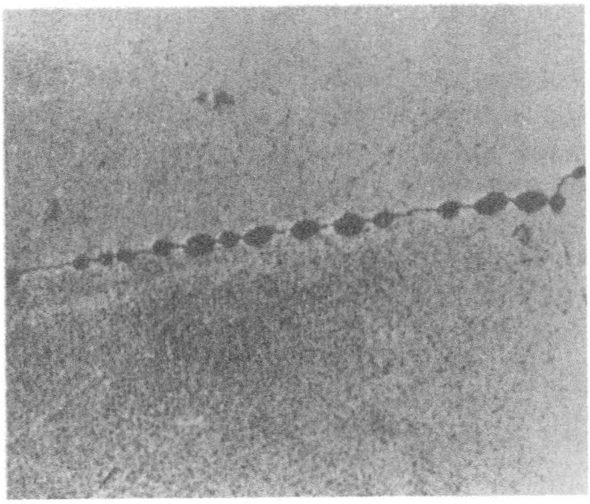

(a) r-type voids on grain boundary of copper [From Needham, Wheatley and Greenwood (1975)]

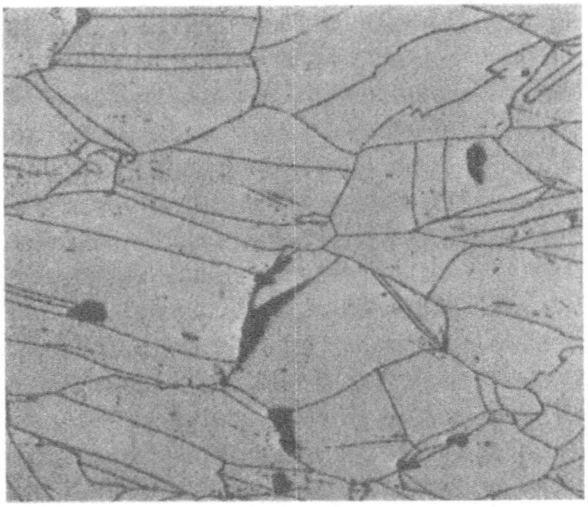

(b) w-type cracks in type 304 stainless steel [From Matera and Rustichelli (1979)]

Fig.2.8 Cavities observed in creep damage

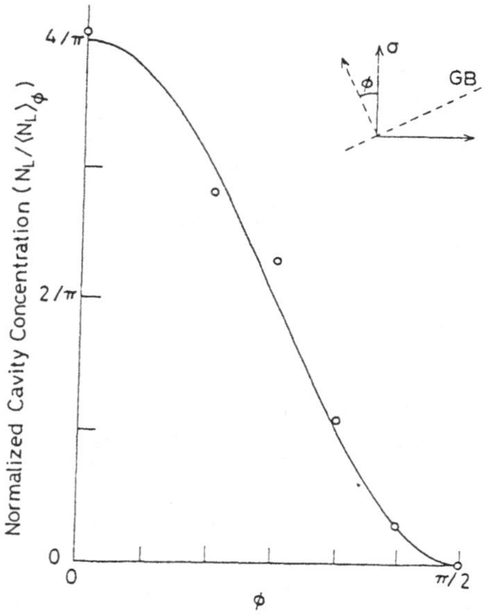

Fig.2.9 Relation between normalized cavity density and the inclination ϕ of grain boundaries in type 304 stainless steel (Stabilized at 775°C for 40 h, tested at 600°C/700°C in vacuum) [From Chen and Argon (1981)]

particles. The growth of these voids occurs by two mechanisms; the growth due to diffusion and condensation of vacancies along the grain boundaries, and the strain-controlled growth as a result of grain boundary sliding.

The w-type cracks, on the other hand, have been observed to initiate at triple points on grain boundaries, as a result of strain concentration due to grain boundary sliding or as a result of linkage of microcavities. These cracks develop mostly on grain boundaries perpendicular to the maximum tensile stress direction due to the grain boundary sliding. Fig.2.9 shows the dependence of cavity densities along grain boundaries on their orientation measured by Chen and Argon (1981) for type 304 stainless steel. This figure ascertains that the cavities develop most markedly on grain boundaries perpendicular to the tensile stress.

Creep damage of ceramic polycrystals evolves also by the nucleation, growth and coalescence of cavities at microstructural heterogeneities, mainly by diffusive mechanisms. The heterogeneities include large grain regions, inclusions and innate cracks. The initial cracks brought about by the cavity coalescence propagate first. But they then blunt usually, and new cracks nucleate continuously. The final creep fracture occurs by the coalescence of these blunted cracks. Fig.2.10 shows a scanning electron micrograph exhibiting the cavities and the crack nucleation in a ceramic polycrystal [Evans, A.G.1982]. Crack growth occurs in the direction perpendicular to the maximum tensile principal stress, and marked anisotropy of damage is again recognized.

2.4 Fatigue Damage

Fatigue damage in polycrystalline metals advances by entirely different mechanisms from those of the preceding types of damage. The process of fatigue damage of metals consists of 1) formation of intrusions or initiation of microcracks in active slip bands in

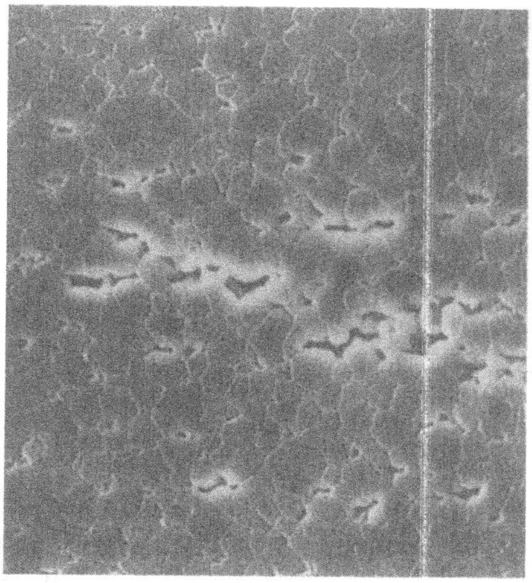

Fig.2.10 Scanning electron micrograph illustrating cavity coalescence and crack nucleation in ceramic polycrystal [From Evans, A.G. (1982)]

favourably oriented grains, as a result of irreversible dislocation glide process, 2) crack growth along slip bands which form angles of about 45° with the maximum tensile stress direction (stage I), and finally of 3) crack growth normal to the direction of maximum tensile stress (stage II) [Knott 1973, Tomkins 1981]. In the case of high cycle fatigue, because of the small amplitudes of the plastic strains, the formation of intrusions occurs only on slip planes of adequately oriented grains, and distributes sparsely in the material. A major part of the total life is

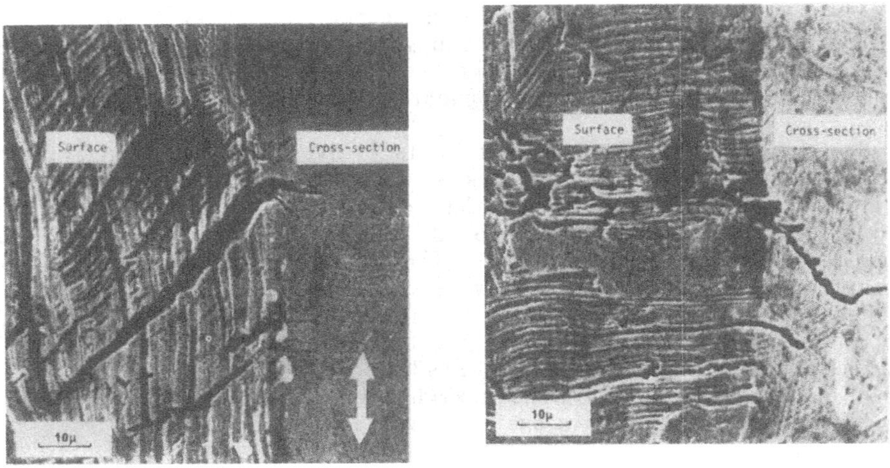

(a) Type A cracking (b) Type B cracking

Fig.2.11 Surface and in-depth growth of stage I fatigue cracks in low cycle fatigue tests of type 316L stainless steel [From Jacquelin, Hourlier and Pineau (1983)]

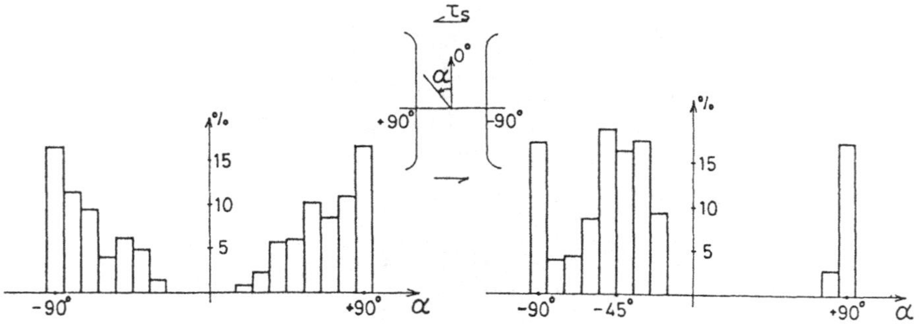

(a) Strain-controlled reversed tension ($\Delta\epsilon_P/2=0.5\%$, $\tau_s=0$)

(b) Strain-controlled reversed tension with superposed steady torque ($\Delta\epsilon_P/2=0.2\%$, $\tau_s=100$ MPa)

Fig.2.12 Distribution of crack orientation in low cycle fatigue tests of type 316L stainless steel [From Jacquelin, Hourlier and Pineau (1983)]

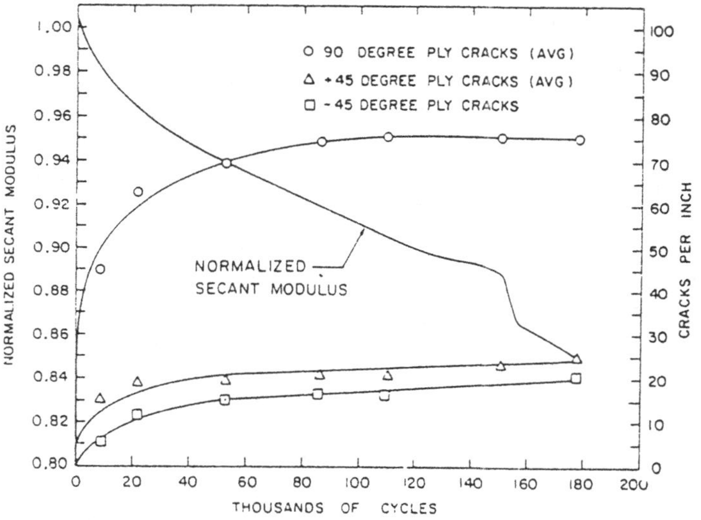

Fig.2.13 Stiffness reduction and crack development in a graphite/epoxy $[0,90,\pm 45]_s$ laminate. [From Jamison, Schulte, Reifsnider and Stinchcomb (1984)]

consumed in the formation of these incipient intrusions. In low cycle fatigue, on the other hand, since it is characterized by high level of stresses, numerous active slip bands appear in many grains, and a large number of distributed microcracks develop in the material. In this case, most of the fatigue life is occupied by the crack growth of stage II.

Fig.2.11 shows the SEM micrographs of the stage I cracks observed on type 316L stainless steel specimens damaged by low cycle fatigue under repeated tensile loads. In this case, type B cracks in the figure were observed more predominantly. However, when steady torque is superimposed to the reversed tension, type A cracks tend to appear more frequently. Fig.2.12 shows the histogram of the distribution of the crack orientation. As observed from these figures, the crack formation generally starts from the material surface, and their orientation and arrangement are closely related to the direction and the state of the applied stresses.

As regards fatigue damage of composite materials, though individual damage modes are identical to those of elastic-plastic damage mentioned in Section 2.1, several characteristic aspects to fatigue are also recognized. In tension-tension fatigue tests of graphite/epoxy laminates, for example, fiber fracture occurs early in the fatigue life and no accelerated fiber fracture rate is observed in the later stage [Jamison, Schulte, Reifsnider and Stinchcomb 1984]. The strong relationship of damage modes to one another is another generic aspect of fatigue in this material. Fig.2.13 illustrates the relation between stiffness reduction, crack density and the number of loading cycles observed in a tension-tension fatigue test (stress ratio R=0.1, maximum stress $\sigma/\sigma_{ult}=0.62$) of graphite/epoxy laminates. Much larger damage of the specimen was observed before the fracture in this case than the case of elastic-plastic damage.

2.5 Creep-Fatigue Damage

When metals and alloys are stressed repeatedly at elevated temperatures, the materials are deteriorated by the combined damage of creep and fatigue. In view of the difference in the mechanisms of these

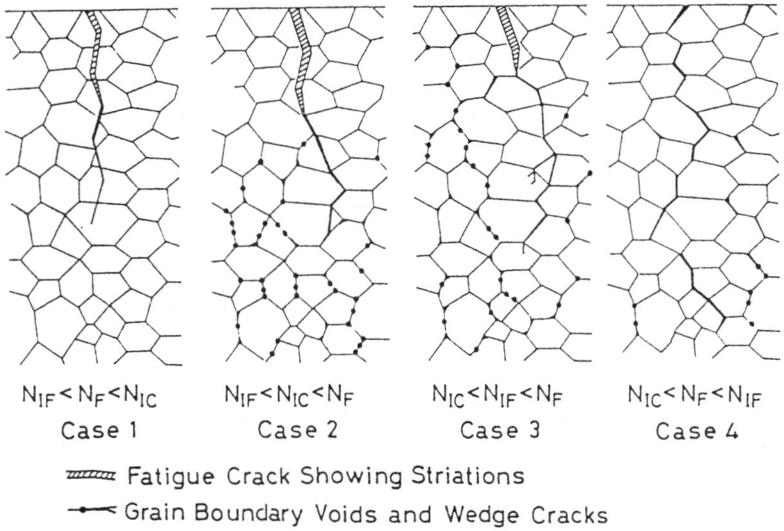

Fig.2.14 Schematic representation of four possible modes of creep-fatigue damage [From Goodall, Hales and Walters (1981)]

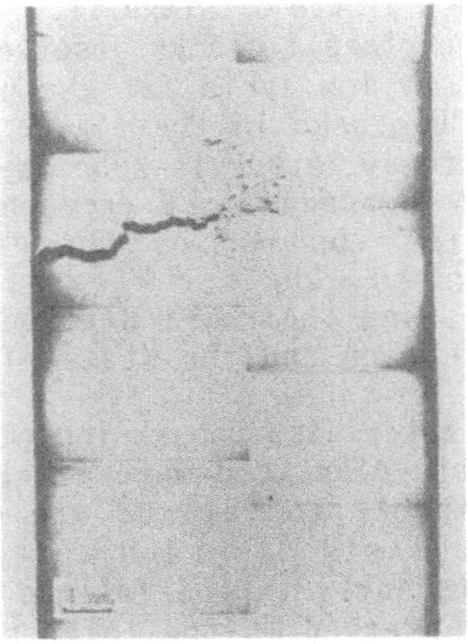

(a) 1.2 % total strain range 500 cycles with 10 min hold time, then 100 cycles with 5 h hold time, and again 55 cycles with 10 min hold time

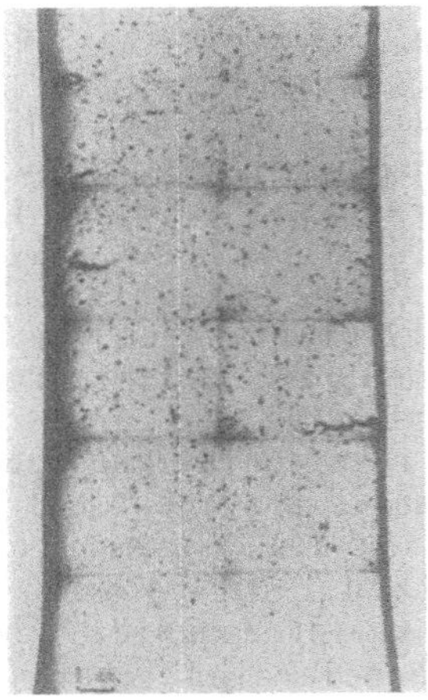

(b) 1.2 % total strain range, 373 cycles with 24 h hold time

Fig.2.15 Creep fatigue damage in type 316L stainless steel at 600°C [From Cailletaud and Levaillant (1984)]

two types of damage mentioned above, two effects of these damages are essentially separable. Then, the extent of the interaction of creep and fatigue damage depends on the loading conditions, namely on the relative effects of each load cycle exerted on these two kinds of damage. Fig.2.14 shows the schematic presentation of four possible modes of creep-fatigue failure, where N_F, N_{IF} and N_{IC} denote the number of cycles to failure, those to the initiation of fatigue cracks and those to the initiation of creep cavities.

Because of the difference in damage mechanisms illustrated in Fig.2.14, the coupling of these damages does not appear in early stage of the damage in general. As regards the later stage of the damage, it has been observed that, while the creep damage enhances the fatigue damage generally, the effect of the fatigue damage on the creep damage is less significant than the reverse case. Fig.2.15 (a) and (b) show the micrographs of the creep fatigue damage of type 316L stainless steel at 600°C, and corresponding to the cases 2 and 3 of Fig.2.14, respectively. It will be observed that the existence of a fatigue crack in Fig.2.15 (a) accelerates the intergranular creep cavities ahead of the crack tip because of enhanced stress level in this region, while the surface cracks in Fig.2.15 (b) have induced no noticeable concentration of creep damage.

2.6 References

Ashby, M.F. and L.M. Brown: Perspectives in Creep Fracture, Pergamon, Oxford (1983).

Ashby, M.F. and R. Raj: Creep Fracture, in: The Mechanics and Physics of Fracture, Metals Society and Institute of Physics, The Metals Society, London 1975, 148-158.

Barbee, T.W., Jr.,L. Seaman, R. Crewdson and D. Curran: Dynamic Fracture Criteria for Ductile and Brittle Metals, J. of Materials, 7 (1972), 393-401.

Butcher, B.M., L.M. Barker, D.E. Munson and C.D. Lundergan: Influence of Stress History on Time-Dependent Spall in Metals", AIAA J., 2 (1964), 977-990.

Cailletaud, G. and C. Levaillant: Creep-Fatigue Life Prediction: What about Initiation?, Nucl. Engng Design, 83 (1984), 279-292.

Chen, I.-W. and A.S. Argon: Creep Cavitation in 304 Stainless Steel, Acta Metallurgica, 29 (1981), 1321-1333.

Cordebois, J.P.and F. Sidoroff: Damage Induced Elastic Anisotropy, in: Mechanical Behavior of Anisotropic Solids (Ed. J.-P. Boehler), Martinus Nijhoff, The Hague 1982, 761-774.

Davison, L., A.L. Stevens and M.E. Kipp: Theory of Spall Damage Accumulation in Ductile Metals, J. Mech. Phys. Solids, 25 (1977), 11-28.

Dragon, A.: On Phenomenological Description of Rock-Like Materials with Account of Kinetics of Brittle Fracture, Archives of Mechanics, 28 (1976), 13-30.

Dragon, A. and Z. Mroz: A Continuum Model for Plastic-Brittle Behaviour of Rock and Concrete, Int. J. Engng Sci., 17 (1979), 121-137.

Dyson, B.F., M.S. Loveday and M.J. Rodgers: Grain Boundary Cavitation Under Various States of Applied Stress, Proc. Roy. Soc. London A, 349 (1976), 245-259.

Evans, A.G.: High Temperature Failure of Ceramics, In: Recent Advances in Creep and Fracture of Engineering Materials and Structures (Ed. B. Wilshire and D.R.J. Owen) Pineridge, Swansea 1982, 53-133.

Evans, H.E.: Mechanisms of Creep Fracture, Elsevier Applied Science,

London (1984).
Garofalo, F.: Fundamentals of Creep and Creep-Rupture in Metals, Macmillan, New York (1965).
Goods, S.H, and L.M. Brown: The Nucleation of Cavity by Plastic Deformation, Acta Metallurgica 27 (1979), 1-15.
Gurland, J.: Observations on the Fracture of Cementite Particles in a Spheroidized 1.05%C Steel Deformed at Room Temperature, Acta Metallurgica, 20 (1972), 735-741.
Hayhurst, D.R., W.A. Trampczynski and F.A. Leckie: Creep-Rupture Under Non-Proportional Loading, Acta Metallurgica, 28 (1980), 1171-1183.
Highsmith, A.L. and K.L. Reifsnider: Stiffness-Reduction Mechanisms in Composite Laminates, in: Damage in Composite Materials, ASTM STP 775, American Society for Testing and Materials (Ed. K.L. Reifsnider), New York 1982, 103-117.
Jacquelin, B., F. Hourlier and A. Pineau: Crack Initiation Under Low Cycle Multiaxial Fatigue in Type 316L Stainless Steel, ASME J. Pressure Vessel Technology, 105 (1983), 138-143.
Jaeger, J.C. and N.G.W. Cook: Fundamentals of Rock Mechanics, Third edition, Chapman and Hall, London 1979.
Jamison, R.D., K. Schulte, K.L. Reifsnider and W.W. Stinchcomb: Characterization and Analysis of Damage Mechanisms in Tension-Torsion Fatigue of Graphite/Epoxy Laminates, in: Effects of Defects in Composite Materials, ASTM STP 836, American Society for Testing and Materials, New York 1984, 21-45.
Janson, J.: Dugdale-Crack in a Material with Continuous Damage Formation, Engng Fracture Mech., 9 (1977), 891-899.
Janson J. and J. Hult: Fracture Mechanics and Damage Mechanics, A Combined Approach, J. Mec. Appl., 1 (1977), 69-84.
Knott, J.F. Fundamentals of Fracture Mechanics, Butterworths, London (1973).
Lagneborg, R.: Creep: Mechanisms and Theories, in: Creep and Fatigue in High Temperature Alloys (Ed. J. Bressers), Elsevier Applied Science, London 1981, 41-71.
Lemaitre, J. and J.L. Chaboche: Aspect Phenomenologique de la Rupture par Endommagement, J. Mec. Appl., 2 (1978), 317-365.
Lemaitre, J. and J.L. Chaboche: Mecanique des Materiaux Solides, Dunod, Paris 1985.
Matera, R. and F. Rustichelli: The Evalution of Creep Damage, in: Creep of Engineering Materials and Structures(Ed. G. Bernasconi and G. Piatti), Applied Science, London 1979, 389-412.
Murakami, S.: Anisotropic Damage in Metals, in: Failure Criteria of Structured Media(Ed. J. -P. Boehler), A.A. Balkema, Rotterdam 1986.
Needham, N.G., J.E. Wheatley and G.W. Greenwood: The Creep Fracture of Copper and Magnesium, Acta Metallurgica, 23 (1975), 23-27.
Seaman,L., D.R. Curran and D.A. Shockey: Computational Models for Ductile and Brittle Fracture, J. Appl. Phys., 47 (1976), 4814-4826.
Tetelman, A.S. and A.J. McEvily, Jr.: Fracture of Structural Materials, John Wiley, New York 1967.
Tomkins, B.: Fatigue: Mechanisms, in: Creep and Fracture in High Temperature Alloys(Ed. J. Bressers), Elsevier Applied Science, London 1981, 111-143.

3 MODELING OF ANISOTROPIC DAMAGE AND DAMAGE VARIABLES

3.1 Damage Measures to Define and to Quantify Material Damage

As observed in the discussion of the previous Section, material damage is characterized by the development of distributed microscopic cavities (more generally, by the microscopic internal structural change) in materials leading to the deterioration of their mechanical properties. Two problems will arise when we describe such damage states in terms of certain mechanical variables. The first problem is how to define the damage variables; i.e., what kind of physical and mathematical (tensorial) nature these mechanical variables should have in order to represent such damage state property. The second problem, on the other hand, is how to quantify the magnitude of these variables. These two problems are usually discussed on the basis of various damage measures shown in Table 3.1. These mechanical variables should have not only definite physical meaning, but also should have proper physical and mathematical properties.

Table 3.2 shows the examples of damage variables so far proposed for anisotropic damage of materials. Though most of these variables have been defined on the basis of the microscopic measures in Table 3.1, i.e., on the basis of the configuration of microscopic cavities, the quantification of them should reply more or less on some macroscopic phenomena influenced by the existence of cavities, i.e., the macroscopic measures of Table 3.1.

In the next Section, we will start with the definition of damage variables in terms of measure 3) of Table 3.1, i.e., on the basis of the reduction of net area due to cavity formation.

3.2 Definition of Damage Variables in Terms of Effective Area Reduction

3.2.1 Modeling by scalar and vector variables

In Section 1-2.2*, it was shown that the damage variables ψ or D can be interpreted as the fraction of net area reduction caused by the development of microscopic cavities. In order to extend the isotropic damage theory of Section 1-2.2 to the case of anisotropic damage, Kachanov (1974) introduced three scalar variables ψ_1, ψ_2 and ψ_3 characterizing the net area fractions of planes normal to the direction of the principal stresses σ_1, σ_2 and σ_3. Then, the effective stresses defined in Section 1-2.2 can be written as

$$S_1 = \sigma_1/\psi_1,\ S_2 = \sigma_2/\psi_2,\ S_3 = \sigma_3/\psi_3 \tag{3.1}$$

Though the notion of this theory is clear and resonable, expression (3.1) is valid only in the cases of coincident principal directions of damage and stress; it can not be applied to the problems of rotating principal stress directions. A simpler theory has been proposed also by Rabotnov [1969] and Leckie and his coworkers [Hayhurst and Leckie 1973, Martin and Leckie 1972]. By assuming that, among the principal stresses σ_1, σ_2 and σ_3, only the maximum principal stress σ_1 is magnified by the effective

* Hereafter, Section 1-2.2 and equation (1-2.1), for example, represent Section 2.2 and equation (2.1) in Part 1 (Introduction and General Overview, by J. Hult).

Table 3.1 Damage measures to define and quantify material damage

	Physical quantities employed as measures
Microscopic measures	1) Number, length, area and volume of cavities 2) Geometry, arrangement and orientation of cavities 3) Net area reduction specified by geometry, arrangement and orientation of cavities
Macroscopic measures	4) Elastic constant, creep rate, stress or strain amplitude 5) Yield stress, tensile strength 6) Endurance limit, creep rupture time 7) Elongation 8) Mass density 9) Electric resistance 10) Velocity of ultrasonic waves 11) Acoustic emission

Table 3.2 Damage variables and their tensorial nature

Damage variables	Reference	Material damage
Scalar	Kachanov (1958)	Creep
	Kachanov (1974)	Creep (anisotropic)
	Rabotnov (1969)	Creep (anisotropic)
	Martin-Leckie (1972), Hayhurst-Leckie (1973)	Creep (anisotropic)
	Davison et al. (1977)	Spalling
	Gurson (1977)	Elastic-plastic
Vector	Kachanov (1974)	Creep
	Hayhurst-Storakers (1976)	Creep
	Davison-Stevens (1973)	Spalling
	Krajcinovic-Fonseka (1981)	Elastic-brittle
	Krajcinovic et al. (1983, 1984)	General, creep
Second rank tensor	Rabotnov (1968)	Creep
	Vakulenko-Kachanov (1971)	Elastic-brittle
	Murakami-Ohno (1978, 1981)	Creep
	M. Kachanov (1980)	Elastic
	Çordebois-Sidoroff (1982a, b)	Elastic, elastic-plastic
	Betten (1983)	General, creep
	Murakami et.al. (1986)	General
Higher rank tensor (fourth rank)	Chaboche (1982)	Creep
(a set of tensors of even rank)	Leckie-Onat (1981)	Creep

area reduction caused by cavities, they defined the following effective stresses in terms of a single damage variable D:

$$S_1 = \sigma_1/(1-D),\ S_2 = \sigma_2,\ S_3 = \sigma_3, \tag{3.2}$$

Thus, the effects of damage anisotropy is represented as the difference in the magnification of principal stresses. This theory is subjected to the limitations similar to those of equation (3.1).

Description of the damage states by means of vector variables also has been attempted [Kachanov 1974]. Let us take an area element of an arbitrary direction $\underset{\sim}{\nu}$. Then, this area element is subjected to normal stress σ_ν, and the damage advances thereby. By characterizing the damage state on this area element by a vector $\underset{\sim}{\psi}_\nu$ of the magnitude ψ_ν and in the direction $\underset{\sim}{\nu}$, Kachanov defined an effective stress

$$S_\nu = \sigma_\nu / \psi_\nu \tag{3.3}$$

Kachanov applied this theory to analyze the creep damage process of materials subjected to non-proportional loading. However, in order to describe completely the damage state at a point in the material, this theory requires an entire distribution of the vector $\underset{\sim}{\psi}$ in all directions through that point, and hence it is rather difficult to discuss a general state of damage by this theory. Another difficulty of vector damage variables is that the combined effect of plane cracks in different orientations can not be represented by a simple addition of the damage vectors corresponding to the cracks of each orientation. In Section 4.3, predictions of equation (3.3) will be compared with those of a tensor damage theory.

3.2.2 Modeling by a second rank tensor variable

The above arguments suggests the limitation of scalar and vector damage variables in describing the anisotropic damage of materials. Now, let us discuss the extension of the notion of equations (I-2.6) and (I-2.8) to the general case of three-dimensional anisotropic damage.

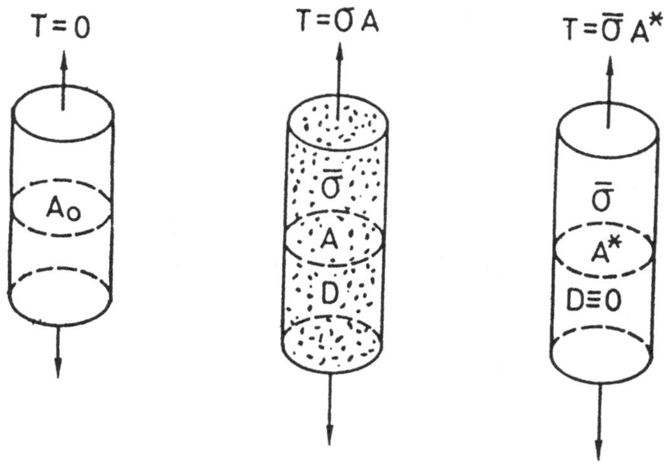

(a) Initial undamaged state (b) Current damaged state (c) Fictitious undamaged state

Fig.3.1 Uniaxial tensile bar

Let us take a tensile bar of Fig.3.1, and denote by A the cross sectional area of the bar at the actual (current) configuration under tensile stress σ [Fig.3.1(b)]. Then, A_n (which will be denoted by A^* hereafter) defined by equation (I-2.8) may be interpreted as the net load carrying area decreased by damage in an equivalent fictitious undamaged material [Fig.3.1(c)]. Thus, the damage variable D can be defined, once the relation between the areas A and A^* of the material in these two configurations can be specified.

In order to apply this idea to the general states of damage, we will first take an area element PQR of an arbitrary orientation in the damaged material as shown in Fig.3.2, and call it as the current (actual) damaged configuration B_t. Then, let us represent the line elements PQ, PR and the area of PQR by the vectors $d\underset{\sim}{x}$, $d\underset{\sim}{y}$ and $\underset{\sim}{\nu}\,dA$. Because of the three-dimensional distribution of microcavities, the load carrying net area of PQR is decreased; we postulate that its net effect is equivalent to an diminished area element $P^*Q^*R^*$ in the fictitious undamaged material, which will be called the <u>fictitious undamaged configuration</u> B_f. The line elements P^*Q^*, P^*R^* and the area of $P^*Q^*R^*$ will be denoted by the vectors $d\underset{\sim}{x}^*$, $d\underset{\sim}{y}^*$ and $\underset{\sim}{\nu}^*dA^*$. Since the net area reduction due to damage results not only in the plane PQR but also in the planes of other orientations, the directions of the vectors $\underset{\sim}{\nu}\,dA$ and $\underset{\sim}{\nu}^*dA^*$ do not always coincide with each other.

The decrease in the net area from PQR to $P^*Q^*R^*$ induced by the cavity distribution can be specified, if we introduce a <u>fictitious deformation</u> at the point P from the current configuration B_t of the damaged material to the corresponding fictitious undamaged configuration B_f. If we represent the fictitious deformation gradient from B_t to B_f by $\underset{\sim}{G}$ the segments $d\underset{\sim}{x}^*$ and $d\underset{\sim}{y}^*$ in B_f are given as follows:

$$d\underset{\sim}{x}^* = \underset{\sim}{G}\,d\underset{\sim}{x}\,, \quad d\underset{\sim}{y}^* = \underset{\sim}{G}\,d\underset{\sim}{y} \tag{3.4}$$

By use of Nanson's theorem, the area vector $\underset{\sim}{\nu}^*dA^*$ in B_f can be related to the vector $\underset{\sim}{\nu}\,dA$ in B_t;

$$\begin{aligned}\underset{\sim}{\nu}^*dA^* &= (1/2)d\underset{\sim}{x}^* \otimes d\underset{\sim}{y}^* = (1/2)(\underset{\sim}{G}d\underset{\sim}{x})\otimes(\underset{\sim}{G}d\underset{\sim}{y})\\ &= K(\underset{\sim}{G}^{-1})^T(\underset{\sim}{\nu}\,dA)\,, \qquad K = \det \underset{\sim}{G}\end{aligned} \tag{3.5}$$

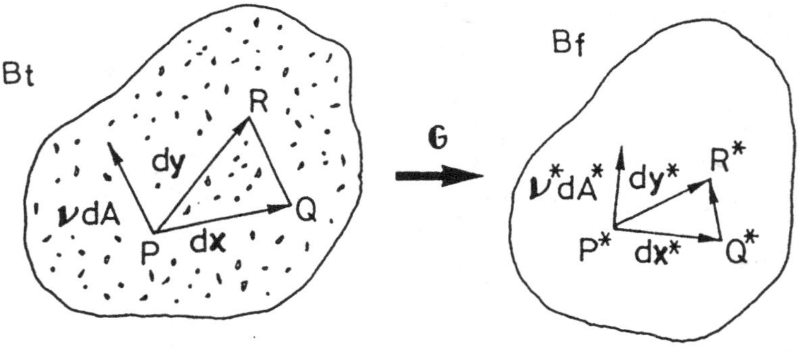

(a) Current (real) damaged configuration (b) Fictitious (reference) undamaged configuration

Fig.3.2 Definition of damage state in terms of fictitious undamaged configuration where ($)^T$ denotes the transpose of the second rank tensors.

According to the above argument, the damage states can be specified by the linear transformation $K(\underset{\sim}{G}^{-1})^T$ of equation (3.5). If we represent $K(\underset{\sim}{G}^{-1})^T$ by a new tensor

$$\underset{\sim}{I} - \underset{\sim}{D} = K(\underset{\sim}{G}^{-1})^T \tag{3.6a}$$

or

$$\underset{\sim}{G} = K[(\underset{\sim}{I}-\underset{\sim}{D})^T]^{-1} = K[(\underset{\sim}{I}-\underset{\sim}{D})^{-1}]^T$$

$$K = [\det(\underset{\sim}{I}-\underset{\sim}{D})]^{1/2} \tag{3.6b}$$

equation (3.5) has the form

$$\underset{\sim}{\nu}^* dA^* = (\underset{\sim}{I}-\underset{\sim}{D})\ (\underset{\sim}{\nu}\, dA) \tag{3.7}$$

where $\underset{\sim}{I}$ is the unit tensor of rank two. The second rank tensor $\underset{\sim}{D}$ of equation (3.7) can be an internal state variable which represents the anisotropic damage states of material, and will be called <u>damage tensor</u> hereafter.

By examining the nature of the tensor $(\underset{\sim}{I}-\underset{\sim}{D})$ of equation (3.7), we can assume the symmetry of the tensor $\underset{\sim}{D}$. Thus, the tensor $\underset{\sim}{D}$ has always three orthogonal principal directions $\underset{\sim}{n}_i$ (i=1,2,3) and the corresponding real principal values D_i, and can be expressed in the canonical form

$$\underset{\sim}{D} = \Sigma\, D_i \underset{\sim}{n}_i \otimes \underset{\sim}{n}_i \tag{3.8}$$

Let us take the principal coordinates $0\text{-}x_1x_2x_3$ and $0^*\text{-}x_1x_2x_3$ which pass through the points P,Q,R and P*,Q*,R*, and form small tetrahedrons OPQR and O*P*Q*R* by the area elements PQR, P*Q*R* and the coordinate planes, as shown in Fig.3.3. Substitution of equation (3.8) into equation (3.4) furnishes

$$\begin{aligned} \underset{\sim}{\nu}^* dA^* &= (\underset{\sim}{I} - \Sigma\, D_i \underset{\sim}{n}_i \otimes \underset{\sim}{n}_i)\ (\underset{\sim}{\nu}\, dA) \\ &= \Sigma (1 - D_i) dA_i \underset{\sim}{n}_i \\ &= \underset{\sim}{n}_1 dA_1^* + \underset{\sim}{n}_2 dA_2^* + \underset{\sim}{n}_3 dA_3^* \end{aligned} \tag{3.9a}$$

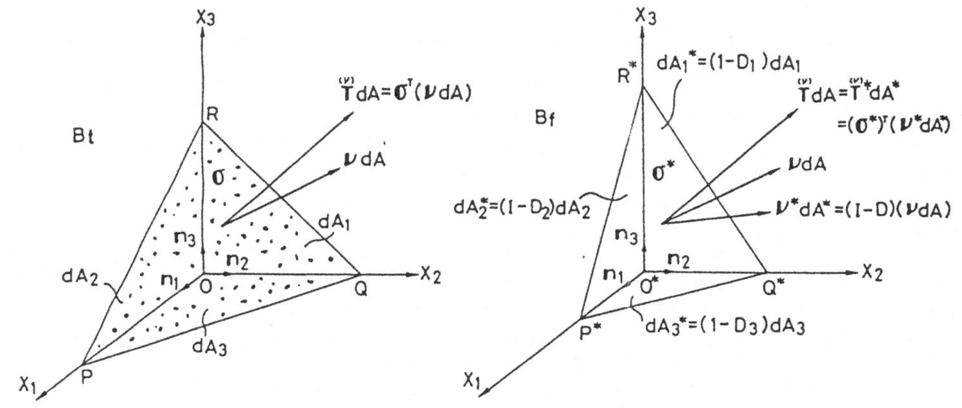

(a) Current damaged configuration (b)Fictitious undamaged configuration

Fig.3.3 Tetrahedrons in the current damaged configuration and in the fictitious undamaged configuration

$$dA_i^* = (1-D_i)dA_i \ (i : \text{no sum}), \ (i=1,2,3) \tag{3.9b}$$

where $dA_i = \nu_i dA$ and $dA_i^* = \nu_i^* dA^*$ (i=1,2,3) signify the facet areas of the tetrahedrons in the coordinate planes. According to equation (3.9) the principal values D_i can be interpreted as the cavity area densities on three principal planes of $\underset{\sim}{D}$, as shown in Fig.3.3.

According to the notion of the effective stress expressed in equation (I-2.8), the effective stress can be defined as the magnified stress σ^* which is induced in the fictitious configuration B_f when the diminished load carrying area $\underset{\sim}{\nu}^* dA^*$ is subjected to the same force vector $T^{(\nu)} dA$ as that of the current configuration B_t (Figure 3.3). Thus this stress corresponds exactly to the Lagrange stress tensor $\underset{\sim}{s}$ with respect to the reference configuration B_f:

$$\underset{\sim}{\sigma}^* = \underset{\sim}{s} = K^{-1} \underset{\sim}{G} \underset{\sim}{\sigma} \tag{3.10a}$$

Substitution of equation (3.6) into this equation gives

$$\underset{\sim}{\sigma}^* = (K^{-1})K[(\underset{\sim}{I}-\underset{\sim}{D})^{-1}]^T \underset{\sim}{\sigma} = [(\underset{\sim}{I}-\underset{\sim}{D})^{-1}] \underset{\sim}{\sigma} \tag{3.10b}$$

The first Piola-Kirchhoff stress tensor $\underset{\sim}{\tau}$ with respect to the reference configuration B_f, on the other hand, is given as follows

$$\underset{\sim}{\tau} = \underset{\sim}{s}^T = \underset{\sim}{\sigma}(\underset{\sim}{I}-\underset{\sim}{D})^{-1} \tag{3.11}$$

The effective stress tensor $\underset{\sim}{\sigma}^*$ and $\underset{\sim}{\tau}$ of equations (3.10),(3.11) are both non-symmetric. Since it is difficult to formulate the evolution equation of damage and the constitutive equations of the damaged material by use of non-symmetric tensors, we need proper symmetrization of these effective stresses. One possible procedure for this purpose is to take the symmetric part of the Cartesian decomposition of $\underset{\sim}{\sigma}^*$ and $\underset{\sim}{\tau}$ of equations (3.10) and (3.11):

$$\underset{\sim}{S} = (1/2)[\underset{\sim}{\sigma}^* + \underset{\sim}{\tau}] = (1/2)[(\underset{\sim}{I}-\underset{\sim}{D})^{-1} \underset{\sim}{\sigma} + \underset{\sim}{\sigma}(\underset{\sim}{I}-\underset{\sim}{D})^{-1}] \tag{3.12}$$

In the particular case when the principal directions of the stress tensor $\underset{\sim}{\sigma}$ and those of the damage tensor $\underset{\sim}{D}$ coincide with each other, we have component forms of equations (3.9) and (3.12) with respect to these principal coordinates

$$dA_1^* = (1-D_1)dA_1, \ dA_2^* = (1-D_2)dA_2, \ dA_3^* = (1-D_3)dA_3 \tag{3.13}$$

$$S_1 = \sigma_1/(1-D_1), \quad S_2 = \sigma_2/(1-D_2), \quad S_3 = \sigma_3/(1-D_3) \tag{3.14}$$

which are in analogous forms to equations (I-2.8) and (I-2.6). Hence, equations (3.7) and (3.12) are the three dimensional version of the effective area and the effective stress in the classical Kachanov-Rabotnov theory. Equation (3.14) is identical to equation (3.1) proposed by Kachanov.

The deformation of the damaged materials depends not only on the effective area reduction caused by cavities but also on their three-dimensional arrangement. Thus, strictly speaking, the effective stress tensor $\underset{\sim}{S}$ of equation (3.12) needs some modifications for the constitutive equations for deformation [Murakami 1986].

Finally, it should be noted that the damage tensor $\underset{\sim}{D}$ of equation (3.8) can not represent the damage states with more complicated symmetry than orthotropy. Nevertheless, as far as the damage states are identified by the net area reduction due to cavity distributions, equation (3.8) provides a general expression of the damage states. Actually M. Kachanov (1980), for example, disoussed the anisotropy of the effective elastic properties of solids with distributed cracks by representing their effects in terms of a crack density tensor of the form of equation (3.8), and confirmed that wide variety of crack arrays not only of small concentration but also of finite concentration can be modeled by equation (3.8) with sufficient accuracy (see also Section 4.1).

3.3 Other Definitions of Anisotropic Damage Variables

3.3.1 Definition of damage variables in terms of elastic constants and other quantities

Material damage usually induces the stiffness reduction of the material. The modulus of elasticity, for example, decreases in the course of the material damage, as shown in Fig.1-5.1. Thus, the damage state can be characterized also by the change of elastic constant, i.e., the measure (4) of Table 3.1. [Lemaitre and Chaboche 1978].

Let us start with the one-dimensional case of Fig.3.4. If we apply the notion of equation (1-5.3) to the elastic deformation of the damaged material with the elastic strain ϵ under stress σ, we have the relation

$$\epsilon = \sigma/\bar{E} = S/E = \sigma/[E(1-D)] \tag{3.15}$$

$$D = 1-(\bar{E}/E), \quad S = (E/\bar{E})\ \sigma \tag{3.16}$$

where E(D) and E are the moduli of elasticity of the damaged and the undamage material, respectively, and S is the effective stress. Equation (3.16) is an alternative definition of the damage variable D.

This notion can be easily extended to the three dimensional case [Chaboche 1982]. If the elasticity tensors of rank four of the damaged and the undamaged materials are denoted by $\bar{\underset{\sim}{E}}(\underset{\sim}{D})$ and $\underset{\sim}{E}$, respectively, equation (3.16) can be written as

$$\underset{\sim}{S} = [\underset{\sim}{E}:\bar{\underset{\sim}{E}}(\underset{\sim}{D})^{-1}]:\underset{\sim}{\sigma} \tag{3.17}$$

where (:) is the contraction of rank two, and $\underset{\sim}{D}$ is a certain damage tensor specifying the damage state of the material. For the definition of the strain energy function for the damaged material, the tensor $\bar{\underset{\sim}{E}}$ is

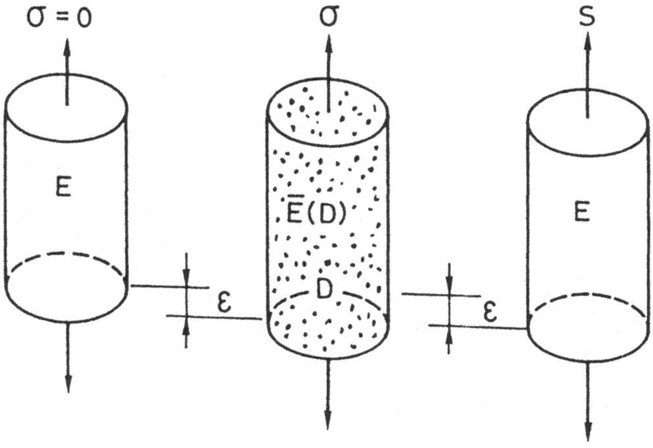

Fig.3.4 Definition of effective stress tensor by the change of elastic constants

subjected to the symmetry condition

$$\bar{E}_{ijkl} = \bar{E}_{klij} \tag{3.18}$$

The transformation $[\underset{\sim}{E}:\bar{\underset{\sim}{E}}(\underset{\sim}{D})^{-1}]$ in equation (3.17) can be identified, if we have the elasticity tensor $\bar{\underset{\sim}{E}}(\underset{\sim}{D})$ of the damaged material. If we restirct our discussion to a class of homogeneous materials damaged by a set of cracks with periodic array and with some planes of symmetry, the result of homogenization theory gives

$$\bar{E}_{ijkl} = [\delta_{ir}\delta_{js} - \{(1-V^*/V)\delta_{ir}\delta_{js} + (1/V)\int_{V^*} b_{ijrs}dV\}]E_{rskl} \tag{3.19}$$

where V and $V^*$are the apparent and the net volume (the apparent volume minus cavity volume) of the material, and b_{ijrs} stands for the matrix of the coefficient of stress concentration. In view of equation (3.19), Chaboche (1982) postulated a fourth rank non-symmetric damage tensor $\underset{\sim}{D}$ which defines the transformation

$$\bar{\underset{\sim}{E}}(\underset{\sim}{D}) = (\underset{\sim}{I}-\underset{\sim}{D}):\underset{\sim}{E} \tag{3.20}$$

where $\underset{\sim}{I}$ denotes the unit tensor of rank four. Then, the damage tensor $\underset{\sim}{D}$ and the effective stress tensor $\underset{\sim}{S}$ are specified as follows:

$$\underset{\sim}{D} = \underset{\sim}{I}-\bar{\underset{\sim}{E}}:\underset{\sim}{E}^{-1}, \qquad \underset{\sim}{S} = (\underset{\sim}{I}-\underset{\sim}{D})^{-1}:\underset{\sim}{\sigma} \tag{3.21}$$

In the case of periodic array of penny shaped cracks, the matrices of $\underset{\sim}{D}$ and $\underset{\sim}{S}$ have the forms

$$D = \begin{bmatrix} D & 0 & 0 & 0 & 0 & 0 \\ (\lambda/\ 1-\lambda)\ D_1 & 0 & 0 & & & \\ (\lambda/\ 1-\lambda)\ D_1 & 0 & 0 & & & \\ 0 & & & 0 & & \\ 0 & & & & D_5 & \\ 0 & & & & & D_5 \end{bmatrix} \tag{3.22}$$

$$\begin{bmatrix} S_{11} \\ S_{22} \\ S_{33} \\ S_{23} \\ S_{31} \\ S_{12} \end{bmatrix} = \begin{bmatrix} 1/(1-D_1) & 0 & 0 & 0 & 0 & 0 \\ (\lambda/\ 1-\lambda)\ (D_1/1-D_1) & 1 & 0 & & & \\ (\lambda/\ 1-\lambda)\ (D_1/1-D_1) & 0 & 1 & & & \\ 0 & & & 1 & & \\ 0 & & & & 1/(1-D_5) & \\ 0 & & & & & 1/(1-D_5) \end{bmatrix} \begin{bmatrix} \sigma_{11} \\ \sigma_{22} \\ \sigma_{33} \\ \sigma_{23} \\ \sigma_{31} \\ \sigma_{12} \end{bmatrix} \tag{3.23}$$

where direction 1 signifies the direction perpendicular to the cracks.

Chaboche's theory outlined above is quite elegant in the sense that it can describe general states of orthotropic damage in terms of three parameters λ, D_1 and D_5. It not only satisfies the requirement of equation (3.18), but also can be incorporated in the framework of the irreversible thermodynamics.

3.3.2 Requirements for effective stress

In the theories of continuum damage mechanics, the mechanical effects of damage on the deformation and on the damage growth of the damaged materials are often represented in terms of a corresponding effective stress. In these cases, according to the notion of the effective stress in the classical Kachanov-Rabotnov theory, the constitutive equations of the damaged materials are assumed to be given by those of the undamaged materials simply by replacing the Cauchy stress in the latter equations by the corresponding effective stress tensor [see equation (I-5.2) and (I-5.3)].

Sidoroff (1981) discussed in detail the conditions so that these notions may be compatible with the thermodynamical requirement. He examined particularly the elastic response of an initially isotropic elastic material, and elucidated that the thermodynamical requirement (3.18) on the elasticity tensor $\bar{E}(D)$ of equation (3.17) imposes a strong restriction on the functional forms of $\bar{E}(D)$ and $S(\sigma, D)$. Then, in order to relax this restriction, Sidoroff abandoned the idea of deriving the constitutive equation of the damaged material directly from that of the undamaged material, and postulated that the free enthalpy function (complementary strain energy function) of the damaged material can be obtained by replacing the stress σ in the function of the undamaged material by the corresponding effective stress S. This idea eliminates the above mentioned restriction, and provides a number of freedom in the definition of the effective stress.

On the basis of this notion, Cordebois and Sidoroff (1982) proposed an effective stress

$$S = (I-D)^{-1/2} \cdot \sigma \cdot (I-D)^{-1/2} \tag{3.24}$$

by use of a second rank symmetric damage tensor D. They then employed this effective stress, together with that of equation (3.12), to model the elastic-plastic damage in some metals, and discussed the validity of their theory by comparing the theoretical predictions of Young's modulus and Poisson's ratio of damaged materials with the corresponding experiments.

3.3.3 Definition of damage variables in terms of microscopic features of cavity

Description of the damage state by means of the measure 2) in Table 3.1 was discussed by Vakulenko and M. Kachanov (1971) and M. Kachanov (1980).

Vakulenko and M. Kachanov (1971) firstly made a comprehensive discussion on the modeling of damaged materials from continuum damage

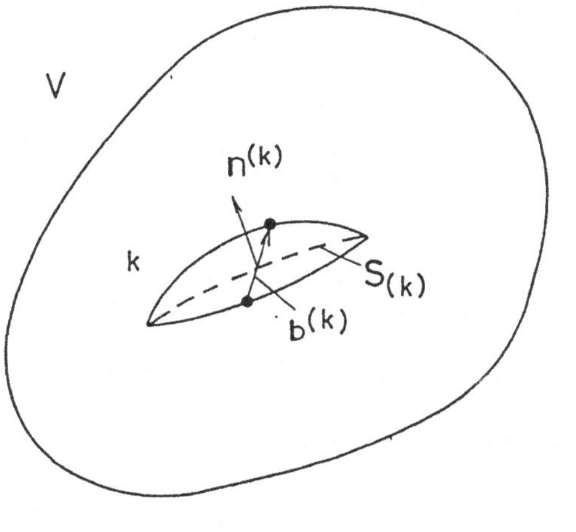

Fig.3.5 Definition of a crack density tensor of rank two

mechanics point of view, and proposed a tensor damage variable of rank two for a cracked elastic material. By noting the similarity between dislocations and microcracks, and taking an arbitrary volume element V including a set of crack (Fig.3.5), they defined a crack density tensor of rank two

$$\underset{\sim}{D} = (1/V) \sum_{k} \int_{S(k)} \underset{\sim}{b}^{(k)} \otimes \underset{\sim}{n}^{(k)} dS_{(k)} \tag{3.25}$$

where $S_{(k)}$ denotes the crack surface of the k-th cracks in V before the crack formation, and $\underset{\sim}{n}^{(k)}$ and $\underset{\sim}{b}^{(k)}$ are the unit normal vector to $S_{(k)}$ and the discontinuity in the displacement of a point on $S_{(k)}$ caused by the crack initiation. In particular, when $\underset{\sim}{b}^{(k)}$ is directed in the direction of $\underset{\sim}{n}^{(k)}$ equation (3.25) stands for a set of cleavage cracks, while the case of $\underset{\sim}{b}^{(k)}$ tangential to $S_{(k)}$ corresponds to the shear cracks.

Though the tensor $\underset{\sim}{D}$ may specify typical features of the crack arrangement in materials, some questions were posed as to its utility as a damage variable. For example, though a plane crack $\underset{\sim}{b}^{(k)}=0$ surely induces a certain material damage, it does not make any contribution to $\underset{\sim}{D}$ of equation (3.25).

In order to eliminate this defficiency, M. Kachanov (1980) modified equation (3.25) by taking $\underset{\sim}{b}^{(k)}$ as the crack opening at the current state but suppressing the influence of the magnitude of crack opening displacement on the damage states:

$$\underset{\sim}{D} = (1/V) \sum_{k} [S_{(k)}]^{1/2} \int_{S(k)} \underset{\sim}{n}^{(k)} \otimes \underset{\sim}{n}^{(k)} dS_{(k)} \tag{3.26}$$

This equation in fact has the same form as equation (3.8). The application of equation (3.26) to evaluate equivalent elastic modulus of cracked media will be discussed in Section 4.1.

When Leckie and Onat (1981) developed a creep damage theory of polycrystalline metals, they discussed the modeling of material damage on the basis of micrographical observations of creep damage process. As already observed in Chapter 2, the creep damage advances as a result of two main mechanisms; i.e., nucleation and growth of the grain boundary cavities. A direct measure to quantify the first mechanism may be the number of cavities per unit volume, while the second mechanism may be quantified by the volume density of an individual cavity. Thus, by denoting the number and the volume density of the cavities in an surface element $dA(\underset{\sim}{n})$ on a grain boundary of orientation $\underset{\sim}{n}$ by $N(\underset{\sim}{n})$ and $V(\underset{\sim}{n})$, Leckie and Onat (1981) constructed moment tensors of $\underset{\sim}{V}$ and $\underset{\sim}{N}$ of even ranks

$$N_0 = \int_S N(\underset{\sim}{n}) dA(\underset{\sim}{n}), \quad N_{ij} = \int_S N(\underset{\sim}{n}) n_i n_j dA(\underset{\sim}{n}), \cdots\cdots \tag{3.27a}$$

$$V_0 = \int_S V(\underset{\sim}{n}) dA(\underset{\sim}{n}), \quad V_{ij} = \int_S V(\underset{\sim}{n}) n_i n_j dA(\underset{\sim}{n}), \cdots\cdots \tag{3.27b}$$

where the integrations are performed on a spherical surface S of unit radius. They postulated that the damage states can be described by these variables.

Characteristic points of Leckie and Onat's theory exist not only in that it is based on the physical mechanisms of creep damage in metals, but also, and more important, is that it gives the lists of proper representation of most damage variables proposed so far.

The modelling of the damage states by means of vector variables related to the local cavity arrangement was discussed by Davison and Stevens (1973), Krajcinovic and Fonseka (1981).

3.4 References

Chaboche, J.L.: The Concept of Effective Stress Applied to Elasticity and

Viscoplasticity in the Presence of Anisotropic Damage, in: Mechanical Behavior of Anisotropic Solids (Ed. J.-P. Boehler), Martinus Nijhoff, The Hague 1982, 737-760.

Cordebois, J.P. and F. Sidoroff: Damage Induced Elastic Anisotropy, in: Mechanical Behavior of Anisotropic Solids (Ed. J.-P. Boehler), Martinus Nijhoff, The Hague 1982, 761-774.

Hayhurst, D.R. and F.A. Leckie: The Effect of Creep Constitutive and Damage Relationships upon the Rupture Time of a Solid Circular Torsion Bar, J. Mech. Phys. Solids, 21 (1973), 431-446.

Hult, J.: CDM-Capabilities, Limitations and Promises, in: Mechanisms of Deformation and Fracture (Ed. K.E. Easterling), Pergamon, Oxford 1979, 233-247.

Kachanov, L.M.: Foundations of Fracture Mechanics, Nauka, Moscow 1974.

Kachanov, M.: Continuum Model of Medium with Cracks, J.Engng Mech. Div., Trans. ASCE, EM5, 106 (1980), 1039-1051.

Krajcinovic, D.: Continuum Damage Mechanics, Applied Mechanics Review, 37 (1984), 1-6.

Krajcinovic, D. and G. Fonseka,: The Continuum Damage Theory of Brittle Materials, Part 1: General Theory, ASME J. Appl. Mech., 48 (1981), 809-815.

Leckie, F.A and E.T. Onat: Tensorial Nature of Damage Measuring Internal Variables, in: Physical Non-Linearities in Structural Analysis (Ed. J. Hult and J. Lemaitre), Springer, Berlin 1981, 140-155.

Lemaitre, J. and J.L. Chaboche: Aspect Phenomenologique de la Rupture par Endommagement, J. Mec. Appl., 2 (1978), 317-365.

Lemaitre, J. and J.L. Chaboche: Mecanique des Materiaux Solides, Dunod, Paris 1985.

Murakami, S.: Anisotropic Damage in Metals, in: Failure Criteria of Structured Media (Ed. J.-P. Boehler), A.A. Balkema, Rotterdam 1986.

Murakami, S. and N. Ohno: A Continuum Theory of Creep and Creep Damage, in: Creep in Structures (Ed. A.R.S. Ponter and D.R. Hayhurst), Springer, Berlin 1981, 422-444.

Rabotnov, Yu.N.: Creep Problem in Structural Members, North-Holland, Amsterdam 1969

Sidoroff,F.: Description of Anisotropic Damage Application to Elasticity, in: Physical Non-Linearities in Structural Analysis (Ed. J. Hult and J. Lemaitre), Springer, Berlin 1981, 237-244.

Vakulenko,A.A. and M.L. Kachanov: Continuum Theory of Cracked Media, Mech. Tverdogo Tiela, NO.4 (1971), 159-166 (in Russian).

4. ANISOTROPIC DAMAGE THEORIES AND THEIR APPLICATIONS

4.1 Elastic Damage and Elastic-Plastic Damage

4.1.1 Deformation of cracked elastic material

As we learned in Section 2.1, nucleation and growth of microcavities in elastic-plastic damage of metals, composites, concretes and rocks are intrinsically directional. Thus, appropriate modeling of these kinds of damage necessitates vector or tensor variables discussed in the preceding Chapter. We will start this Chapter with the elastic response of an elastic-brittle material damaged by a set of distributed microscopic cracks [M. Kachanov 1972,1980].

Now, let us consider an elastic material with a number of microscopic penny-shaped cracks, and let assume that the damage state of the material can be described by the second rank symmetric tensor $\underset{\sim}{D}$ defined by equation (3.26). If we assume that the elastic potential (complementary strain energy density) of the damaged material is represented by

$$G = G(\underset{\sim}{\sigma}, \underset{\sim}{D}) \tag{4.1}$$

then the elastic strain is specified as follows:

$$\underset{\sim}{\varepsilon} = \partial G/\partial \underset{\sim}{\sigma} \tag{4.2}$$

When the material without cracks (material matrix) is isotropic, the function G must be invariant with respect to arbitrary orthogonal transformations; i.e., G is an isotropic tensor function of symmetric second rank argument tensors $\underset{\sim}{\sigma}$ and $\underset{\sim}{D}$ [Truesdell and Noll 1965, Jaunzemis 1967, Leigh 1968, Malvern 1969]. Then, the most general expression of G can be given as a function of the ten basic invariants of $\underset{\sim}{\sigma}$ and $\underset{\sim}{D}$:

$$\begin{aligned}&\mathrm{tr}\,\underset{\sim}{\sigma},\ \mathrm{tr}\underset{\sim}{\sigma}^2,\ \mathrm{tr}\,\underset{\sim}{\sigma}^3,\ \mathrm{tr}\underset{\sim}{D},\ \mathrm{tr}\underset{\sim}{D}^2,\ \mathrm{tr}\underset{\sim}{D}^3,\\ &\mathrm{tr}(\underset{\sim}{\sigma}\underset{\sim}{D}),\ \ \mathrm{tr}(\underset{\sim}{\sigma}^2\underset{\sim}{D}),\ \mathrm{tr}(\underset{\sim}{\sigma}\underset{\sim}{D}^2),\ \mathrm{tr}(\underset{\sim}{\sigma}^2\underset{\sim}{D}^2)\end{aligned} \tag{4.3}$$

If the material matrix is linear elastic, the function G must be a quadratic function of stress $\underset{\sim}{\sigma}$, and thus G has a form

$$\begin{aligned}G = {} & a_1(\mathrm{tr}\,\underset{\sim}{\sigma})^2 + a_2\mathrm{tr}\,\underset{\sim}{\sigma}^2 + b_1\mathrm{tr}\underset{\sim}{\sigma}\ \mathrm{tr}(\underset{\sim}{\sigma}\underset{\sim}{D}) + b_2\mathrm{tr}\,\underset{\sim}{\sigma}\ \mathrm{tr}(\underset{\sim}{\sigma}\underset{\sim}{D}^2)\\ & + b_3[\mathrm{tr}(\underset{\sim}{\sigma}\underset{\sim}{D})]^2 + b_4\mathrm{tr}(\underset{\sim}{\sigma}\underset{\sim}{D})\ \mathrm{tr}(\underset{\sim}{\sigma}\underset{\sim}{D}^2) + b_5[\mathrm{tr}(\underset{\sim}{\sigma}\underset{\sim}{D}^2)]^2\\ & + b_6\mathrm{tr}(\underset{\sim}{\sigma}^2\underset{\sim}{D}) + b_7\mathrm{tr}(\underset{\sim}{\sigma}^2\underset{\sim}{D}^2)\end{aligned} \tag{4.4}$$

where $a_1, a_2, b_1, \cdots\cdots, b_7$ are functions of the invariants of $\underset{\sim}{D}$.

In the particular case of vanishing cracks ($\underset{\sim}{D}=\underset{\sim}{0}$), G is reduced to the elastic potential of an isotropic linear elastic material

$$(G)_{\underset{\sim}{D}=\underset{\sim}{0}} = (-\ \nu_0/2E_0)\ (\mathrm{tr}\underset{\sim}{\sigma})^2 + [(1+\ \nu_0)/(2E_0)]\ \mathrm{tr}\underset{\sim}{\sigma}^2 \tag{4.5}$$

where E_0 and ν_0 are Young's modulus and Poisson's ratio of the undamaged material. Comparison between equations (4.4) and (4.5) suggests that the first two terms in equations (4.4) represent the isotropic elastic response. In the case of damaged material, if a_1 and a_2 are functions of the invariants of tensor $\underset{\sim}{D}$, they must be symmetric functions of the principal values $D_i (i=1,2,3)$ of $\underset{\sim}{D}$. However, the effect of the parallel plane cracks on Young's modulus in the direction parallel to the cracks is negligibly small. The coefficients a_1 and a_2, therefore, are material constants independent of $\underset{\sim}{D}$ also for the damaged material, and should be identical to the coefficients of equation (4.5):

$$a_1 = -\nu_0/(2E_0), \qquad a_2 = (1+\nu_0)/(2E_0) \tag{4.6}$$

We already learned in Section 3.2 that the second rank symmetric tensor $\underset{\sim}{D}$ represents orthotropic damage states characterized by the three principal values and the corresponding principal directions. Hence, G of equation (4.4) represents the elastic potential of orthotropic elastic materials. When the material has only one set of parallel cracks ($\underset{\sim}{D}=D\underset{\sim}{n}\otimes\underset{\sim}{n}$) equation (4.4) specifies the elastic response of a transverse isotropic material, while it represents an isotropic response if the material has distributed cracks of completely random directions ($\underset{\sim}{D}=D\underset{\sim}{I}$).

It is not easy to determine the other coefficients $b_1,\cdots,b_7$ of equation (4.4). However, in the case of small crack density, the potential of equation (4.4) can be linearized with respect to $\underset{\sim}{D}$. Then, equations (4.4) and (4.6) lead to a simplified form

$$G = -(\nu_0/2E_0)\,(\mathrm{tr}\,\underset{\sim}{\sigma})^2 + [(1+\nu_0)/(2E_0)]\;\mathrm{tr}\,\underset{\sim}{\sigma}^2 + b_1\mathrm{tr}\,\underset{\sim}{\sigma}\;\mathrm{tr}(\underset{\sim}{\sigma}\underset{\sim}{D}) + b_6\mathrm{tr}(\underset{\sim}{\sigma}^2\underset{\sim}{D}) \tag{4.7}$$

From this relation, we can calculate Young's modulus E_i, modulus of rigidity G_{ij} and Poisson's ratio ν_{ij} (with respect to the principal axis x_i of damage), and have [M. Kachanov 1980]

$$E_i = E_0/[1+2E_0D_i(b_1+b_6)], \qquad G_{ij} = E_0/[2(1+\nu_0)+2E_0(D_i+D_j)b_6]$$
$$\nu_{ij} = [\nu_0-E_0(D_i+D_j)b_1]/[1+2E_0D_i(b_1+b_6)] \tag{4.8}$$

These elastic-constants identically satisfies Saint-Venant's relation

$$1/G_{ij} = (1+\nu_{ij})/E_i+(1+\nu_{ji})/E_j \tag{4.9}$$

Thus, the elastic potential of equation (4.7) is characterized by seven unknown constants E_0, ν_0, b_1, b_6, D_1, D_2 and D_3. If E_0 and ν_0 are known, in particular, the coefficients b_1 and b_6 can be determined from equations (4.7) and (4.8) by performing two kinds of tests for a given value of $\underset{\sim}{D}$.

The above theory in which we represent the effects of microscopic plane cracks in terms of the second rank symmetric tensor $\underset{\sim}{D}$ and describe the elastic response of the cracked material by means of the orthotropic elastic potential of equation (4.4) is an approxiate model for damaged materials. M. Kachanov (1980) discussed the validity of the above theory as regards two cases; the case of small crack density and the case of large crack density with significant interaction between cracks. He showed that, in both cases, the damaged material with distributed microcacks can be modeled by an orthotropic material with sufficient accuracy, and therefore equation (4.4) describes well the elastic response of such material. He also discussed the case when the material matrix (material without cracks) is anisotropic; in this case G is an anisotropic tensor function of two tensors, $\underset{\sim}{\sigma}$ and $\underset{\sim}{D}$. This will lead to a much larger number of terms in G. Such expressions were later explicitly given by Talreja [1985].

4.1.2 Thermodynamical theory of elastic-plastic damage

Though Kachanov's theory in the preceding Subsection elucidated in detail the anisotropic elastic behaviour of materials with distributed microcracks, it does not give any information on the development of material damage induced by the increase of stress and strain, or on the evolution of the damage variable. In order to describe the progression of such elastic-plastic damage, Dougill (1975,1976), Dragon and Mroz (1979), for example, extended the notion of the loading surface of plasticity theory, and developed damage rules by specifying fracture surface in the stress or strain space.

We now take a more systematic approach to this problem by means of the thermodynamical theory discussed in Chapter II-2 (Part II Foundation and Identification of Damage Constitutive Equations, by J. Lemaitre), and show the unified derivation of the constitutive and the evolution equations for quasi-static anisotropic elastic-plastic damage [Krajcinovic 1983, Krajcinovic and Selvaraj 1984].

Let us assume that the deformation is small, and the strain tensor $\underset{\sim}{\epsilon}$ can be decomposed into a sum of elastic part $\underset{\sim}{\epsilon}^e$ and plastic part $\underset{\sim}{\epsilon}^p$

$$\underset{\sim}{\epsilon} = \underset{\sim}{\epsilon}^e + \underset{\sim}{\epsilon}^p \tag{4.10}$$

The evolution of internal states of materials advances mainly by two classes of microstructural change; one is strain hardening related to the change of dislocation structures (or other relevant structures) and the other is internal damage characterized by the formation of distributed microcavities. Though the former state is usually represented by a scalar parameter κ for isotropic hardening together with a second rank tensor $\underset{\sim}{\alpha}$ for kinematic hardening, various damage variables (Table 3.2) have been proposed to describe the later state. Then, if we postulate a second rank symmetric tensor $\underset{\sim}{D}$ for the damage variable, we have the Helmholtz free energy ψ for a damaged material

$$\psi = \psi(\underset{\sim}{\epsilon}^e, \kappa, \underset{\sim}{\alpha}, \underset{\sim}{D}, T) \tag{4.11}$$

where T stands for temperature. In fact, all the following discussion holds also for other choice of damage variable.

The process of inelastic deformation of the materials with internal damage should satisfy the following Clausius-Duhem inequality [Truesdell and Noll 1965, Jaunzemis 1967, Leigh 1968, Malvern 1969]

$$\sigma_{ij}\dot{\epsilon}_{ij} - \rho(\dot{\psi} + \eta\dot{T}) - (1/T)\underset{\sim}{q}\cdot\mathrm{grad}T \geq 0 \tag{4.12}$$

where ρ, η and $\underset{\sim}{q}$ are mass density, entropy and heat flux vector, respectively. Substitution of equations (4.10) and (4.11) into equation (4.12) furnishes

$$\begin{aligned}(\sigma_{ij} - \rho\partial\psi/\partial\epsilon_{ij}^e)\dot{\epsilon}_{ij}^e - \rho(\partial\psi/\partial T + \eta)\dot{T} \\ + \sigma_{ij}\dot{\epsilon}_{ij}^p - \rho(\partial\psi/\partial\kappa)\dot{\kappa} - \rho(\partial\psi/\partial\alpha_{ij})\dot{\alpha}_{ij} \\ - \rho(\partial\psi/\partial D_{ij})\dot{D}_{ij} - (1/T)\mathrm{grad}T\cdot\underset{\sim}{q} \geq 0\end{aligned} \tag{4.13}$$

The requirement that this inequality is identically satisfied independently of the elastic strain rate $\dot{\epsilon}_{ij}^e$ and the rate of temperature $\dot{T}$ entails

$$\sigma_{ij} = \rho(\partial\psi/\partial\epsilon_{ij}^e), \qquad \eta = -\partial\psi/\partial T \tag{4.14}$$

Let us now introduce thermodynamical forces conjugate to $\dot{\kappa}$, $\dot{\underset{\sim}{\alpha}}$ and $\dot{\underset{\sim}{D}}$ in equation (4.13) as follows

$$K = -\rho(\partial\psi/\partial\kappa), \quad A_{ij} = -\rho(\partial\psi/\partial\alpha_{ij}), \quad R_{ij} = -\rho(\partial\psi/\partial D_{ij}) \tag{4.15}$$

In terms of these variables, if we define the thermodynamical flux vector $\underset{\sim}{J}$ and the conjugate thermodynamical force vector $\underset{\sim}{X}$ as follows

$$\underset{\sim}{J} = \{\dot{\epsilon}_{ij}^p, \dot{\kappa}, \dot{\alpha}_{ij}, \dot{D}_{ij}, q_i\}^T \tag{4.16}$$

$$\underset{\sim}{X} = \{ \sigma_{ij}, K, A_{ij}, R_{ij}, (-1/T)\mathrm{grad}T\} \tag{4.17}$$

then the Clausius-Duhem inequality (4.13) can be written in the form of entropy production rate inequality per unit volume

$$\rho \vartheta = X_m J_m \geq 0 \tag{4.18}$$

The constitutive equations for thermodynamical flux vector $\underset{\sim}{J}$ of equation (4.16) are generally given as functions of the thermodynamical force $\underset{\sim}{X}$. However, under certain assumptions regarding the functional dependence of $\underset{\sim}{J}$ on $\underset{\sim}{X}$, a flow potential $F(\underset{\sim}{X})$ related to the entropy production rate ϑ of equation (4.18) can exists; namely, the flux vector $\underset{\sim}{J}$ can be obtained from the potential F [Krajcinovic 1983, Krajcinovic and Selvaraj 1984];

$$J_m = \partial F/\partial X_m \tag{4.19a}$$

or

$$\dot{\epsilon}_{ij}{}^{p} = \partial F/\partial \sigma_{ij} \qquad \dot{\kappa} = \partial F/\partial K,$$

$$\dot{\alpha}_{ij} = \partial F/\partial A_{ij}, \qquad \dot{D}_i = \partial F/\partial R_{ij} \tag{4.19b}$$

Thus, if the Helmholtz free energy ψ and the flow potential F can be specified appropriately, theelastic-plastic constitutive equation and the evolution equation of damaged materials are furnished by equations(4.14) and (4.19).

In the majority of damage theories developed hitherto, evolution equations for damage have been formulated in some a priori ways independently of the corresponding constitutive equations. In the present theory, however, specification of a single potential F is sufficient to provide the inelastic constitutive equations and the damage evolution equations, and the mathematical procedure is formally identical to the theory of plasticity of undamaged materials.

Krajcinovic et al. [Krajcinovic 1983, Krajcinovic and Selvaraj 1983] applied the above theory to discuss the damage behaviour of

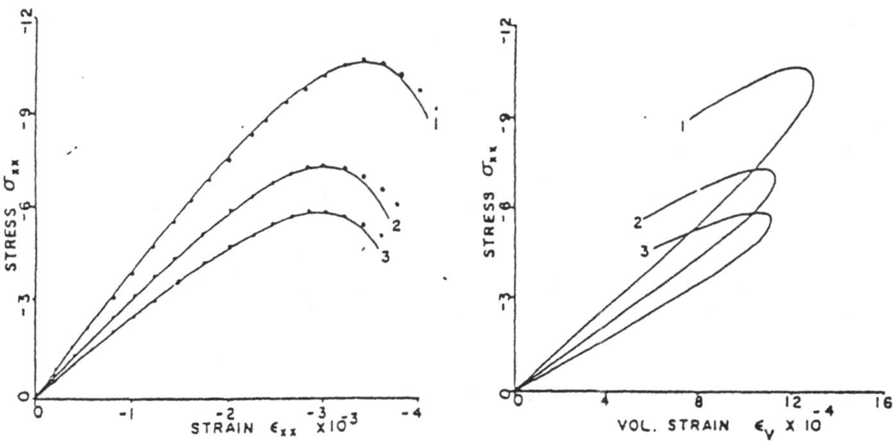

(a) Stress-strain curves (b) Change of volumetric strain

Fig.4.1 Stress-strain curves and change of volumetric strains in uniaxial compression tests of concretes [From Krajcinovic and Selvaraj (1983)] elastic-brittle materials caused by plane cracks by employing a vector damage variable $\underset{\sim}{\omega}$ and assuming the free energy function and the flow potential of the following forms:

$$\psi = \psi(\underset{\sim}{\epsilon}^e, \underset{\sim}{\omega}), \quad F = F(\underset{\sim}{\epsilon}^e, \underset{\sim}{R}, T) = L_{mn}R_mR_n - R_0^2 \tag{4.20}$$

where $R_0 = R_0(\underset{\sim}{\omega}, T)$ and $L_{mn} = L_{mn}(T)$. They established the explicit forms of equation (4.20), and analysed uniaxial compression tests of three different grades of concretes. Fig.4.1 shows the predictions for stress-strain curves and the stress-volumetric strain curves. The curves in Fig.4.1(a) describe well the corresponding experimental results entered by small circles. The curves in Fig.4.1(b), on the other hand, coincide well with the trends of experimental results reported in literature.

An analogous theory based on the second rank symmetric damage tensor $\underset{\sim}{D}$ has been developed also by Cordebois and Sidoroff (1982). Their theory is for an isothermal elastic-plastic damage mainly of metals, and corresponds to specify the Helmholtz free energy and flow potential of the following form in equations (4.14) and (4.19):

$$\psi = \psi(\underset{\sim}{\epsilon}^e, \kappa, \underset{\sim}{D}, \lambda) \tag{4.21}$$

$$F = F_1(\underset{\sim}{\sigma}, K ; \underset{\sim}{D}) + F_2(\underset{\sim}{R}, L) \tag{4.22}$$

where λ in equation (4.21) stands for a scalar variable which plays the similar role as the strain hardening parameter κ for plastic deformation. The symbols $K, \underset{\sim}{R}$ and L in equation (4.22) are the thermodynamical forces conjugate to the thermodynamical flux κ, $\underset{\sim}{D}$ and λ, and can be derived from equation (4.21) by means of an analogousrelation to equation (4.15). The stress $\underset{\sim}{\sigma}$ in equation (4.22), furthermore, is given by equations (4.14) and (4.21). In this case, $\underset{\sim}{J}$ and $\underset{\sim}{X}$ of equation (4.16) have the form

$$\underset{\sim}{J} = \{\dot{\epsilon}_{ij}^p, \dot{\kappa}, \dot{D}_{ij}, \dot{\lambda}\}^T, \quad \underset{\sim}{X} = \{\sigma_{ij}, K, R_{ij}, L\} \tag{4.23}$$

and the expressions for the components of $\underset{\sim}{J}$ are given by equations (4.19) and (4.22). Cordebois et al. derived the explicit expressions of equations (4.21) and (4.22), and discussed the identification of the functions $K(\kappa)$, $L(\lambda)$ and $\underset{\sim}{R}(\underset{\sim}{D})$ by means of uniaxial tension tests. However, the experimental evaluation of this theory remains an open problems.

The elastic-plastic damage of metals involves not only the nucleation and growth of cavities, but also their coupling with the strain-hardening of the material matrix or with the strain localization due to large deformation. Because of the complexity of these problems, though the process of the elastic-plastic damage of metals has been analysed in some detail by assuming damage isotropy [Gurson 1979, Perzyna 1984, Lemaitre 1985], the work on anisotropic features of these problem is still lacking.

In the recent monograph of Lemaitre and Chaboche (1985), thermodynamical theories of the elastic-brittle and elastic-plastic damage have been developed on the basis of the scalar and the tensor damage variables, and will be useful for interested readers.

4.2 Spall Damage

Besides the elastic-brittle damage and the elastic-plastic damage due to quasi-static loading, the material damage induced by intensive stress waves (i.e., the spall damage) also has been objectives of some theoretical papers. Davison et al., first of all, formulated a thermodynamical theory of spall damage by use of a vector damage variable [Davison and Stevens 1973] and a scalar damage variable [Davison, Stevens and Kipp 1977]. Then, they analysed the growth of damage and the effects of damage on the stress distribution, particle velocity and on the heat generation in the impact of various discs. Fig.4.2 shows the velocity history and the stress distribution in a 1100-0 aluminium target disc of 6.4mm thick impacted by a fused-quartz disc of 3.2mm at the impact velocity of $v=142ms^{-1}$. The solid lines and the dashed lines in the figure reprsent the numerical results with and without regards to the effects of spall damage, while small symbols are the corresponding experimental results. As observed in Fig.4.2(a), while the solid lines describe well the experimental results, the dashed lines show significant deviation from the solid lines and the experimental results for the stage after about 2.4 μs after impact; this is a manifestation of the accumulation of internal damage. Namely, the experimental results and the solid lines in Fig.4.2(a) exhibit a temporary increase of the free surface velocity after about 2.4μs, and this characteristic feature is accounted for by the arrival at the back surface of the stress-release wave produced by the development of spall damage within the material. Fig.4.2(b), furthermore, shows the plots of the calculated axial stress distribution in the target at 2.3 μs after the impact. The stress relief observed in the solid line is attributed again to the damage accumulation.

Analogous problems were investigated also by Seaman, Shockey and Curran and others [Seaman, Curran and Shockey 1976, Shockey, Seaman, Dao and Curran 1980]. They performed a series of experiments by impacting

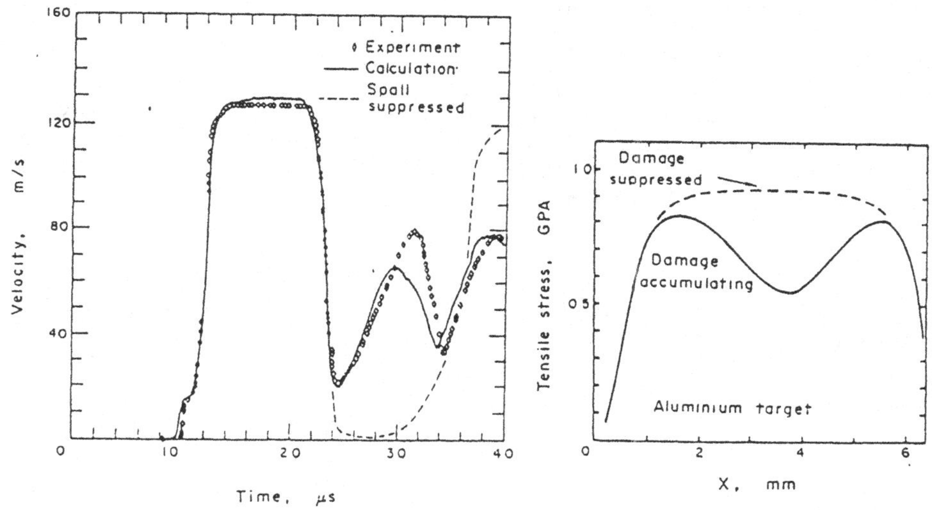

(a) Calculated and observed velocity histories of the free surface of the target

(b) Calculated stress distributions in the target at 2.3 μs after impact

Fig.4.2 Spall damage in a 1100-0 aluminium target disc of 6.4 mm thick impacted by a fused-quartz disc of 3.2 mm thick (impact velocity v=142 m/s) [From Davison, Stevens and Kipp (1977)]

flyer discs on target discs of various materials, and proposed a mechanical model to describe the density, size and volume ratio of the cavities observed in the ductile and the brittle spall damage. Then, they incorporated this model into computational code of one- and two-dimensional wave propagation, and calculated the damage growth in the targets.

Fig.4.3 shows the comparison between the calculated and the experimental results on damage distribution in XAR30 steel specimens caused by the impact by a tapered flyer of OFHC copper. A number of microcracks are coalesced into a single spall crack, and these features are simulated very well by the calculated result.

4.3 Creep Damage

4.3.1 Modeling of creep damage by means of a second rank damage tensor

Because of the intrinsic anisotropy in the cavity formation due to creep damage, a number of anisotropic creep damage theories have been proposed by employing various damage variables listed in Table 3.2. Now, we will develop a creep damage theory on the basis of a second rank symmetric tensor $\underset{\sim}{D}$ of equation (3.8), and apply it to the creep damage analysis of thin-walled tubes under non-proportional loading [Murakami and Ohno 1978,1981]. The deformation is assumed to be small.

We first assume that the creep damage state in polycrystalline metals can be characterized by the three dimensional cavity area density and represented by the second rank symmetric damage tensor $\underset{\sim}{D}$ of equation (3.8). Then, because of the net area reduction due to such cavity distribution, the effects of the Cauchy stress $\underset{\sim}{\sigma}$ are magnified to the effective stress tensor $\underset{\sim}{S}$ of equation (3.12)

$$\underset{\sim}{S} = (1/2)(\underset{\sim}{\Phi}\,\underset{\sim}{\sigma} + \underset{\sim}{\sigma}\,\underset{\sim}{\Phi}), \qquad \underset{\sim}{\Phi} = (\underset{\sim}{I}-\underset{\sim}{D})^{-1} \tag{4.24}$$

where $\underset{\sim}{\Phi}$ will be called damage effect tensor.

The rate of damage growth depends mainly on the current states of material damage and stress. If the effects of stress and damage appears only through $\underset{\sim}{S}$ and $\underset{\sim}{\Phi}$ of equation (4.24), the evolution equation of damage can be specified as follows:

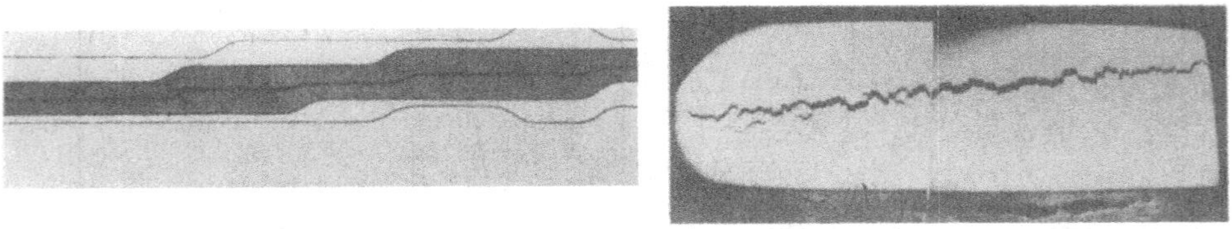

(a) Computed damage on cross section of XAR30 target

(b) Cross section of XAR30 target showing partial spall

Fig.4.3 Comparison of computed and observed spall damage in a XAR30 steel target disc (thickness and diameter are 10.16 mm and 38.1 mm, respectively) impacted by a tapered OFHC copper flyer disc (thickness and diameter are 2.55-6.25 mm and 38.1 mm)(impact velocity v=306.5 m/s) [From Seaman, Curran and Shockey (1976)]

$$\dot{\underset{\sim}{D}} = \underset{\sim}{H}(\underset{\sim}{S}, \underset{\sim}{\Phi}, \kappa) \tag{4.25}$$

where κ is a work-hardening parameter of the material matrix, and $(\cdot)$ signifies the material time derivative.

In Section 4.1, we mentioned two possibilities to establish the damage equations; one is to introduce the damage surfaces in the stress or in the strain space, while the other is to establish a dissipation potential on the basis of the thermodynamical theory of inelasticity. As an alternative method, we can specify the evolution equations of damage by taking account of the micromechanisms of the damage growth or of their mechanical effects.

In view of the mechanisms of creep damage in polycrystalline metals mentioned in Section 2.3 together with the fact that $\underset{\sim}{D}$ in equation (3.8) stands for the increase in the three dimensional cavity area density caused by the damage, equation (4.25) can be expressed in a sufficiently general form [Murakami and Ohno 1981]

$$\dot{\underset{\sim}{D}} = \gamma \underset{\sim}{I} + \sum_i \underset{\sim}{M}^{(i)} : [\underset{\sim}{\nu}^{(i)} \otimes \underset{\sim}{\nu}^{(i)}] + \sum_j \underset{\sim}{N}^{(j)} : [\underset{\sim}{\nu}_D^{(j)} \otimes \underset{\sim}{\nu}_D^{(j)}] \tag{4.26}$$

where γ, $\underset{\sim}{M}^{(i)}$ and $\underset{\sim}{N}^{(j)}$ are a scalar function and fourth rank tensor functions, respectively, of $\underset{\sim}{S}$, $\underset{\sim}{\Phi}$ and κ. The symbols $\underset{\sim}{\nu}^{(i)}$ and $\underset{\sim}{\nu}_D^{(j)}$ in equations (4.26), furthermore, are the principal vectors corresponding to the positive principal values of $\underset{\sim}{S}$ and its deviator $\underset{\sim}{S}_D$, whereas $(:)$ signifies the contraction of the second rank. While the second term on the right-hand side of equation (4.26) represents the increase in the cavity area fraction on planes perpendicular to the positive principal stress, the first and the third term imply the isotropic increase of cavity area fraction and the possibility of the damage growth under compressive stress.

One of the simple forms of equation (4.26) capable of describing the anisotropic creep damage can be expressed as follows:

$$\dot{\underset{\sim}{D}} = B[(1-\zeta)\{(3/2)\mathrm{tr}\underset{\sim}{S}_D^2\}^{1/2} + \zeta S^{(1)}]^k [(1-\eta)\underset{\sim}{I} + \eta \underset{\sim}{\nu}^{(1)} \otimes \underset{\sim}{\nu}^{(1)}]$$

$$\underset{\sim}{S}_D = \underset{\sim}{S} - (1/3)(\mathrm{tr}\underset{\sim}{S})\underset{\sim}{I} \tag{4.27}$$

where B, k, ζ and η are material constants, and $S^{(1)}$ and $\underset{\sim}{\nu}^{(1)}$ are the maximum positive principal value of the tensor $\underset{\sim}{S}$ and the corresponding principal direction.

When the creep rate $\dot{\underset{\sim}{\epsilon}}^c$ of damaged materials depends on the stress $\underset{\sim}{\sigma}$ only through $\underset{\sim}{S}$, and the effect of damage tensor $\underset{\sim}{D}$ on $\dot{\underset{\sim}{\epsilon}}^c$ is represented by $\underset{\sim}{\Phi}$, the constitutive equation of creep has a form

$$\dot{\underset{\sim}{\epsilon}}^c = \underset{\sim}{G}(\underset{\sim}{S}, \underset{\sim}{\Phi}, \kappa) \tag{4.28}$$

The most general expression of this equation can be given by the representation theorem for isotropic tensor functions [Truesdell and Noll 1965, Jaunzemis 1967, Leigh 1968, Malvern 1969], and expressed as a tensor polynomial of $\underset{\sim}{S}$ and $\underset{\sim}{\Phi}$.

In the case of undamaged material, however, creep analyses of structural components are often performed by use of the three-dimensional version of the Bailey-Norton equation combined with the strain-hardening hypothesis:

$$\dot{\underset{\sim}{\epsilon}}^c = (3/2) A^{1/m} \kappa^{(m-1)/m} [(3/2)\mathrm{tr}(\underset{\sim}{\sigma}_D^2)]^{(n-m)/2m} \underset{\sim}{\sigma}_D^2$$

$$\kappa = \int (2/3)\mathrm{tr}[(\dot{\underset{\sim}{\epsilon}}^c)^2]dt, \qquad \underset{\sim}{\sigma}_D = \underset{\sim}{\sigma} - (1/3)(\mathrm{tr}\underset{\sim}{\sigma})\underset{\sim}{I} \tag{4.29a}$$

or, in a particular case of uniaxial constant stress

$$\epsilon^c = A \sigma^n t^m \tag{4.29b}$$

where A, n and m are material constants. Hence, one of the simplest expression of equation (4.28) for damaged material which is reduced to equation (4.29a) for vanishing damage can be obtained by replacing $\underset{\sim}{\sigma}$ in equation (4.29a) by the corresponding net stress tensor $\underset{\sim}{S}$ of equation (3.14b):*

$$\dot{\epsilon}^{C} = (3/2)\ A^{1/m}\kappa^{(m-1)/m}[(3/2)\mathrm{tr}(\underset{\sim}{S}_D{}^2)]^{(n-m)/2m}\ \underset{\sim}{S}_D \tag{4.30}$$

Fig.4.4 shows the results of the application of equation (4.27) to the creep rupture analysis of thin walled tubes under the stress history entered in the figure [Murakami and Ohno 1978,1984], where t_R, t^* and t_σ denotes, respectively, the rupture time, the time of stress change and the rupture time for constant uniaxial tension σ defined by

$$t_\sigma = [B(k+1)\sigma^k]^{-1} \tag{4.31}$$

The solid line represents the result of equation (4.27) with $\zeta = \eta = 1$:

$$\dot{\underset{\sim}{D}} = B[S^{(1)}]^k(\ \underset{\sim}{\nu}^{(1)} \otimes \underset{\sim}{\nu}^{(1)}) \tag{4.32}$$

while the dashed line is Kachanov's result calculated from the following equation (Kachanov 1974):

$$\dot{\Psi}_y = -\ B(\ \sigma_y/\Psi_y)^k \tag{4.33}$$

where $\sigma_y/\Psi_y = S_y$ is the effective stress defined by equation (3.3).

As will be seen from Fig.4.4(a), the damage process of torsion due to Kachanov's theory for larger values of k proceeds almost independently of the damage accumulated in the preceding process of tension. This is

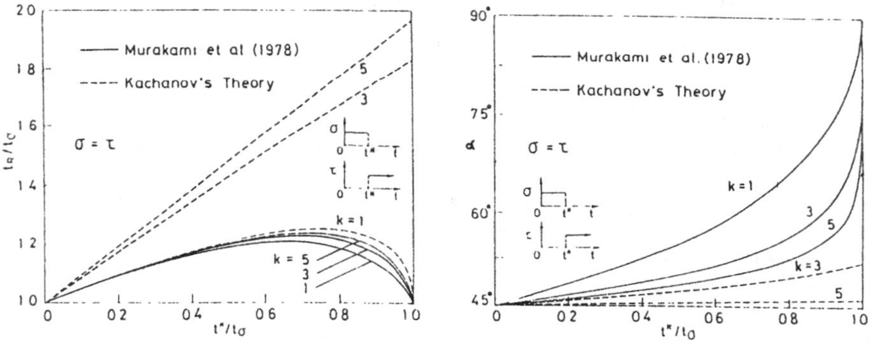

(a) Relation between the rupture time and the time of stress change

(b) Relation between the direction of fracture surface and the time of stress change

Fig.4.4 Creep rupture of thin-walled tubes under tension followed by torsion [From Murakami and Ohno (1978)]

---

* Besides equation (4.30), we can incorporate also more elaborate constitutive equations of creep [e.g., Murakami, Ohno and Tagami 1986, Chaboche 1977] into the present creep damage theory.

obviously attributable to the simplified assumption in equation (4.33) that the rate of damage on the plane $\underset{\sim}{\nu}$ does not depend on the damage state of other planes.

The orientation of the rupture plane measured from the axial direction in relation to the time of stress change is shown in Fig.4.4(b). It will be noticed that the rupture plane predicted by Kachanov's theory is closer to that of torsion (ϕ=45°) than that of the present theory.

In order to elucidate the effects of anisotropic creep damage in more detail, Murakami and Ohno (1981) further calculated the creep rupture time of thin-walled tubes subjected to the combined constant tension and reversed or cyclic torsion as shown in Fig.4.5. The material was tough pitch copper at 250 °C, and equation (4.32) and material constants

$$B = 1.40\times 10^{-13} h^{-1} (\mathrm{MPa})^{-5.6}, \qquad k = 5.6 \tag{4.34}$$

were employed to describe the evolution of creep damage in this material.

The curves A and B in Fig.4.5 exhibit the rupture time t_R corresponding to the stress histories entered in the figure. The symbol t is the rupture time under constant uniaxial tension of $\sigma^{(1)}$=46.8MPa.

The rupture times of the curves A and B increase with τ/σ. This is due to the fact that, when the damage growth is directional, the change of the principal stress direction induce the change of the plane of significant accumulated damage, and thus creep cavities develop mainly on two different planes. This trend has been confirmed by the micrograph of Fig.4.6, which was observed by Trampczynski et al. (1981) for the creep damage of copper under non-proportional loading. Since the isotropic damage theory entered by a chain line in the figure cannot describe these features, it predicts a constant rupture time t_R/t =1 independently of the stress ratio τ/σ. The analysis without regards to the damage

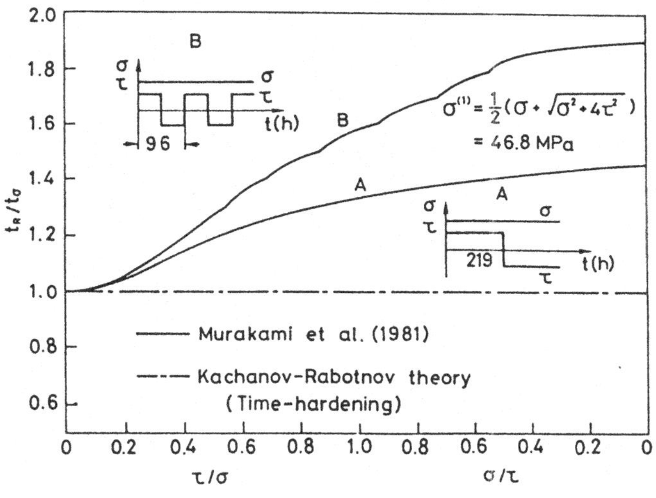

Fig.4.5 Effect of material damage on the rupture time of thin-walled tubes under non-proportional loading: curve A, constant tension and reversed torsion; curve B, constant tension and cyclic torsion [From Murakami and Ohno (1981)]

anisotropic gives the rupture times 1/1.5 ~1/2 shorter than the anisotropic theory.

Recently, equation (4.27) was applied also to the analysis of creep crack bifurcation under non-proportional loading [Murakami, Kawai and Rong 1987].

4.3.2 Other anisotropic theories of creep damage

Microscopic mechanisms of creep damage are simpler than other types of damage, and the modeling of creep damage is relatively easy. Therefore, several anisotropic creep damage theories have been formulated on the basis of various damage variables listed in Table 3.2. We already learned that early anisotropic creep damage theories based on the scalar or vector variables were not applicable to the general histories of loading. Thus, we will now outline three tensorial creep damage theories; Chaboche's theory based on the damage tensor of fourth rank, Leckie and Onat's theory postulating the second rank moment tensor of cavity volume density, and Betten's theory for creep constitutive equations of damaged material of initial anisotropy.

By characterizing the damage state by the change of elastic constant, Chaboche (1982,1984) formulated another creep damage theory on the basis of the damage tensor $\underset{\sim}{D}$ and the corresponding net stress tensor $\underset{\sim}{S}$ of equations (3.22) and (3.23).

Now, we assume that the cavities observed in the actual process of creep damage can be simulated by a combination of the plane cracks specified by equation (3.22) and a set of isotropic voids. Then, the evolution equations of creep damage can be given as

$$\dot{\underset{\sim}{D}} = \underset{\sim}{Q}D, \quad \underset{\sim}{Q} = \gamma \underset{\sim}{I} + (1-\gamma)\underset{\sim}{\Gamma} \tag{4.35a}$$

Fig.4.6 Grain boundary cracks in a copper tube subjected to creep under combined cyclic torsion and constant tension (250℃) [From Trampczynski, Hayhurst & Leckie (1981)]

$$\underset{\sim}{\Gamma} = \begin{bmatrix} 1 & 0 & 0 & 0 & 0 & 0 \\ \lambda/(1-\lambda) & 0 & 0 & & & \\ \lambda/(1-\lambda) & 0 & 0 & & & \\ 0 & & & 0 & & \\ 0 & & & & \mu & \\ 0 & & & & & \mu \end{bmatrix} \tag{4.35b}$$

$$\dot{D} = A[\alpha S^{(1)} + \beta \,\mathrm{tr}S + (1-\alpha-\beta)\, S_{EQ}]^{k}(1-D)^{-1} \tag{4.35c}$$

where $\underset{\sim}{I}$ denotes a unit tensor of rank four, whereas γ, λ, μ, α, β, A,k and l are material constants.

As regards the creep constitutive equation of the damaged materials, he assumed von Mises flow rule and the isotropic creep potential expressed in terms of the effective stress tensor $\underset{\sim}{S}$ of equation (3.23), and obtained as follows

$$\dot{\underset{\sim}{\varepsilon}}^{c} = (3/2)A[(1/2)\mathrm{tr}(\underset{\sim}{S}_D{}^2)]^{n-1}\ \kappa^{-n/m}[(\underset{\sim}{I}-\underset{\sim}{D})^{-1}:\underset{\sim}{S}_D]$$
$$\dot{\kappa} = A[(1/2)\mathrm{tr}(\underset{\sim}{S}_D)^2]^{n}\ \kappa^{-n/m} \tag{4.36}$$

Chaboche (1982,1984) analyzed by this theory the creep rupture times of thin-walled tubes under combined tension and torsion and the damage process of thin plates simulated by perforated plates. He compared the predictions of these analyses with those of Murakami and Ohno's theory mentioned in Subsection 4.3.1, but no marked difference was observed between them.

Leckie and Onat (1981), on the other hand, proposed a continuum theory of creep damage in terms of the damage variables discussed in Section 3.3. Among damage variables of equation (3.27), they assumed the second moment of the cavity volume density V_{ij}, and postulated the constitutive equation, the evolution equation and the rupture criterion as follows:

$$\dot{\underset{\sim}{\varepsilon}}^{c} = f[(1/2)\mathrm{tr}\ \underset{\sim}{\sigma}_D{}^2,\ \mathrm{tr}\underset{\sim}{V})\ \underset{\sim}{\sigma}_D \tag{4.37a}$$

$$\underset{\sim}{V} = g[\sigma^{(1)}]\underset{\sim}{n}^{(1)}\otimes \underset{\sim}{n}^{(1)} \tag{4.37b}$$

$$R[\sigma^{(1)},\ \mathrm{tr}\underset{\sim}{V},\ \underset{\sim}{V}\underset{\sim}{n}^{(1)}\cdot\underset{\sim}{n}^{(1)}] = 0 \tag{4.37c}$$

where $\underset{\sim}{\sigma}_D$ denotes the deviatoric stress tensor, and $\sigma^{(1)}$ and $\underset{\sim}{n}^{(1)}$ are the maximum tensile principal stress and the corresponding principal direction. Though this theory is characteristic because of its close relation to the metallurgical observations, the concrete forms of the above equations have not been proposed yet.

Creep damage theories discussed so far are mainly concerned with initially isotropic materials. Betten (1983) developed a creep constitutive equation of damaged materials of initial anisotropy on the basis of a tensor function theory of non-linear algebra. By characterizing the initial anisotropy of the material by means of a fourth rank tensor $\underset{\sim}{A}$, he first assumed a creep constitutive equation of the form

$$\dot{\underset{\sim}{\varepsilon}}^{c} = \underset{\sim}{f}(\underset{\sim}{\sigma},\ \underset{\sim}{\Phi},\ \underset{\sim}{A}) \tag{4.38a}$$

$$\underset{\sim}{\Phi} = (\underset{\sim}{I}-\underset{\sim}{D})^{-1} \tag{4.38b}$$

where $\underset{\sim}{D}$ is the symmetric second order damage tensor of equation (3.8). Then, he derived the general representation of the tensor function $\underset{\sim}{f}$ in terms of the irreducible set of tensor generators and the integrity basis of the argument tensors. In stead of the argument of equation (4.38a), he then derived two pseudo-tensors $\underset{\sim}{t}$, $\underset{\sim}{\tau}$ constructed by $\underset{\sim}{\sigma}$, $\underset{\sim}{\Phi}$ and $\underset{\sim}{\sigma}$, $\underset{\sim}{A}$, respectively, and developed a few simplified theories in terms of these pseudo-tensors.

4.4 Influence of Prior Plastic Damage on Creep Damage

The plastic damage discussed in Section 4.1 has significant influence also on the subsequent creep rates and the creep rupture time. According to the observations on Nickel base alloys, for example, plastic deformation at room temperature brings about profuse distributed cavities in the material. These cavities deteriorate largely the subsequent creep strength at elevated temperature, and their effects have marked anisotropy [Dyson, Loveday and Rodgers 1976]. Let us show that the damage mechanics is applicable successfully also to this problem [Murakami and Sanomura 1986].

The plastic damage of polycrystalline metals at the temperature range of significant slip or grain boundary sliding usually develops as a results of the cavity formation on grain boundary. As we learned already, the creep damage also is governed by the development of grain boundary cavities. Thus, if we postulate that the damage state of the plastic damage can be characterized by the second rank symmetric tensor of equation (3.8), the damage rate $\overset{\circ}{\underset{\sim}{D}}$ may be decomposed into the sum of the rates of plastic damage $\overset{\circ}{\underset{\sim}{D}}{}^{p}$ and that of creep damage $\overset{\circ}{\underset{\sim}{D}}{}^{c}$:

$$\overset{\circ}{\underset{\sim}{D}} = \overset{\circ}{\underset{\sim}{D}}{}^{p} + \overset{\circ}{\underset{\sim}{D}}{}^{c}$$
$$\overset{\circ}{(\ )} = \dot{(\ )} + (\ )\underset{\sim}{W} - \underset{\sim}{W}(\ ) \tag{4.39}$$

where $\overset{\circ}{(\ )}$, $\dot{(\ )}$ and $\underset{\sim}{W}$ are the Jaumann derivative, material time derivative and spin tensor, respectively.

Dyson et al. applied plastic prestrains of tension, compression and torsion to Nimonic 80A specimens at room temperature, annealed them for two hours at 750°C, and then observed quantitatively the cavities formed in the specimens by use of a high voltage electron microscope. The characteristic features of this observation are:

1) Cavities due to plastic damage were observed mainly on grain boundaries parallel to the direction of maximum principal stress,
2) The volume density of the grain boundary cavities increases with the equivalent plastic strain,
3) The magnitude of the density is larger for tensile prestrains than that for torsion of the same value of equivalent plastic strain.

A simple evolution equation for the plastic damage conforming to these observations can be expressed as follows:

$$\overset{\circ}{\underset{\sim}{D}}{}^{p} = C[\{K_2{}^3 + 4\lambda K_3{}^2\}/\{1+(4/27)\lambda\}]^{l}[2\underset{\sim}{I} - \underset{\sim}{\nu}^{p(1)} \otimes \underset{\sim}{\nu}^{p(1)}]$$
$$\times [(2/3)\mathrm{tr}(\underset{\sim}{d}^{p})^2]^{1/2}$$
$$K_2 = (1/2)\int \mathrm{tr}[\underset{\sim}{d}^{p}(\tau)]^2 d\tau,\ K_3 = (1/3)\int \mathrm{tr}[\underset{\sim}{d}^{p}(\tau)]^3 d\tau \tag{4.40}$$

where $\underset{\sim}{\nu}^{p(1)}$ is the unit vector in the direction of the maximum positive value of the plastic rate of deformation tensor $\underset{\sim}{d}^{p}$, and C, l and λ are material constants.

As regards the evolution equation of creep damage of Nimonic 80A at elevated temperature, we assume that equation (4.27) with $\eta=1$ holds

$$\overset{\circ}{\underset{\sim}{D}}{}^{c} = B[\ \zeta S^{(1)} + (1-\zeta)S_{EQ}]^{k}\ (\underset{\sim}{\nu}^{c(1)} \otimes \underset{\sim}{\nu}^{c(1)}) \tag{4.41a}$$

$$S_{EQ} = [(3/2)\mathrm{tr}(\underset{\sim}{S}_D{}^2)]^{1/2}, \quad \underset{\sim}{S}_D = \underset{\sim}{S} - (1/3)(\mathrm{tr}\underset{\sim}{S})\underset{\sim}{I} \tag{4.41b}$$

The creep curves of this material under constant tension and under constant torsion have no apparent primary creep stage, but show the secondary creep stage from the beginning, and then lead to the tertiary

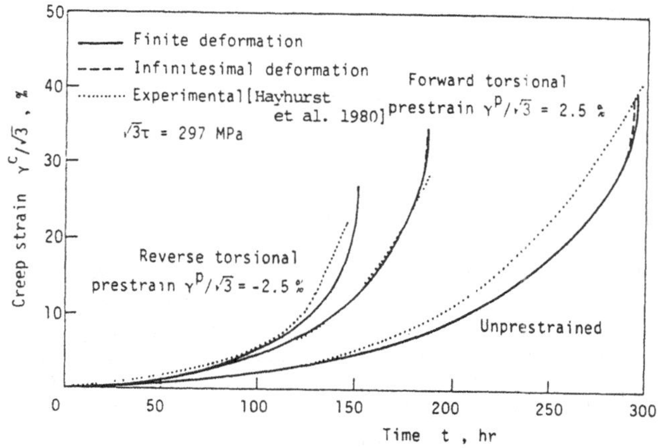

(a) Torsion creep curves after plastic pretorsion

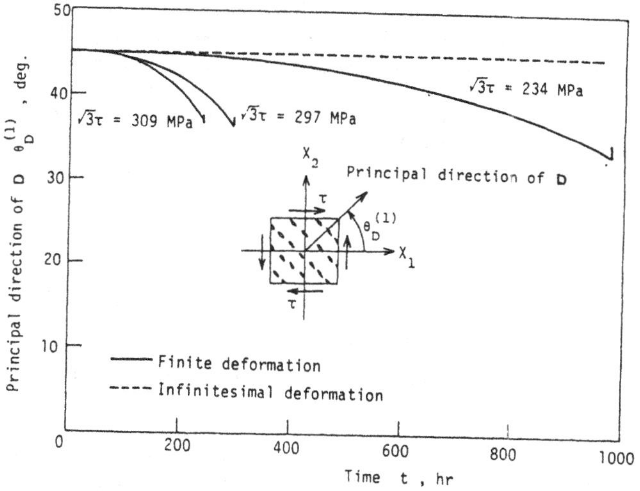

(b) The change of the principal direction $\theta_D^{(1)}$ of the damage tensor $\underset{\sim}{D}$ (unprestrained material)

Fig.4.7 Creep curves and the rotation of the principal direction of damage tensor in Nimonic 80A at 750°C . [From Murakami and Sanomura 1986] stage. As regards this tertiary stage, besides the acceleration induced by the cavity formation as a result of creep damage, material softening due to the dislocation networks at particle interfaces has been pointed out. If we assumed that the change in such dislocation structures in the tertiary creep can be represented by a scalar internal state variable R, the creep constitutive equation of this material can be represented as follows:

$$\underset{\sim}{d}^{c} = (3/2)A[S_{EQ}/R]^{n-1}(\underset{\sim}{S}_{D}/R) \tag{4.42a}$$

$$\dot{R} = P(q-R)[(2/3)tr(\underset{\sim}{d}^{c})^{2}]^{1/2} \tag{4.42b}$$

where A,n,P and q are material constants. The symbols $\underset{\sim}{S}_D$ and S_{EQ} in equation (4.42) are given in equation (4.41).

By use of the above equations and on the basis of finite deformation theory, Murakami and Sanomura (1986) analysed the effects of plastic prestrains at room temperature on the subsequent creep damage process of Nimonic 80A at 750℃.

Fig.4.7 exhibits the predictions of equations (4.39)-(4.42) and the results of the corresponding experiments performed by Hayhurst, Trampczynski and Leckie (1980). Fig.4.7(a), to begin with, shows the torsional creep curves subjected to the torsional plastic prestrain of $\gamma^{P}/\sqrt{3} = \pm 2.5$ percent, together with those for unprestrained material. The solid and the dashed lines in the figure are the numerical prediction of the above equations and the experimental results, while dashed lines are the predictions of the infinitesimal deformation theory. The marked aspects observed in this figure is the significant dependence of the succeeding creep behaviour not only on the prior plastic torsion itself, but also on the direction of the pretorsion. Such dependence of creep rupture times on the direction of pretorsion is attributable to the anisotropic features of the plastic and the creep damage, and cannot be accounted for by the isotropic damage theories reported hitherto. The solid and the dashed lines in Fig.4.7(a) almost coincide with one another. These feature were confirmed also for the cases of plastic pretorsion of other conditions as well as for plastic pretension of some magnitudes. Fig.4.7(b), on the other hand, shows the effect of the rotation of the cavity arrangement (i.e., the rotation of the principal direction of the damage tensor) due to material spin in creep torsion of the unprestrained material.

4.5 References

Betten,J.: Damage Tensors in Continuum Mechanics, J. Mec. Theor. Appl., 2 (1983), 13-32.

Chaboche,J.L.: Viscoplastic Constitutive Equations for the Description of Cyclic and Anisotropic Behaviour of Metals, Bull. de l'Acad. Polonaise des Sciences, 25 (1977), 33-42.

Chaboche,J.L.: The Concept of Effective Stress Applied to Elasticity and Viscoplasticity in the Presence of Anisotropic Damage, in: Mechanical Behavior of Anisotropic Solids (Ed. J.-P.Boehler), Martinus Nijhoff, The Hague (1982), 737-760.

Chaboche,J.L.: Anisotropic Creep Damage in the Framework of Continuum Damage Mechanics, Nuclear Engng Design, 79 (1984), 309-319.

Cordebois,J.P. and F.Sidoroff: Endommagement Anisotrope en Elasticite et Plasticite, J. Mec. Theor. Appl., Numero Special (1982), 45-60.

Davison,L. and A.L.Stevens: Thermodynamical Constitution of Spalling Elastic Bodies, J. Appl. Phys., 44 (1973), 668-674.

Davison,L., A.L.Stevens and M.E. Kipp: Theory of Spall Damage Accumulation in Ductile Metals, J. Mech. Phys. Solids, 25 (1977), 11-28.

Dougill,J.W.: Some Remarks on Path Independence in the Small in Plasticity, Quart. Appl. Math., 33 (1975), 233-243.

Dougill,J.W.: On Stable Progressively Fracturing Solids, J. Appl. Math. Phys. (ZAMP), 27 (1976), 423-437.

Dragon,A. and Z. Mroz: A Continuum Model for Plastic-Brittle Behaviour of Rock and Concrete, Int. J. Engng Sci., 17 (1979), 121-137.

Dyson,B.F., M.S. Lovely and M.J. Rodgers: Grain Boundary Cavitation Under Various States of Applied Stress, Proc. Roy. Soc. London A, 349 (1976), 245-259.

Gurson,A.L.: Continuum Theory of Ductile Rupture by Void Nucleation and Growth, ASME J. Engng Mat. Tech., 99 (1977), 2-15.

Hayhurst,D.R. and F.A. Leckie: The Effect of Creep Constitutive and Damage Relationships upon the Rupture Time of a Solid Circular Torsion Bar, J. Mech. Phys. Solids, 21 (1973), 431-446.

Hayhurst,D.R., W.A. Trampczynski and F.A. Leckie: Creep-Rupture Under Non-Proportional Loading, Acta Metallurgica, 28 (1980), 1171-1183.

Jaunzemis,W.: Continuum Mechanics, Mcmillan, New York (1967).

Kachanov,L.M.: Foundations of Fracture Mechanics, Nauka, Moscow (1974).

Kachanov,M.L.: On the Continuum Theory of Cracked Media, Mekh. Tverdogo Tiela, NO.2 (1972), 54-59 (in Russian).

Kachanov,M.: Continuum Model of Medium with Cracks, J. Engng Mech. Div., Trans. ASCE, EM 5, 106 (1980), 1039-1051.

Krajcinovic,D.: Constitutive Equations for Damaging Materials, ASME J. Appl. Mech., 50 (1983), 355-360.

Krajcinovic,D. and S. Selvaraj: Constitutive Equations for Concrete, in: Constitutive Laws for Engineering Materials, Theory and Application (Ed.C.S.Desai and R.H.Gallagher), University of Arizona, Tucson (1983), 399-406.

Krajcinovic,D. and S. Selvaraj: Creep Rupture of Metals - An Analytical Model, ASME J. Engng Mat. Tech., 106 (1984), 405-409.

Leckie,F.A. and E.T. Onat: Tensorial Nature of Damage Measuring Internal Variables, in: Physical Non-Linearities in Structural Analysis (Ed.J.Hult and J. Lemaitre), Springer, Berlin (1981), 140-155.

Leigh,D.C.: Nonlinear Continuum Mechanics, McGraw-Hill, New York 1968.

Lemaitre,J.: A Continuous Damage Mechanics Model for Ductile Fracture, ASME J. Engng Mat. Tech., 107 (1985), 83-89.

Lemaitre,J. and J.L. Chaboche: Mecanique des Materiaux Solides, Dunod, Paris 1985.

Malvern,L.E.: Introduction to the Mechanics of a Continuous Medium, Prentice-Hall, Englewood Cliffs 1969.

Murakami,S. and N. Ohno: A Continuum Theory of Creep and Creep Damage, in: Creep in Structures (Ed.A.R.S. Ponter and D.R. Hayhurst), Springer, Berlin 1981, 422-444.

Murakami, S., M. Kawai and H. Rong: Finite Element Analysis of Creep Damage Process Based on Anisotropic Creep Damage Theory, Trans. JSME(1987) (in press)

Murakami,S. and N. Ohno: Constitutive Equations of Creep and Creep Damage in Polycrystalline Metals, Memoirs of the Faculty of Engineering, Nagoya University, 36 (1984), 179-190.

Murakami,S., N. Ohno and H. Tagami: Experimental Evaluation of Creep Constitutive Equations for Type 304 Strainless Steel Under Non-Steady Multiaxial States of Stress, ASME J. Engng Mat. Tech., 108 (1986), 119-126..

Murakami,S. and Y. Sanomura: Analysis of the Coupled Effect of Plastic Damage and Creep Damage in Nimonic 80A at Finite Deformation", Engng Fracture Mech., 17 (1986).

Perzyna,P.: Constitutive Modelling of Dissipative Solids for Postcritical Behaviour and Fracture, ASME J. Engng Mat. Tech., 106 (1984), 410-419.

Seaman,L., D.R. Curran and D.A. Shockey: Computational Models for Ductile and Brittle Fracture, J. Appl. Phys., 47 (1976), 4814-4826.

Shockey,D.A., L. Seaman, K.C. Dao and D.R. Curran: Kinetics of Void Development in Fracturing A533B Tensile Bars, ASME J. Engng Mat. Tech., 102 (1980), 14-21.

Trampczynski,W.A., D.R. Hayhurst and F.A. Leckie: Creep Rupture of Copper and Aluminium Under Non-Proportional Loading, J. Mech. Phys. Solids, 29 (1981), 353-374.

Truesdell,C. and W. Noll: The Non-Linear Field Theories of Mechanics, in: Encyclopdia of Physics, III/3 (Ed.S.Flugge), Springer, Berlin 1965.

5. CONCLUSIONS

The present lecture was concerned with the anisotropic features of material damage caused by the distributed internal defects, and with the mechanical description of the anisotropic damage process in engineering materials.

As described in Chapter 2, wide variety of fracture process ranging from quasi-static creep rupture to spalling due to impulsive load is brought about by the nucleation and the growth of distributed microcavities, and such cavity formation is intrisically directional. These damage processes not only lead to final fracture, but also precede the nucleation of discrete macroscopic cracks, or induce the deterioration of the mechanical properties of materials; the damage was observed in metals, polymers, geological materials, ceramics, composites etc.

Mechanical modeling of such anisotropic damage were discussed in Chapter 3. For this purpose, three kinds of measure, i.e., the reduction of load carrying effective area, the change of elastic constants in the damaged materials, and the microscopic character of cavity configuration, were employed to quantify the states of material damage, and showed the possibility of describing such damage states by means of mechanical variables with proper mathematical property (tensorial property).

Finally in Chapter 4, formulation of anisotropic damage theories on elastic-plastic damage, spall damage, creep damage and the coupling of plastic damage and creep damage, and comparison of their predictions with experiments were discussed. As observed in this Chapter, in spite of their utility, continuum mechanics theories for these problems are still in their beginning. As regards creep damage, in paticular, several three-dimensional theories incorporating oriented features of internal deterioration have been proposed so far, and achieved considerable success.

The damage mechanics theories of elastic-brittle and elastic-plastic damage, however, are less developed than the creep damage. This may be attributable mainly to the complicated mechanisms of such damage, and to the lack of experimental results under more general states of stress, mainly as a result of technical difficulty associated with large deformation and strain localization. Damage mechanics theories for fatigue damage and creep-fatigue damage have much more utilities and versatilities in comparison with the conventional ad hoc methods of life prediction, they are limited to one-dimensional theories or three-dimensional theories postulating material isotropy. The difficulties of formulating more elaborate theories for these problems will consists mainly in the complexity of the microscopic mechanisms involved in these kinds of damage. Development of appropriate three-dimensional theories neccessitates intensive future efforts from metallurgical, mechanical and engineering points of view.

MICROMECHANICS OF THE DAMAGE PROCESSES

Dusan Krajcinovic - Dragoslav Sumarac
University of Illinois at Chicago

ABSTRACT

The presented study focuses on the formulation of the damage models based on the actual mesostructural geometry and the kinetics of its irreversible changes. Assuming that the process is sufficiently well defined by the volume averages of the state and internal variables, the overall compliance tensor is derived using both Taylor's and self-consistent approximations. The kinetic equations are derived from the hierarchy of toughnesses at the mesoscale in conjunction with the Griffith's criterion.

INTRODUCTION

The elastic response of a typical engineering material is almost entirely dependent on the primary interatomic forces bonding the atoms together. In contrast, the nonlinear response of a solid, and ultimately its mechanical strength, is attributable to the change in the concentration, distribution, orientation and type of defects in its structure. The increasing demands which the modern technology places on the materials brought in its wake a strong interest in micromechanics as the only rational avenue for a realistic assessment of the material behavior in the nonlinear regime.

Limited by space and time, this particular study will be restricted to microcracks being only one of the many classes of microdefects influencing the mechanical response of solids. The difference between the ductile and brittle behavior in view of the underlying modes of microstructural rearrangements was at some length discussed in [1]. Omitting the details of the discussion, it is important to emphasize the fact that the theory of plasticity provides an almost uniquely inappropriate framework for the development of analytical models for the analysis of processes dominated by the nucleation, growth and coalescence of microcracks. From a purely physical standpoint, the microcracking is characterized by a net loss of atomic bonds while the plastic flow reflects the slip of matter through the crystalline lattice during which the number of bonds remains to a large extent constant. The essential difference between the two processes is emphasized in unloading. In fact, as shown in [2], the difference between the slopes of the initial segment of the stress-strain curve in loading and the unloading path is an appropriate way of measuring the accumulated damage.

The relation between the continuum damage mechanics and the fracture mechanics is of more insidious nature and is, in essence, a question of scale. The prominent role of scale can be clarified by the following discussion of the energies needed to propagate a crack in an elastic solid. According to Stroh [3], the energy needed to initiate a crack exceeds the energy needed to propagate the same crack through a homogeneous and isotropic elastic solid. Thus excluding the dynamic (inertial) effects, a crack nucleated in a homogeneous, elastic solid could be arrested only if the level of the externally supplied energy is lowered. Consequently, if the a microcrack is nucleated, the ultimate rupture is imminent if the external load is ever so slightly increased. However, even a cursory glance at one of the many micrographs available in the published literature demonstrates a reasonably large population of microcracks in most brittle solids showing no serious intentions of falling apart.

The answer to the apparent paradox of the Stroh's model is in fact very simple. A microcrack, being the size of a grain or any second-phase particle, sees the surrounding medium as strongly inhomogeneous and fortified by a network of energy barriers (such as grain boundaries or any other surfaces or volumes of higher toughness). These crack arresting barriers can be penetrated only at the expense of the additional externally supplied energy. On the other hand, a macrocrack traversing many grain boundaries can be assumed to be surrounded by a homogeneous (in a statistical sense) medium.

The problem of scale is in many ways a product of the choice of analytical models and their structure. The strong preference for continuum models dealing with homogeneous continua and gradually and smoothly varying parameters and variables is, of course, fully justified by their relatively simple mathematical structure and the attendant computational efficiency. However, the range of the application of such a theory is limited to the volumes of matter which contain a statistically large number of samples (defects, grains, etc.) rendering the averaging meaningful.

At some danger of oversimplification [1,4], the entire spectrum of sizes can be divided into three distinct scales (Table 1).

Table 1: Volume Scales and Their Relation to Analytical Models

Scale	Material	Defects	Models
Micro	Atoms, molecule chains	Vacancies and dislocations	Physics and material sciences
Meso	Ensemble of grains	Microcracks, pores, slips	Micromechanics
Macro	Test specimen structure	Macrocracks, shear band	Continuum theories

All analytical models used in engineering describe the events on the two latter scales. This includes the Continuum Damage Mechanics (CDM) as well. The phenomenological models, alluring in their simplicity, are primarily focused on the macroevents, i.e., average values of state and internal variables. However, since the kinetics of a microcrack depend on the mesostructure formulation of theories based solely on the observations of macroscale events, it is at the very best a perilous venture. In view of the simplicity of the continuum theories and the physical foundation of micromechanical models, a promising strategy would consist of combining the best features of both models. Specifically, a particular mode of structural rearrangement is observed and analyzed on the mesoscale and a homogenization (or averaging) procedure is subsequently applied providing for a transition to a continuum model containing only the salient features of the physical process.

Some of the recent controversies related to the applicability of the fracture mechanics to a material such as concrete [4], rephrased occasionally as a question of whether the stress intensity factor or a J integral are material parameters, can be resolved introducing the scale effect. In the case of notched specimens, the propagation of the major crack is a dominant energy sink. Thus an appropriately modified stress intensity factor (using, for example, the crack layer model [5]) would take care of the problem. However, in the case of the unnotched specimens, most of the energy is dissipated in creating new and propagating already existing microcracks [6,7]. Thus a model based on a single macrocrack will miss most of the action. In other words, in the case of unnotched specimens responding in a brittle mode (characterized by a dominant role of microcracks in energy dissipation), the CDM presents a rational framework for the development of an analytical model.

BASIC CONCEPTS AND DEFINITIONS

A certain degree of confusion related to the basic concepts and definitions is typical of the adolescence of any developing theory. The CDM is by no means an exception having its share of different models based on different interpretations of physical reality. Thus, a cursory discussion of the terminology to be used in the sequel might prove to be useful.

The structure of a solid is, for present purposes, defined by its macroparameters (such as elastic moduli), distance D separating energy barriers and the characteristic lengths, and a defining the size of the defect. An energy barrier is defined simply as a region or plane of higher toughness (crack resistance) capable of arresting the crack. In the case of polycrystalline solids, a typical energy barrier is a grain boundary (unless wetted as in some ceramics), while the same role is played by aggregates in concrete. Thus the size of a grain will, in fact, be defined in polycrystalline solids as the distance D which is in most cases a random variable.

For the purpose of this discussion, assume that the geometry of an inclusion (defect) can be approximated by an ellipsoid with semi-axes $a_1 > a_2 > a_3$. If its stiffness (i.e., ability to transmit stresses) is vanishingly small, a defect can be classified either as a void or a crack. A defect is considered to be a crack if $a_3 \to 0$ (high aspect ratio). Finally, a crack will be considered to be a microcrack if $a_1 \ll D/2$.

As already mentioned above, the distinction between the microcrack and a macrocrack is essential from the analytical standpoint. A macrocrack is assumed to be embedded into a homogeneous solid and is cognizant of the mesostructure of the material only in a statistical (average) sense. Consequently, the pattern of its growth is defined solely by the state of stress and the macro geometry of the specimen. In contrast, the size of the microcrack and the pattern of its growth is strongly influenced by the distribution of the energy barriers.

A crack will further be called active if the displacement field across its plane is discontinuous. Thus the damage is not in itself an internal variable but just a consequence of the status of the existing cracks (i.e., if the cracks are present but none are active, the material behaves as undamaged). Furthermore, in order to distinguish between different classes of active cracks, it is convenient to speak of open cracks (across which the normal stress is non-negative) and closed cracks (which are compressed across its own plane but allow the displacement discontinuity along its surface).

The consideration of a solid weakened by an ensemble of a large number of cracks arranged into irregular patterns represents a very difficult analytical problem which, in general, does not admit a closed-form analytical solution. A common strategy in dealing with these types of problems is to assume that a considered crack is surrounded by some effective medium which in some appropriately smeared (averaged or homogenized) sense reflects the influence of adjacent cracks.

The simplest strategy, used originally by Taylor in the context of the slip theory, is to neglect totally the influence of all other cracks and analyze each crack as being embedded into the virgin, undamaged material. This method obviously overestimates the stiffness of the medium and its applicability is, strictly speaking, restricted to small microcrack concentrations.

The more realistic model is provided by the so-called effective field theories. Every crack is considered to be embedded in isolation into a homogeneous medium, the moduli of which depend not only on the matrix but on the crack distribution and concentration as well. In the case of the self-consistent theories, the effective field is assumed to be constant and identical for all cracks. Notably, the actual distribution of microcracks is in this case assumed to be irrelevant. In other words, the microcrack pattern is assumed to be well defined by the volume average (and orientation) of the microcracks. Thus this type of a theory should be applicable in the case of modest levels of crack concentrations (weak interaction). An even better, but much more complicated, model is obtained introducing higher statistical momenta of the stress and strain field (see Kunin [8]). This type of model, involving a significant increase in computational effort, avoids some of the problems inherent to self-consistent models for the case of large concentrations of microcracks. As such this model, which will not be pursued in the sequel, should in all probability be used during the phase preceding the onset of localization (i.e., the process during which the microcracks coalesce into a macrocrack). More importantly, a model of this type will provide not only the expected values but also the dispersion and fluctuation of all involved variables and parameters.

In summary, the continuum damage models to be discussed in this study should provide a transition between the theory of elasticity (during which no dissipation takes place) and the fracture mechanics (during which a macrocrack is the dominant source of dissipation). In many cases, however, the role of the fracture mechanics becomes insignificant since the coalescence of microcracks into a large macrocrack is followed almost instantaneously by a global loss of integrity.

COMPLIANCE TENSOR

General Considerations

Consider a relatively large sample of a perfectly elastic and brittle solid of total volume V bounded by the exterior surface A. The sample contains a large number of crack-like defects. Each defect (or better, slit) in the elastic matrix is assumed to be elliptical in shape with area $A^{(k)}$ and negligible volume. The shape of each defect is fully determined by two scalars, say the length a_i of two semi-axes of the ellipse. The position of each defect is determined by three scalars, i.e. coordinates $\underset{\sim}{x}$ measured along the axes of a global rectangular and cartesian coordinate system with unit vectors e_i (i=1, 2 3). In addition, define N local cartesian coordinate systems $\underset{\sim}{x}'$ embedded into each of the N cracks and oriented in such a manner that the normal to the crack surface $n^{(k)}(x)$ is colinear with the axis e_3'. The axes e_1' and e_2' are colinear with the principal axes of the ellipse. The relative orientation between the two cartesian coordinate systems is fully defined by three Euler angles ϕ, θ and ψ. Thus the geometry of the crack pattern within the sample is defined by 8N scalars (two radii of the crack surface, three coordinates of the crack center and three Euler angles).

The number of variables needed to describe the microcrack pattern is often reduced by 2N considering the cracks to be circular in planform. This assumption is justified by computations performed by Budiansky and O'Connell [8] according to which the ellipticity of cracks has a negligible influence on the overall material compliance.

The precise information related to the details of the microcrack pattern is further complicated by a fairly irregular pattern reflecting the effects of the inhomogeneities in the mesostructure of the solid. A rather random disposition of energy barriers and domains of inferior toughnesses (such as wetted grain boundaries, cleavage planes in crystals or aggregate-cement interfaces) or cohesion have almost as much influence on the eventual crack configuration as the state of the stress itself.

Additionally, the exact disposition of cracks and the fine details of their random arrangement are not reproducible from one experiment to the other performed on virtually identical (in the macroscopic sense) specimens. Since the macroscopically similar specimens subjected to identical force and displacement boundary conditions respond in a virtually identical manner, it is often concluded that the finer details of the microcrack pattern and the mesostructure as a whole have a limited influence on the macro-response. This is extremely encouraging since the exact analysis based on a model containing 8N random variables could be hardly considered appealing for practical applications.

In summary, the 'exact' theory of an elastic solid weakened by an ensemble of microdefects is inherently non-local. The distance separating the cracks is obviously the additional scale parameter which must be introduced to enable consideration of the 'strong' interaction. However, for reasonably dilute concentrations, the zeroth approximation (see, for example, Kunin [8]) becomes an appealing alternative. According to this approximation, the so-called effective tensor of elastic moduli is considered to be sufficient to capture the salient properties of the inhomogeneous (and often discontinuous) medium. The underlying premise is that the external field for a given crack weakly depends on the surrounding defects which can, therefore, be smeared out into an effective medium.

Consequently, in the sense of the self-consistent theory [8,9], every microcrack can be considered as an isolated defect embedded into an effective, elastic and anisotropic continuum. The properties of this effective continuum depend on the crack density and configuration in some averaged sense. The overall external field is obtained by a simple juxtaposition of the perturbations associated with each of the existing cracks considered independently.

Derivation of the Overall Compliance Tensor

The considerations within this study will be restricted to specimens with:

- a perfectly elastic matrix,
- a reasonably dilute concentration of planar microcracks, and time independent, quasi-static processes under isothermal conditions.

Assuming the response to be perfectly brittle, the strains can be considered as being infinitesimal. In view of the listed restrictions, emphasizing microcracking as the sole, or at least dominant, energy sink, the present study implicitly focuses on the materials such as rocks, concrete, ceramics, etc., in unconfined conditions.

According to the computations performed by Delameter and Herrmann [11] and Kunin [9], the strong interaction of cracks may be considered significant when the size of the crack exceeds the distance separating the adjacent cracks. In the case of polycrystalline solids, this would occur after the cracks start penetrating the grain boundaries. At that point, the crack may be assumed to be a macrocrack and the rupture is probably imminent. The similar situation occurs in concrete when the original interfacial crack starts propagating through the cement paste in an essentially runaway fashion. Consequently, the strong interaction becomes, in all probability, important just before the onset of the global failure which may occur either through the coalescence of several neighboring microcracks or as a result of a runaway propagation of a single, preferentially oriented and situated single microcrack.

As a result of the restriction to weak interaction, every microcrack in the specimen can be considered independently of the others. The influence of the adjacent microcracks will be reflected, in an appropriately smoothed sense, through the effective moduli determined on the basis of some physically acceptable considerations (such as the self-consistent method). The macro-response is then obtained through the juxtaposition of the contributions (external fields) of all individual active cracks. The question of superposition and the attendant errors was examined in detail by Margolin [11]. Since the number of cracks N is typically very large, the sum evolves into an integral taken over the domain of all active cracks. The argument of the integral contains probability density functions of the crack density, size and orientation consistent with the mesostructure of the solid. The fields determined in this manner are, consequently, the averages over the entire ensemble of realizations.

Consider a specimen defined above subjected to a homogeneous far field stress $\underset{\sim}{\sigma}^{o}$ in equilibrium with a given set of externally applied tractions $F_i(x \in A)$, such that in the absence of body forces

$$\overset{o}{\sigma}_{ij,j} = 0 \qquad x \in V_M \tag{1}$$

and

$$\overset{o}{\sigma}_{ij} n_j = F_i \qquad x \in A \tag{2}$$

where $\underset{\sim}{n}(x)$ is a normal to the external surface A.

The corresponding strain tensor is defined by the mapping

$$\overset{o}{\varepsilon}_{ij} = \overset{o}{S}_{ijkm} \overset{o}{\sigma}_{km} \tag{3}$$

where $\underset{\sim}{S}^{o}(\underset{\sim}{x})$ is the elastic compliance tensor reflecting all symmetries and other properties of the elastic matrix (such as possible anisotropy related to the orientation of mesostructure and grain boundaries).

The presence of the crack-like microdefects introduces disturbances $\underset{\sim}{\sigma}$ and $\underset{\sim}{\varepsilon}$ into the stress and strain field. The total stresses and strains in the matrix must satisfy the relationship

$$\underset{\sim}{\varepsilon}^{o} + \underset{\sim}{\varepsilon} = \underset{\sim}{S}^{o} : (\underset{\sim}{\sigma}^{o} + \underset{\sim}{\sigma}) \qquad x \in V_M \tag{4}$$

In view of (1) and (2), the stress disturbance must satisfy the following conditions

$$\sigma_{ij,j} = 0 \qquad \underset{\sim}{x} \in V_M \tag{5}$$

$$\sigma_{ij} n_j = 0 \qquad \underset{\sim}{x} \in A \tag{6}$$

The tractions on each of the surfaces of an open (and dry) crack must vanish,

$$(\overset{o}{\sigma}_{ij} + \sigma_{ij}) n_j^{(k)} = 0 \qquad \underset{\sim}{x} \in A^{(k)} \tag{7}$$

while the displacement across the open crack is, by definition, discontinuous,

$$[\overset{o}{u}_i + u_i]^{(k)} = b_i^{(k)} \qquad \underset{\sim}{x} \in A^{(k)} \tag{8}$$

where $\underset{\sim}{b}$ is the displacement discontinuity vector, while the brackets signify the jump in the magnitude of the bracketed variable (here the displacement $\underset{\sim}{u}$) across the crack.

The next step in the derivation consists of the determination of the relation existing between the crack opening displacement $\underset{\sim}{b}$ and the far field stress σ^o. Since a crack is a limiting case of an ellipsoidal inclusion (of vanishing stiffness and thickness), it is possible to use the Eshelby's [13] result according to which the eigenstrains and eigenstresses within the inclusion are homogeneous providing that the far field stress $\underset{\sim}{\sigma}^o$ is homogeneous as well. Orienting the axis e_3' along the normal to the crack, the transition from the ellipsoid to the elliptical crack requires (see, for example, Mura [10], Ch. 5) that

$$\lim_{a_3 \to 0} a_3 \varepsilon_{3i}^* = b_i' n_3' \qquad \text{(no summation)} \tag{9}$$

where the asterisk denotes eigenstrain, while a_3 is the length of the semi-axis of the ellipsoid in the direction of e_3' and n_3' the normal to the crack.

Since, according to [10], the eigenstrain $\underset{\sim}{\varepsilon}^*$ is constant within the inclusion, the ellipsoid deforms into another ellipsoid (Hoenig [14]). Consequently, the displacement discontinuity must be proportional to $\{1-(x_i'/a_i)^2\}^{1/2}$. Furthermore, since the problem is linear, b_i must be proportional to $B_{ik}\sigma_{3k}^{o\prime}$, where B_{ik} is a matrix of influence coefficients which can be shown to be symmetric (Hoenig [14]).
Consequently,

$$b_i'^{(k)} = \left\{1 - (x_j'/a_j)^2\right\}^{1/2} B_{ik}' \sigma_{3k}^{o\prime} a_1 \tag{10}$$

where $x_j' \leq a_j$ is measured along the coordinate axis e_j'.

Formally simple, expression (10) conceals the fact that the determination of the crack opening displacement in an anisotropic solid represents a non-trivial computational problem. As shown in Mura [10] (Sect. 28), in the absence of a closed-form for the Green's function for the displacements in an anistropic solid, the determination of the crack opening displacement $\underset{\sim}{b}$ involves delicate numerical quadratures of double integrals with complicated arguments.

The two final steps leading to the determination of all tensors in the constitutive equation (4) consists of deriving the functional relationship between the eigenstrains $\underset{\sim}{\varepsilon}^*$ and the stress and strain disturbances $\underset{\sim}{\sigma}$ and $\underset{\sim}{\varepsilon}$ in the exterior of the inclusion (crack). This task, for much the same reasons mentioned above, involves arduous computations involving Eshelby tensor and elliptic integrals [10].

Thus, in conclusion, knowing the far field stress $\underset{\sim}{\sigma}^o$ and the compliance tensor of the elastic matrix $\underset{\sim}{S}^o(x)$ it is, at least in principle, possible to determine the stress and strain fields in every point of an anisotropic solid weakened by a single crack-like defect. The solution, however, involves non-trivial computational effort. The stress and strain fields will be strongly inhomogeneous with peaks coinciding with the position of cracks distributed in a rather irregular fashion.

However, as indicated before, there is a valid reason to suspect that the macro-response of solid is dominated by the local averages of the stress and strain field. The local fluctuations are often believed to have a limited influence on the salient and reproducible features of the macroscopic behavior. Assuming, therefore, that the stress and strain disturbances are a second order effect (prior to the onset of localization), the problem reduces to the determination of the constitutive law relating the averages of the stresses and strains.

Since $x_{j,k} = \delta_{jk}$ in a cartesian coordinate system, the total stress averaged over the volume of the unit cell is

$$V\bar{\sigma}_{ij} = \int_{V_M} \sigma_{ij}\, dV = \int_{V_M} \sigma_{ik} x_{j,k}\, dV$$

Integrating the expression above by parts

$$V\bar{\sigma}_{ij} = \int_{A} \sigma_{ik} n_k x_j\, dA - \int_{V_M} \sigma_{ik,k} x_j\, dV$$

and using (5) and (6), it directly follows that the volume average of the stress disturbance

$$\int_{V_M} \overset{*}{\sigma}_{ij}\, dV = 0 \tag{11}$$

vanishes as expected. Consequently, the far field stress is at the same time the expected value of the stress

$$\frac{1}{V}\int_{V_M} (\overset{o}{\sigma}_{ij} + \overset{*}{\sigma}_{ij})\, dV = \overset{o}{\sigma}_{ij} = \bar{\sigma}_{ij} \tag{12}$$

While the stress exists only in the matrix, the strain must be averaged over the entire volume consisting of the elastic matrix and N voids (cracks). The expected value of the strain in the matrix is obtained directly from (1) and (12). Since the eigenstrains within the void are homogeneous, in conjunction with expression (9), the final expression for the expected value of the strain in the volume V is

$$\frac{1}{V}\int_V (\overset{o}{\varepsilon}_{ij}+\varepsilon_{ij})\, dV = \overset{o}{S}_{ijkm}\,\frac{1}{V}\int_{V_M}(\overset{o}{\sigma}_{km}+\sigma_{km})\, dV + \frac{1}{V}\sum_k \int_{V^{(k)}} \varepsilon^{*}_{ij}\, dV$$

i.e.,

$$\bar{\varepsilon}_{ij} = \overset{o}{S}_{ijkm}\bar{\sigma}_{km} + \frac{1}{V}\sum_k V^{(k)}\varepsilon^{*(k)}_{ij} \tag{13}$$

where the sum extends over the entire ensemble of active cracks in the unit cell.

In the case of a flat crack, the quantity $\underset{\sim}{\varepsilon}^*(x_1,x_2;a_3)$ can be interpreted [9] as the coefficient of the principal term of the expansion of the function $\underset{\sim}{\varepsilon}^*(\underset{\sim}{x})\, dV$ into multipoles concentrated on the crack surface $dA^{(k)}$ Thus,

$$\varepsilon^{(k)*}_{ij}(\underset{\sim}{x}')A(\underset{\sim}{x}') = \varepsilon^{(k)*}_{ij}(\underset{\sim}{x}';a_3)\delta(A) + \ldots \tag{14}$$

where, by definition,

$$\int_{V^{(k)}} \delta(A)\, d\underset{\sim}{x}' = \int_{A^{(k)}} dA \tag{15}$$

Consequently, from (9) and (14), since in the case of flat cracks $\underset{\sim}{n} \neq \underset{\sim}{n}(A)$ (i.e., all normals to the crack surface are parallel)

$$a_3\bar{\varepsilon}^{(k)*}_{ij} = \frac{1}{2}(n_i\bar{b}_j + n_j\bar{b}_i)^{(k)} \tag{16}$$

where

$$A^{(k)}\bar{b}^{(k)}_i = \int_{A^{(k)}} b_i\, dA \qquad x' \in A^{(k)} \tag{17}$$

is the crack opening displacement averaged over the entire surface of the crack.

The subsequent steps leading to the derivation of the overall compliance involves simply the substitution of the expressions (10) and (16) into (13). The ensuing expression will relate the expected (or macro) values of stresses and strains. In the process, the expression (10) for the displacement discontinuity vector must first be transformed from the local to the global coordinate system by means of the transformation

$$e'_i = g_{ij}e_j \tag{18}$$

where the components of the coordinate transformation (rotation) matrix are trigonometric functions of the Euler angles. For the case of circular cracks (Fig. 1), the matrix [g] reads

$$[g] = \begin{bmatrix} \cos\phi & \sin\phi & 0 \\ -\cos\theta\sin\phi & \cos\theta\cos\phi & \sin\theta \\ \sin\theta\sin\phi & -\sin\theta\cos\phi & \cos\theta \end{bmatrix} \tag{19}$$

From (18)

$$b_i' = g_{ij} b_j \quad \text{and} \quad \sigma_{ij}' = g_{im} g_{jn} \sigma_{mn} \tag{20}$$

The transformation (18) is obviously orthogonal such that

$$g_{im} g_{mj} = \delta_{ij} \tag{21}$$

and

$$n_i = g_{i3} \tag{22}$$

Mindful of the fact that the expressions for the second order tensor $\underset{\sim}{B}$ in (10) (i.e., the crack opening displacements) have their simplest form in the local coordinate system, the final expression for the eigenstrain can be written in a concise form [1,15] as

$$\varepsilon_{ij}^{(k)*} = a_{(k)}^{3} b_{ijmn}^{(k)} \bar{\sigma}_{mn} \tag{23}$$

where the fourth order tensor $\underset{\sim}{b}$ has the form

$$b_{ijmn} = \beta_{ijmn} + \beta_{ijnm} \tag{24}$$

with

$$\beta_{ijmn} = \frac{\pi}{3} B_{pq}' (g_{pi} g_{qn} n_m n_j + g_{pj} g_{qn} n_m n_i) \tag{25}$$

Consequently, the final expression mapping the macrostresses into macrostrains is obtained introducing (23) into (13)

$$\bar{\varepsilon}_{ij} = \left(\overset{o}{S}_{ijmn} + \sum_k \frac{a_{(k)}^3}{V} b_{ijmn}^{(k)} \right) \bar{\sigma}_{mn} \tag{26}$$

or

$$\bar{\varepsilon}_{ij} = (\overset{o}{S}_{ijmn} + \overset{*}{S}_{ijmn}) \bar{\sigma}_{mn} = \bar{S}_{ijmn} \bar{\sigma}_{mn} \tag{27}$$

The overall compliance tensor $\underset{\sim}{\bar{S}}$ is comprised of two terms: the first reflects the elastic properties of the undamaged matrix and the second term provides an estimate of the influence

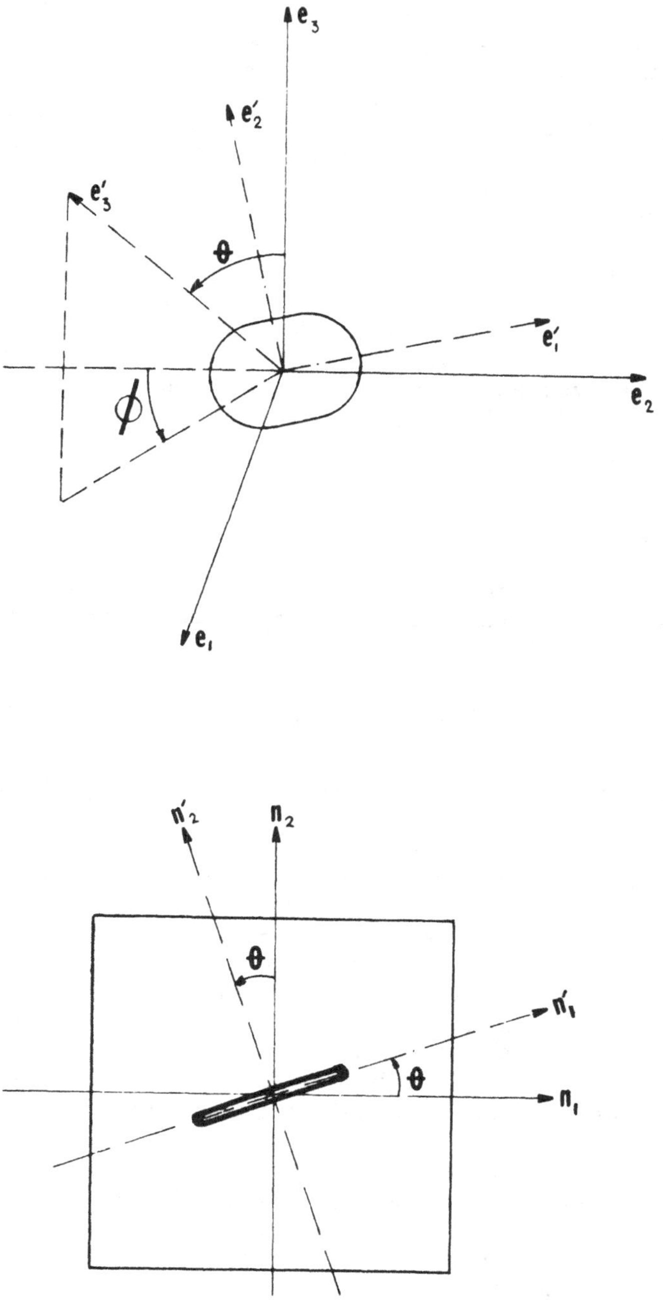

Fig. 1. The crack (primed) and global cartesian coordinate systems for the three- and two-dimensional cases.

of the displacement discontinuities across the cracks on the macro response of the specimen.

The expressions (24) and (25) are valid in the case of open cracks characterized by tensile stress $\sigma'_{33} > 0$ in the direction along the normal to the crack surface. Since the materials such as rocks or concrete are commonly subjected to loading conditions emphasizing compressive stresses, it is of considerable interest to examine the case of closed cracks ($\sigma'_{33} < 0$). A closed crack is active only if the Coulomb's condition

$$\left|\bar{\sigma}'_{31}\right| \geq -\mu\bar{\sigma}'_{33} \tag{28}$$

is satisfied. In (28), μ is the coefficient of sliding friction (regarded as a positive constant) while $\bar{\sigma}_{13}$ is the shear stress resolved on the plane of the crack. Thus, the net stress causing the discontinuity in the displacements across the crack is

$$\tilde{\sigma}'_{3j} = \bar{\sigma}'_{3j} + \mu\bar{\sigma}'_{33} \qquad (j = 1,2) \tag{29}$$

In view of the transformation (20), the fourth order tensor (24) can be derived, following transformations identical to ones used in the case of open cracks, in the form of [15,16]

$$b_{ikmn} = \frac{\pi}{3} B_{pj}(n_k g_{pi} + n_i g_{pk})(g_{jn} - \mu\delta_{2j}n_n)n_m \tag{30}$$

The overall compliance (27) is obtained substituting the expression (30) into (26).

As already indicated, the major problem in the computation of the elements of the overall compliance $\bar{\underset{\sim}{S}}$ (27) consists of the determination of the crack opening displacement $\underset{\sim}{b}$, i.e., components of the tensor $\underset{\sim}{B}$, as a function of the far field stresses $\underset{\sim}{\sigma}^o$. In fact, a closed-form expression relating $\underset{\sim}{b}$ and $\underset{\sim}{\sigma}^o$ is available in the literature only for the case of a penny-shaped or elliptic crack in an isotropic and homogeneous elastic solid and for the two-dimensional case of a slit in an anisotropic body. In all other cases, the determination of the crack opening displacement requires considerable numerical effort.

The most radical approximation leading to the simplest model is based on the assumption that the behavior of each crack is not influenced by the other cracks at all. In other words, implicit to the application of this model is the assumption that the microcrack concentration is too dilute to have a strong effect on the crack opening displacement of any crack. In analogy to the slip theory, this model will be referred to in the sequel as the Taylor model. Thus assuming that $\underset{\sim}{B}(\underset{\sim}{S}) \approx \underset{\sim}{B}(\underset{\sim}{S}^o)$, using the well-known expressions for the crack opening displacements for a penny-shaped crack embedded into an isotropic and homogeneous elastic solid [17], the non-vanishing components of the tensor B are

$$\begin{Bmatrix} B'_{11} \\ B'_{22} \\ B'_{33} \end{Bmatrix} = B \begin{Bmatrix} 2/(2-\nu_0) \\ 2/(2-\nu_0) \\ 1 \end{Bmatrix}$$

where (31)

$$B = 8\,\frac{1-\nu_0^2}{\pi E_0}$$

The subscript 'o' in (31) indicates the reference to the undamaged (virgin) isotropic material. This assumption, alluring in its simplicity, neglects the already accumulated damage and, consequently, overestimates the material stiffness and underestimates the strains attributable to the presence of cracks. As such, the Taylor type of a solution presents a lower bound on the strains. Substitution of (31) into (30) leads to the following final expression for the compliance tensor $\underset{\sim}{S}^*$ in Taylor's approximation

$$S^*_{ijmn} = \sum \frac{a^3}{V}\,\frac{8(1-\nu_0^2)}{3E_0(2-\nu_0)}\,(\delta_{im}n_jn_n + \delta_{in}n_jn_m$$

$$+ \delta_{jm}n_nn_i + \delta_{jn}n_in_m - 2\nu_0 n_in_jn_mn_n) \tag{32}$$

The upper bound on the solution may be obtained assuming that the largest concentration of the damage is isotropically distributed in all planes (akin to the phenomenological model suggested by Lemaitre and Chaboche [2]). In this case, the compliance tensors $\underset{\sim}{S}^*$ and $\underset{\sim}{S}^0$ are proportional and the overall compliance can be written in the form

$$\bar{\underset{\sim}{S}} = (1+\omega)\underset{\sim}{S}^0 \tag{33}$$

where ω is an appropriately selected scalar measure of the accumulated damage (in this case, the 'damage' in the plane perpendicular to the principal tensile stress). Thus, if $\omega \ll 1$, the overall stiffness tensor obtained by the inversion of (33) is of the familiar form

$$\bar{\underset{\sim}{C}} = (1+\omega)^{-1}\underset{\sim}{C}^0 \tag{34}$$

proposed in many papers dealing with the continuum damage theory [2,18,19].

If the microcrack concentration is not dilute enough to allow for the application of a Taylor type of model, it is often necessary to apply the so-called self-consistent model (SCM). According to this model (see, for example, Budiansky and O'Connell [8], Mura [10], Horii and Nemat-Nasser [20], etc.), the influence of the adjacent cracks is reflected in a smeared sense through the material parameters of an effective medium. Since the crack distribution is rather random, the effective medium is, consequently, anisotropic. Moreover, the crack opening displacements are both explicit and implicit functions of the components of the overall compliance tensor. Thus, the involved computations are by no means simple involving by necessity solutions of a system of algebraic equations in conjunction with a predictor-corrector numerical scheme.

The analyses in cases of strong interaction are, as a rule, rather complex. Even after the introduction of a string of simplifying assumptions, the 'exact' formulations of Kanaun [21] and Kunin [9] are too complicated for practical application. The approximations suggested by M. Kachanov [22] seem rather interesting but have as yet to be incorporated into a comprehensive CDM model.

Leaving the determination of the compliance tensor according to specific models to the latter part of this section, it appears necessary to discuss first the general form of the compliance tensor $\underset{\sim}{S}^*$ in the case of a large number of cracks. In the case of a substantial number of cracks, typical to the application of rocks, concrete, ceramics and some brittle metals, it is computationally advantageous to use integration over the region of all active cracks instead of summation over all active cracks (M. Kachanov [23], Krajcinovic and Fanella [24], Krajcinovic [1], etc.). The tensor S^* in (27) is simply recast into the form of a triple integral

$$\underset{\sim}{S}^* = \frac{N}{V}\int_H a^3 \underset{\sim}{b}(\phi,\theta)\,p(a)\,p(\phi)\,p(\theta)\sin\theta\;da\,d\phi\,d\theta \tag{35}$$

where N is the number of all active cracks, $p(a)$, $p(\phi)$ and $p(\theta)$ the probability density functions defining the microcrack radius and orientation. The probability density functions on the right-hand side of expression (35) are assumed to be statistically independent of each other. This assumption is justified by the fact that the microcrack size is by necessity equal to the distance separating the adjacent energy barriers (for example, grain boundaries in case of the polycrystalline solids). Since the distribution of the energy barriers is typically independent of orientation, the probability density functions in the integrand on the right-hand side of (35) are independent of each other as well. In most cases, $p(\phi)$ and $p(\theta)$ are uniform over the entire hemisphere of orientations. Conversely, $p(a)$ depends on the size of grains or aggregates (sieve grading in case of concrete). The domain of integration

H in the space of (a,ϕ,θ) is defined as a locus of all active cracks.

Thus, the problem of the determination of the compliance of a specimen weakened by an ensemble of many microcracks well diffused over a large part of its volume is reduced to the quadrature of the integral (35). The completion of this task is contingent on the knowledge of all involved probability density functions and the domain of integration H. The determination of the domain of integration will be discussed in detail in the following section of this study.

While the constitutive law, i.e., expressions for the overall compliance (27) in conjunction with (35) has an enticingly simple form the entire string of assumptions contributing to its simplicity must be kept in mind. This is especially the case since each simplifying assumption at the same time represents a limitation on the applicability of the model.

KINETIC EQUATIONS

While the so-called non-process models are restricted to the assessment of the influence which the mesostructural defects have on the relations between macroscopic stresses and strains, the formulation of a process model requires in addition the establishment of the kinetic equations defining the rate of the nucleation and growth of defects as a function of the externally supplied energy. The formulation of kinetic equations governing the nucleation and growth of microcracks (damage evolution law) consists of two different, but related, tasks:

- establishment of a mathematical expression defining the condition under which a particular crack becomes unstable and starts growing, and
- definition of a incremental relationship between the fluxes and affinities governing the rate of the damage growth.

Since the second law of thermodynamics in most cases places only trivial constraints on the analytical models, the purely phenomenological theories are often susceptible to arbitrary interpretations of the phenomenon having but a tenuous relationship to the physical reality. The never-ending flood of flow rules in the plasticity theory, a century old and showing no signs of abatement, is a vivid testimony to the fragile relationship between the great majority of phenomenological models and the underlying physical phenomenon.

Since a microcrack, as a result of its size, sees the surrounding material as inhomogeneous and isotropic, a rational formulation of the kinetic law requires appropriate description of the solid structure on the mesoscale. In particular, most of the experimental evidence clearly indicates that the pattern of the microcrack growth depends not only on the state of stress, but also on the distribution of planes characterized by lower toughness or inferior cohesion. Moreover, once destabilized, a microcrack can get arrested as it impinges an energy barrier of superior toughness. Consequently, a rational formulation of the kinetic equations inherently involves a geometric description of the distribution of toughnesses on the mesoscale. Such a description is, in most cases, of stochastic nature.

It can be said that, in general, the process of gradual loss of integrity of a solid attributable to the increasing concentration of microcracks depends on:

- the existence of initial cracks associated with the manufacturing process or the previous loading histories,
- highly localized deformations and the attendant stress concentrations,
- substantial differences in fracture toughnesses of the constituent phases, and
- large fluctuations of the stresses attributable to the inhomogeneity of the mesoscale structure.

Due to the vast disparity in the mesoscale structure of various materials it is, of course, not a reasonable thing to expect that a single model leading to a general kinetic equation applicable over the entire spectrum of solids is possible. For present purposes, the discussion will be focused exclusively on the 'cleavage 1' solids in Ashby's [25] classification. This type of process, typical of unconfined rock, concrete and ceramic specimen, is characterized by the growth of the initially existing defects in an almost total absence of plastic deformation. In other words, the number of cracks N remains constant while the size of cracks measured by some characteristic length a (in the present case radius) increases.

In tune with the preceding derivation, consider again an isolated crack embedded into an effective (homogenized) perfectly elastic and brittle continuum. The three commonly used criteria defining the circumstances leading to the loss of stability of a crack and causing to change its size are: the maximum stress criterion, the strain density criterion and the maximum energy release rate criterion.

The two latter criteria, suggested originally to predict the onset of instability in a mixed mode crack propagation, are in most cases associated with a serious computational effort. For example, the maximum energy release rate criterion [26] is

$$G_I + G_{II} + G_{III} = 2\gamma \tag{36}$$

where G's are the energy release rates in the three basic fracture modes while γ is the effective surface energy of the fracture.

The energy release rates are functions of the stress intensity factors and the elastic moduli of the anisotropic solid [17]. Determination of these energy release rates and the application of this conceptually appealing expression to the case of anisotropic solids involves a substantial computational effort. The most significant problem stems from the fact that the anisotropy of the effective medium surrounding the crack is not known a priori. Thus, the solution will again, for the determination of the stability criterion as well, involve a predictor-corrector iteration routine. The two other prominent sources of computational problems associated with the application of the energy stability criteria are:

- determination of the stress intensity factors for a general anisotropic medium, and
- the occurrence of the mixed mode response (i.e., out-of-plane propagation of the crack).

In the absence of the Green's function for the displacements in an anisotropic solid, the determination of the stress intensity factors involves (see Mura [10], Sect. 28) numerical quadratures of double integrals with complicated arguments. This problem will be dealt with at a later point in this study.

In the case of mixed mode instability, a crack will generally propagate out of its own plane. Since the pattern in which the microcracks grow strongly depends on the mesostructure of the solid (weak interfaces, or cleavage planes in the crystal), the out-of-plane growth is far from being an experimental reality. In other words, it appears more likely that a crack will follow a plane of inferior toughness irrespective of the criteria. Thus it appears reasonable to simplify the discussion assuming that the fracture modes are decoupled and the growth of each crack self-similar. Consequently, the maximum stress criteria will be adopted assuming that a crack will become unstable when the first of the three independent conditions

$$f_N = K_N - K_{NC} = 0 \qquad (N = I, II, III) \qquad (37)$$

is satisfied. The critical stress intensity factors K_{NC} are material parameters which are available for most frequently used materials. It will be further assumed that the condition (37) will govern in the case of a crack in an anisotropic (effective) solid as well.

The stress intensity factor K_I for a penny-shaped crack with radius a embedded in a plane with normal directed along the local coordinate axis e_3' can be written in the form

$$K_I = C\sigma_{33}^{o\prime} a^{1/2} \qquad (38)$$

where C is a parameter depending on the geometry and the material parameters while $\sigma_{33}^{o\prime}$ is the component of the far field stress normal to the plane of the crack. The two other stress intensity factors (for second and third mode) can be written in a similar fashion. Thus, in view of the transformation (20), the fracture criterion (37)

$$f_N(a, \phi, \theta, \underline{F}) = 0 \qquad (N = I, II, III) \qquad (39)$$

represents a hyperplane in the space of crack radius a, Euler angles ϕ and θ, and the externally applied tractions $\underline{F}$ from which the far field stresses are computed (2). The criterion (37) is typical of continuum theories since it involves only the expected values of stresses without requiring detailed knowledge of the stress and strain fluctuations. This is consistent with the derivation of the compliance tensor and the restriction of the model to the expected values over the entire ensemble of possible realizations.

As an illustration, consider the case of a prismatic brittle specimen in uniaxial tension. The externally applied tractions F and the macro stresses are then defined by

$$\underline{F} = q\underline{e}_3 \quad \text{or} \quad \overset{o}{\sigma}_{ij} = q\delta_{i3}\delta_{j3} \tag{40}$$

where q is the magnitude of the externally applied tension in the direction of the global axis e_3. In conjunction with the coordinate transformation (19) and (20), the first of criteria (39) can be written in the form

$$f_I = Cq(\cos^2\theta)a^{1/2} - \overset{o}{K}_{IC} = 0 \tag{41}$$

where $\overset{o}{K}_{IC}$ is the critical intensity factor of the initial energy barrier.

For any combination of a, θ and q satisfying (41), a given crack will become unstable and start propagating through the medium in a runaway fashion. Disregarding, for simplicity, the effect of inertia, a destabilized crack can get arrested only if it runs into a domain of higher toughness (energy barrier) or lesser stress magnitude for which f_N becomes negative.

Assuming that the initial crack radius is defined by a random variable a_o and that the distance separating adjacent energy barriers is defined by another random variable D, the entire ensemble of cracks defined by a volume in the space (a,θ,q) is separated by the hypersurface (41) into two domains (Fig. 2). The cracks which were still not destabilized and which consequently remain of original length a_o constitute a set H

$$\left\{ q, a_o, \theta;\ Cq(\cos^2\theta)a^{1/2} < \overset{o}{K}_{IC} \right\} \in H_o \tag{42}$$

Complementary to the first set is a set of all microcracks which were already destabilized and subsequently arrested by an energy barrier with toughness $K^B_{IC} > \overset{o}{K}_{IC}$

$$\left\{ q,\ a = \frac{D}{2},\ \theta;\ \overset{o}{K}_{IC} < Cq(\cos^2\theta)a^{1/2} < \overset{B}{K}_{IC} \right\} \in H_a \tag{43}$$

In the case of a 'cleavage 1' process, it is important to keep in mind that the total number of cracks N remains constant. However, the number of already destabilized cracks N_a increases at the expense of the cracks N_o which are still of

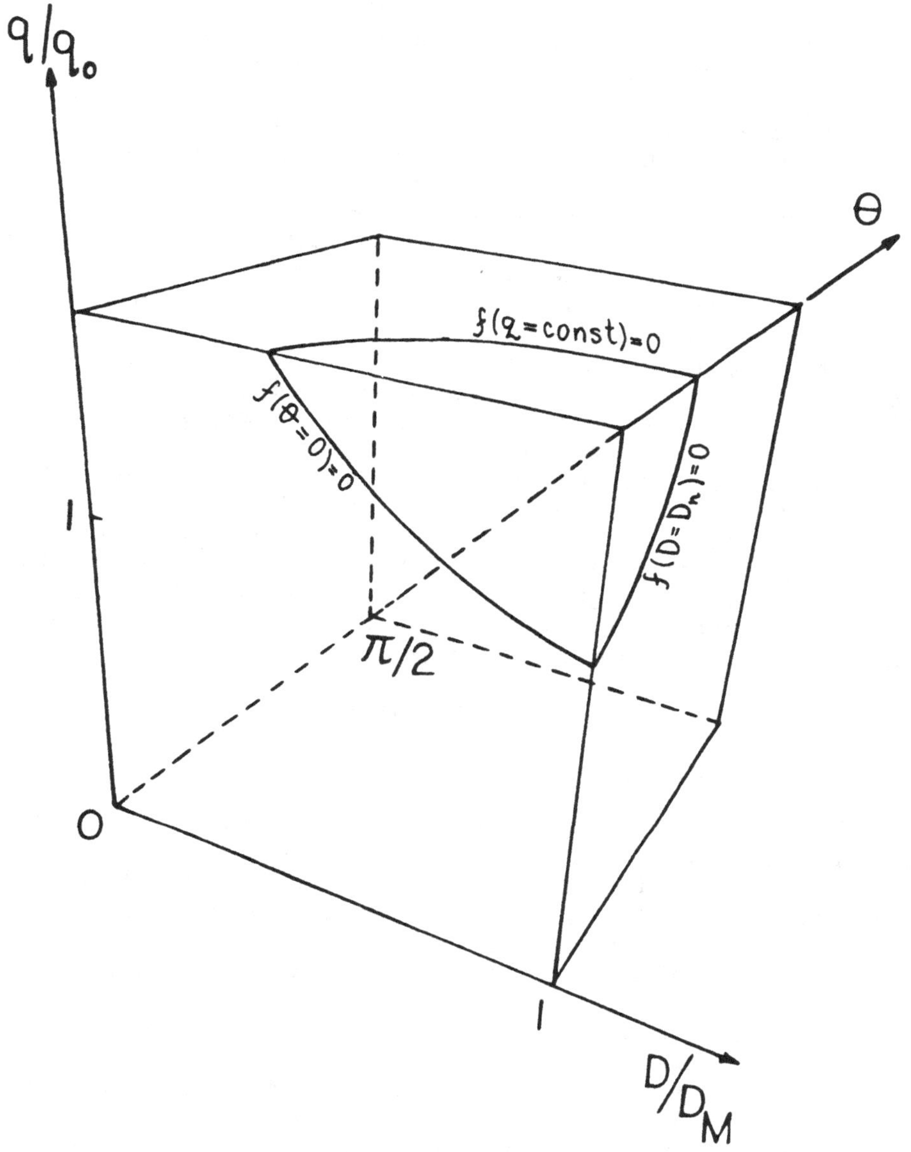

Fig. 2. The hypersurface f = 0 separating the space of all cracks into domains H_o and H_a (tip of the cube).

original length. The question of determining the total number of cracks will be relegated to a more appropriate point.

The described process can be synoptically described in the following manner:

- At a certain threshold level of the externally applied tension the most preferentially oriented large initial crack will become destabilized and subsequently arrested after changing its radius from a_o to D/2.
- As the external load is further incremented, some less preferentially oriented and smaller cracks will destabilize as well increasing the size of the region H_a at the expense of the region H_o.
- At some even higher value of the external load q, one of the arrested cracks will penetrate through the energy barrier with toughness K^B_{IC} and substantially increase in size.

The last event will, in all probability, result in the nucleation of a macrocrack which will in a perfectly brittle solid (unless the tension q is reduced) precede the loss of the integrity of the specimen. It is important to realize that the described analytical model closely reflects the actual process. The acoustic emission tests (see, for example, Holcomb [27,28]) clearly indicate that a seemingly gradual process on the macroscale consists of a series of spasmodic events. Each burst of acoustic energy is related to a loss of stability of a single microcrack, i.e., a change in the position of the hyperplane (39) in the sense of the proposed analytical model.

COMPUTATIONAL MODELS

General Considerations

The derivation of the overall compliance tensor $\bar{S}$ (27), in conjunction with (35), and the kinetic equation in the form of the stability criterion (41) defining the integration domain, represent, in principle, all necessary ingredients for the establishment of the desired process model. The apparent simplicity of this model conceals a host of computational hurdles which must be discussed in some detail. The major computational problem resides in the absence of the closed-form solution for the Green's function for the displacements of a general anisotropic solid. As a consequence, the stress intensity factors in (41) and the crack opening displacements needed to determine the components of the overall compliance tensor can be calculated only after a non-trivial computational effort. Moreover, the proposed scheme by necessity involves iteration for every increment of the external load. Since such a numerical scheme lacks a practical appeal, it is apparent that some additional effort must be made to assess the error in applying the Taylor model which seems to be the simplest possible alternative which is still physically acceptable.

A very fortunate circumstance allowing a reasonably painless accomplishment of this task is that the problem is much less complicated in a two-dimensional case. As shown by Sih, et al. [29], in a two-dimensional case, the stress intensity factor is unaffected by the anisotropy and the crack opening displacements can be written in a reasonably simple analytical form. This suggests the following strategy. The first task would consist of solving the two-dimensional problem using both the Taylor's and self-consistent approximations. The obtained results will be used to assess the error inherent to the use of the Taylor model. The three-dimensional problem will be solved only in Taylor's approximation hoping that the error will not be substantially different from that in the two-dimensional case.

Two-Dimensional Models

In many cases, the geometry of the solid and the loading and boundary conditions justify the approximation of general three-dimensional fields of stress and strain by the corresponding two-dimensional fields. In the present case, such an approximation is justified by purely heuristic reasons as well since the determination of the compliance tensor in a general three-dimensional case involves significant computational complexities.

Introducing the Voigt notation [20] $e_1 = \varepsilon_{11}$, $e_2 = \varepsilon_{22}$, $e_6 = 2\varepsilon_{12}$, $\tau_1 = \sigma_{11}$, $\tau_2 = \sigma_{22}$ and $\tau_6 = \sigma_{12}$, the stress-strain relation may be rewritten in a simpler (matrix) form

$$\bar{e}_i = \bar{S}_{ij}\bar{\tau}_j \tag{44}$$

where $\bar{\underset{\sim}{S}}$ is the overall compliance tensor.

The orthogonal transformation relating the compliance matrices $\underset{\sim}{S}$ in global and local (primed) coordinate systems are:

$$S'_{ij} = g'_{im}g'_{jn}S_{mn} \quad \text{and} \quad S_{ij} = g''_{mi}g''_{nj}S'_{mn} \tag{45}$$

where

$$[g'] = \begin{bmatrix} \cos^2\theta & \sin^2\theta & \frac{1}{2}\sin2\theta \\ \sin^2\theta & \cos^2\theta & -\frac{1}{2}\sin2\theta \\ -\sin2\theta & \sin2\theta & \cos2\theta \end{bmatrix} \tag{46}$$

and

$$[g''] = \begin{bmatrix} \cos^2\theta & \sin^2\theta & \sin 2\theta \\ \sin^2\theta & \cos^2\theta & -\sin 2\theta \\ -\frac{1}{2}\sin 2\theta & \frac{1}{2}\sin 2\theta & \cos 2\theta \end{bmatrix} \qquad \text{(46) Cont'd}$$

The eigenstrains in each void (oriented in such a way that the normal to the slit is colinear with the axis e_2') can now be derived directly from (16) (see [20]) as

$$\begin{aligned} A\bar{e}_1^{*(k)'} &= \int b_1' n_1' \, dx' = S_{1j}^{*(k)'} \bar{\tau}_j' \\ A\bar{e}_2^{*(k)'} &= \int b_2' n_2' \, dx' = S_{2j}^{*(k)'} \bar{\tau}_j' \\ A\bar{e}_6^{*(k)'} &= 2\int \frac{1}{2} (b_1' n_2' + b_2' n_1') \, dx' = S_{6j}^{*(k)'} \bar{\tau}_j' \end{aligned} \qquad (47)$$

where A is the crack area while the integration domain coincides with the length of the k-th slit (microcrack).

Self-Consistent Model

Initially, prior to the first loading the specimen is considered to be macroscopically homogeneous and isotropic. In other words, the initial cracks (associated typically with some manufacturing process) are distributed in a perfectly random fashion, with respect to size and orientation, over the entire solid. Consequently, the compliance of the isotropic and homogeneous elastic matrix is in the case of plane strain defined by a symmetric matrix of rank three

$$[S^0] = \frac{1+\nu_0}{E_0} \begin{bmatrix} 1-\nu_o & -\nu_o & 0 \\ -\nu_o & 1-\nu_o & 0 \\ 0 & 0 & 2 \end{bmatrix} \qquad (48)$$

where E_0 and ν_0 are the elastic modulus and the Poisson's ratio of the elastic matrix in its initial state which is commonly regarded as virgin.

The next step consists of the determination of the displacement discontinuity across the crack as a function of the far field stresses. In the case of an isolated slit embedded into a two-dimensional anisotropic and homogeneous elastic solid, the displacement discontinuities for a slit of length 2a with a normal in the direction of the e_2' axis are, according to Sih, et al. [29] given by the expressions

$$b_1' = [u_1'] = 2\sqrt{a^2-x^2}\,\bar{S}_{11}'[(r_1s_2+r_2s_1)\bar{\tau}_2' + (s_1+s_2)\bar{\tau}_6']$$

$$b_2' = [u_2'] = 2\sqrt{a^2-x^2}\,\bar{S}_{22}'\left[\left(\frac{s_1}{r_1^2+s_1^2} + \frac{s_2}{r_2^2+s_2^2}\right)\bar{\tau}_2' + \frac{r_1s_2+r_2s_1}{(r_1^2+s_1^2)(r_2^2+s_2^2)}\bar{\tau}_6'\right] \tag{49}$$

where $\lambda_j = r_j + is_j$ $(j = 1,2)$, with both parameters s taken as non-negative, are the roots of the characteristic equation [20,29,30]

$$\bar{S}_{11}'\lambda^4 - 2\bar{S}_{16}'\lambda^3 + (2\bar{S}_{12}'+\bar{S}_{66}')\lambda^2 - 2\bar{S}_{26}'\lambda + \bar{S}_{22}' = 0 \tag{50}$$

The coefficients by λ in the fourth order algebraic equation (50) are the elements of the overall compliance of the solid (44) reflecting the existing microcrack distribution in the sense of the self-consistent model.

Since $n_1' = 0$ from the first of equations (47), it directly follows that

$$S_{11}^{*(k)\prime} = S_{12}^{*(k)\prime} = S_{16}^{*(k)\prime} = 0 \tag{51}$$

Furthermore, since

$$2\int_{-a}^{a}\sqrt{a^2-x^2}\,dx = \pi a^2$$

from the second of equations (47) it follows that

$$\pi a^2\bar{S}_{22}'\left[\left(\frac{s_1}{r_1^2+s_1^2} + \frac{s_2}{r_2^2+s_2^2}\right)\bar{\tau}_2' + \frac{r_1s_1+r_2s_2}{(r_1^2+s_1^2)(r_2^2+s_2^2)}\bar{\tau}_6'\right]$$

$$= S_{21}^{*(k)\prime}\bar{\tau}_1' + S_{22}^{*(k)\prime}\bar{\tau}_2' + S_{62}^{*(k)\prime}\bar{\tau}_6'$$

Thus, as in [20], the relation between the overall compliances and the compliances reflecting the existing crack distribution is given, rearranging the expression above, by

$$S^{*(k)'}_{22} = \pi a^2 \left(\frac{s_1}{r_1^2+s_1^2} + \frac{s_2}{r_2^2+s_2^2} \right) \bar{S}'_{22} = h_1(r,s)\bar{S}'_{22} \tag{52}$$

In an identical manner from the third of equations (47)

$$S^{*(k)'}_{62} = \pi a^2 (r_1 s_2 + r_2 s_1)\bar{S}'_{11} = h_2(r,s)\bar{S}'_{11} = S^{*(k)'}_{26} \tag{53}$$

and

$$S^{*(k)'}_{66} = \pi a^2 (s_1 + s_2)\bar{S}'_{11} = h_3(r,s)\bar{S}'_{11} \tag{54}$$

The form of the parameters h being implicit functions of the overall compliance $\bar{S}$ (through parameters r and s) is easily deducible from expressions (52) to (54). For convenience, primes above the parameters r and s in (49) to (54) are omitted.

The expressions (51) to (54), derived for a single crack in its local coordinate system, must be summed (actually integrated) over the entire domain of active cracks in the (a,θ) domain. Thus, the compliance associated with the presence of the microcracks has the following form

$$S^*_{ij} = \frac{N}{A} \int_H S^{*(k)'}_{mn}(r,s) g''_{mi} g''_{nj} \, dH \tag{55}$$

where N is the number of active cracks per unit surface (two-dimensional unit cell in the sense of Hill [31]) of the area A.

The integrand in (55) is both an implicit and explicit function of the as yet unknown components of the overall compliance tensor. The implicit dependence through the parameters r and s represents a rather serious problem since the characteristic equation (50) in general does not admit an analytical solution. In other words, the only possible solution is numerical.

The problem can be significantly simplified in the case of uniaxial and biaxial tension and compression characterized by a special type of crack-induced orthotropy for which the compliance matrix, written in principal coordinates, acquires the following form:

$$[\bar{S}] = \begin{bmatrix} \bar{S}_{11} & \bar{S}_{12} & 0 \\ \bar{S}_{12} & \bar{S}_{22} & 0 \\ 0 & 0 & 2(\bar{S}_{11}-\bar{S}_{12}) \end{bmatrix} \tag{56}$$

The corresponding characteristic equation (50) in the case of a materal with the orthotropy defined by (56) reduces to a biquadratic form

$$\lambda^4 + 2\lambda^2 + m = 0 \tag{57}$$

where

$$m = \frac{\bar{S}_{22}}{\bar{S}_{11}} \tag{58}$$

The solution of the biquadratic algebraic equation (57) is:

$$(\lambda^2)_{1,2} = -1 \pm \sqrt{1-m} \tag{59}$$

For $m < 1$, the roots (59) are purely imaginary and

$$r_1 = r_2 = 0$$

while

$$s_1 = \sqrt{1 - \sqrt{1-m}} > 0 \qquad s_2 = \sqrt{1 + \sqrt{1-m}} > 0 \tag{60}$$

For $m > 1$, the solutions of the characteristic equation (57) must be complex conjugate. After some straightforward manipulations, it can be shown that

$$\begin{aligned} \lambda_1 &= r_1 \pm is_1 \\ \lambda_2 &= r_2 \pm is_2 \end{aligned} \tag{61}$$

The parameters r and s, obtained substituting (61) into (57):

$$r_1 = -r_2 = r = \sqrt{\frac{\sqrt{m} - 1}{2}} \qquad s_1 = s_2 = s = \sqrt{\frac{\sqrt{m} + 1}{2}} \tag{62}$$

However, the parameters r and s derived in the simple form of expressions (60) and (62) are valid only for the principal planes. For all other planes, the matrix (56) is full and the roots of the characteristic equation, having the form of (50), can be determined only numerically. However, an alternate route was provided by Lekhnitsky [30] who demonstrated that the roots of the characteristic equation associated with a plane subtending an angle θ with the direction of the isotropy are given by

$$\lambda_k' = \frac{\lambda_k\cos\theta - \sin\theta}{\cos\theta + \lambda_k\sin\theta} \qquad \bar{\lambda}_k' = \frac{\bar{\lambda}_k\cos\theta - \sin\theta}{\cos\theta + \bar{\lambda}_k\sin\theta} \tag{63}$$

where λ_k are the roots of (59) and $\bar{\lambda}_k$ their complex conjugates. Consequently, the parameters r and s for an arbitrary plane can be written in the form

$$
\begin{aligned}
r_1' &= W_1 r(\cos 2\theta + r \sin 2\theta) \\
s_1' &= s W_1 \\
r_2' &= W_2 r(-\cos 2\theta + r \sin 2\theta) \\
s_2' &= s W_2
\end{aligned} \tag{64}
$$

where r and s are given by (62) and

$$
W_{1,2} = (\sqrt{m}\,\sin^2\theta + \cos^2\theta \pm r \sin 2\theta)^{-1}
$$

The parameters h_i in expressions (52) to (54) are now functions of the angle θ which the observed plane subtends with the plane of isotropy. Substitution of (64) into the expressions for h_i, which can be easily deduced from equations (52) to (54), leads, after some straightforward but tedious manipulations, to

$$
\begin{aligned}
h_1(\theta) &= \pi a^2 \left[\frac{s_1'}{(r_1')^2 + (s_1')^2} + \frac{s_2'}{(r_2')^2 + (s_2')^2} \right] \\
h_2(\theta) &= 2\pi a^2 s \, \frac{r^2 \sin 2\theta}{1 + (m-1)\sin^4\theta} \\
h_3(\theta) &= 2\pi a^2 s \, \frac{2r^2 \sin^2\theta + 1}{1 + (m-1)\sin^4\theta}
\end{aligned} \tag{65}
$$

where s is defined by (62).

It is possible now to express the components of the compliance tensor $\underset{\sim}{S}^*$, reflecting the presence of cracks, as a function of the angle θ defining the plane into which the crack is embedded and the overall compliance moduli. For example,

$$
\begin{aligned}
S_{11}^* = \frac{N_a}{2\pi} \int_{\theta_1}^{\theta_2} \Big\{ & \sin^2\theta [h_1 \sin^2\theta - h_2 \sin\theta\cos\theta + h_3 \cos^2\theta] \bar{S}_{11} \\
& + \sin^4\theta [h_1 \cos^4\theta - 2h_2 \sin^3\theta\cos\theta + h_3 \sin^2\theta\cos^2\theta] (\bar{S}_{22} - \bar{S}_{11}) \Big\} d\theta
\end{aligned} \tag{66}
$$

i.e.,

$$
S_{11}^* = c_1 \bar{S}_{11} + c_2 \bar{S}_{22} \tag{67}
$$

where the parameters c are functions of the overall moduli (through the parameters h and m) and the limits of integration and θ_i (defining the domain of active cracks).

Arranging the non-vanishing elements of the tensor $\underset{\sim}{S}^*$ into a vector, the components of the tensor of overall compliances can be obtained (in the sense of the adopted self-consistent model) from equation (27), after integration of expressions similar to (66), written in matrix form

$$\left\{\bar{S}\right\} = \left\{S^{0}\right\} + [c]\left\{\bar{S}\right\} \tag{68}$$

Thus, the components of the overall compliance matrix (i.e., vector in present notation) are obtained solving the matrix equation (68) as

$$\left\{\bar{S}\right\} = [I-c]^{-1}\left\{S^{0}\right\} \tag{69}$$

Since the elements of the matrix [c] are implicit functions of the overall compliance moduli, the solution of equation (69) involves an iterative process in which the parameter m (58) must be predicted and then corrected until a satisfactory accuracy is achieved.

In order to emphasize the basic features of the model unencumbered by purely mathematical complexities, consider the uniaxial tension of a perfectly brittle specimen with a perfectly random distribution of the initial damage. Thus, in its initial state the material behaves as a linear, homogeneous and isotropic elastic solid. For purely heuristic purposes, consider also that both the distances separating the adjacent energy barriers D and the sizes of initial cracks $2a_o$ to be constant. A more realistic case characterized by the unequal size of the grains or aggregates will be considered later in the study. The only random variable is, therefore, the Euler angle θ which is assumed to be uniformly distributed $p(\theta) = 1/\pi$ within the domain $0 < \theta < \pi$.

The stress intensity factors for a slit of length $2a_o$ in a homogeneous anisotropic (and also isotropic) infinitely extended elastic medium subjected are

$$K_I = \bar{\tau}_2' \sqrt{\pi a_o} \qquad K_{II} = \bar{\tau}_6' \sqrt{\pi a_o} \tag{70}$$

where, as before, $\bar{\tau}_2'$ and $\bar{\tau}_6'$ are the components of the macrostress normal and tangential to the crack. Since, from (40)

$$\bar{\tau}_2' = q\cos^2\theta \quad \text{and} \quad \bar{\tau}_6' = q\sin\theta\cos\theta \tag{71}$$

the stability criterion (41) can be in the case of a Mode I crack rewritten as

$$q\sqrt{\pi a_o}\cos^2\theta - K_{IC} = 0 \tag{72}$$

where K_{IC} is the toughness of the energy barrier at the tip of the initial crack. A similar expression can be written for the case of Mode II crack instability.

The minimum value of $q = q_0$ which satisfies the stability criterion (72) is obviously obtained for $\theta = 0$ and $a = a_0$

$$q_0 = \frac{K_{IC}}{\sqrt{\pi a_0}} \tag{73}$$

In other words, the crack in a plane normal to the tensile axis will become unstable at $q = q_0$ and change its length from $2a_0$ to D. As the load is further incremented, the cracks in planes within a fan $[-\theta_1, \theta_1]$ where from (72)

$$\theta_1 = \pm \cos^{-1}\left(\sqrt{\frac{K_{IC}}{q\sqrt{\pi a_0}}}\right) = \pm \cos^{-1}\sqrt{\frac{q_0}{q}} \tag{74}$$

will also change their size from $2a_0$ to D. At this point it is assumed that the energy barriers arresting the crack are of infinite toughness.

In a 'cleavage 1' process, the total number of cracks N is constant. Moreover, in the considered case, all of the cracks are open. Since their initial orientation is perfectly random, the number of cracks with size D is

$$N_a = \frac{2\theta_1}{\pi} N \tag{75}$$

while the complementary set of cracks which are still of the original size $2a_0$ consists of

$$N_0 = N - N_a = \left(1 - \frac{2\theta_1}{\pi}\right) N \tag{76}$$

cracks.

Consequently, the scenario describing the damage evolution (i.e., the sequence in which the microcracks increase their size) is in the considered case of monotonically increasing tensile load quite straightforward:

(i) $q < q_0$. The material is homogeneous and isotropic and the deformation process is linear and fully reversible (non-dissipative). All cracks are of initial size. The material constants can be determined using a simple analysis which will be discussed in the section on 'isotropic damage'. All cracks have the length $2a_0$.

(ii) $q > q_0$. The cracks within the fan $[-\theta_1, \theta_1]$ will change their length from $2a_0$ to D. The rest of the cracks will retain their original length $2a_0$. The material consequently becomes anisotropic (in the particular case considered here actually transversely isotropic). The response is nonlinear and the energy is dissipated on the creation of new surfaces as the cracks increase their length sequentially.

(iii) $q = q_u$. A global loss of integrity of some type (to be considered later) takes place.

Since all of the cracks are open and since the cracks within the fan $[-\theta_1, \theta_1]$ are of size $2a = D$ while all the other cracks are still of the initial size $2a_o$, the components of the compliance tensor are from (55)

$$S^*_{ij} = \frac{ND^2}{2}\left\{2\,\frac{\theta_1}{\pi}\int_0^{\theta_1} S^{*(k)\,'}_{mn}\left(m,\theta,\bar{\underset{\sim}{S}};\ a = \frac{D}{2}\right) g''_{mi}g''_{nj}\,d\theta\right.$$

$$\left. + \left(1-2\,\frac{\theta_1}{\pi}\right)\rho^2\int_{\theta_1}^{\pi/2} S^{*(k)\,'}_{mn}(m,\theta,\bar{\underset{\sim}{S}};\ a=a_o)g''_{mi}g''_{mj}\,d\theta\right\} \tag{77}$$

where $\rho = 2a_o/D$.

As indicated before (67), the components of the compliance tensor $\underset{\sim}{S}*$ can be written in the form

$$S^*_{ij} = c_{(ij)1}\bar{S}_{11} + c_{(ij)2}\bar{S}_{22} \tag{78}$$

where

$$c_{(ij)1} = \hat{c}_1(m,\theta_1)\ , \qquad c_{(ij)2} = \hat{c}_2(m,\theta_1) \tag{79}$$

The overall moduli $\bar{\underset{\sim}{S}}$ are then, for the given value of the parameter m (58), derived solving the matrix equation (69).

Since the arguments of the integrals on the right-hand side of (77) and, consequently, the elements of the matrix [c] in (69) are implicit functions of the unknown overall moduli (through the parameter m) the determination of the components of the overall compliance tensor, by its very nature, involves an iterative computational scheme. The determination of the overall compliance tensor $\bar{\underset{\sim}{S}}$ during the nonlinear phase (ii) involves the following iterative scheme:

- at some load level $q_i = q_{i-1} + \Delta q$ (where Δq is a 'small' load increment), guess (predict) an appropriate value for m given by (58),
- determine θ_1 from (74),
- compute components of the tensor S^*_{ij}, i.e., coefficients $c_{(ij)1}$ and $c_{(ij)2}$, integrating (77),
- solve matrix equation (69) for the unknown components of the overall compliance tensor $\bar{S}_{ij}$,
- compute m substituting the values for $\bar{S}_{11}$ and $\bar{S}_{22}$ found in the preceding step and compare with the initial guess (predicted value).

If the discrepancy between the predicted and corrected value exceeds some previously established criteria of acceptable errors, the procedure is simply repeated to obtain results of desired accuracy. This computational procedure is repeated for every load increment within the entire range of the externally applied tension.

Taylor's Model

Even though the computational effort attendant to the application of the above discussed self-consistent scheme are by all standards minor, exploration of simplifying conditions leading to a model admitting a closed-form analytical solution is, like always, an intriguing proposition worth pursuing. In particular, it appears reasonable to derive the expressions for the overall compliance tensor within the context of the Taylor model, i.e., neglecting the influence of the adjacent cracks on the crack opening displacement of the observed crack. Consequently, the compliance tensor can be derived directly using the expressions for the crack opening displacements of a crack embedded into an infinitely extended homogeneous, isotropic (undamaged), elastic material

$$\begin{aligned} b_1' &= [u_1'] = \left(4\bar{\tau}_6' \sqrt{a^2-x^2}\right)/E_o \\ b_2' &= [u_2'] = \left(4\bar{\tau}_2' \sqrt{a^2-x^2}\right)/E_o \end{aligned} \tag{80}$$

The expressions (80) can also be derived from (49) substituting, in view of (58), $m = 1$, $r_1 = r_2 = r = 0$ and $s_1 = s_2 = s = 1$.

The non-vanishing components of the compliance tensor S^* can now be computed from (47) and (77) in a closed, analytical form as

$$\begin{aligned} S_{11}^{*T} &= \frac{\omega_D}{E_o}\left\{\left(1-2\,\frac{\theta_1}{\pi}\right)\rho^2 + \frac{1}{\pi}\left[2\,\frac{\theta_1}{\pi} - \rho^2\left(1-2\,\frac{\theta_1}{\pi}\right)\right](2\theta_1-\sin 2\theta_1)\right\} \\ S_{22}^{*T} &= \frac{\omega_D}{E_o}\left\{\left(1-2\,\frac{\theta_1}{\pi}\right)\rho^2 + \frac{1}{\pi}\left[2\,\frac{\theta_1}{\pi} - \rho^2\left(1-2\,\frac{\theta_1}{\pi}\right)\right](2\theta_1+\sin 2\theta_1)\right\} \end{aligned} \tag{81}$$

where the superscript 'T' denotes the Taylor model, and

$$\omega_D = \frac{1}{4}\pi N D^2 \quad \text{and} \quad \rho = 2a_o/D \tag{82}$$

while θ_1 can be easily determined from (74) for a given value of the externally applied tension $q > q_o$.

The overall compliances are, according to (27), given by

$$\bar{S}_{ij}^{T} = \overset{o}{S}_{ij} + S_{ij}^{*T} \tag{83}$$

where the components of $\overset{o}{S}_{ij}$ are given by (48) with E_o and

ν_0 corresponding to the undamaged matrix material. Thus in the case of plane stress

$$E_0\bar{S}^T_{11} = 1+\omega_D\left\{\left(1-2\,\frac{\theta_1}{\pi}\right)\rho^2 + \frac{1}{\pi}\left[2\,\frac{\theta_1}{\pi} - \rho^2\left(1-2\,\frac{\theta_1}{\pi}\right)\right](2\theta_1-\sin2\theta_1)\right\}$$

$$E_0\bar{S}^T_{22} = 1+\omega_D\left\{\left(1-2\,\frac{\theta_1}{\pi}\right)\rho^2 + \frac{1}{\pi}\left[2\,\frac{\theta_1}{\pi} - \rho^2\left(1-2\,\frac{\theta_1}{\pi}\right)\right](2\theta_1+\sin2\theta_1)\right\} \quad (84)$$

In the case of plane strain, the expressions (84) are simply premultiplied by $(1-\nu_0^2)$.

It should be pointed out that the Taylor model, neglecting the influence of other cracks in the matrix, overestimates its stiffness and, therefore, underestimates the total strain.

Isotropic Damage

It is often of interest to consider a simple case characterized by a completely isotropic distribution of damage during the entire process of deformation. In this case, the change in compliance can be defined by a single scalar variable (see Lemaitre and Chaboche [2,32]) and the introduction of the 'effective stress' leads to a simple but interesting analytical model. While such a model would, strictly speaking, apply only in hydrostatic tension or at very low damage levels, it also in some fashion represents the simplest extension of the one-dimensional damage model to the three-dimensional state of stress.

In the case of 'isotropic damage', defined by a perfectly random distribution of active cracks of equal length $2a$, the parameter m in (58) becomes equal to unity, while from (54)

$$h_1 = h_3 = 2\pi a^2 \quad \text{and} \quad h_2 = 0 \qquad (85)$$

Consequently, the arguments of the integrals on the right-hand side of expressions (55) are reducible to a very simple form

$$\tilde{S}^{*(k)}_{11} = \tilde{S}^{*(k)}_{66}\sin^2\theta \qquad \tilde{S}^{*(k)}_{12} = 0$$

$$\tilde{S}^{*(k)}_{16} = -\tilde{S}^{*(k)}_{66}\sin\theta\cos\theta \qquad \tilde{S}^{*(k)}_{22} = S^{*(k)}_{66}\cos^2\theta \qquad (86)$$

$$\tilde{S}^{*(k)}_{26} = \frac{1}{2}\,\tilde{S}^{*(k)}_{66}(2-3\sin^2\theta)\sin2\theta \qquad \tilde{S}^{*(k)}_{66} = 2\pi a^2\,\frac{1-\bar{\nu}^2}{\bar{E}}$$

The overall compliances, in view of (27) and (35), are

$$\bar{S}_{ij} = S^o_{ij} + \frac{N}{\pi}\int_0^{\pi}\tilde{S}^{*(k)}_{ij}\,d\theta \qquad (87)$$

The non-vanishing elements of the compliance tensor for the case of plane stress are

$$\bar{S}_{11} = \bar{S}_{22} = \frac{1}{E_0(1-\omega)} \tag{88}$$

and

$$\bar{S}_{12} = -\frac{\nu_0}{E_0}$$

where E_0 and ν_0 are, as before, the elastic moduli of the undamaged matrix and

$$\omega = \pi N a^2 \tag{89}$$

is an appropriate measure of damage introduced originally by Budiansky and O'Connell [8]. The relation of this volumetric damage measure to the damage variable characteristic of most phenomenological CDM models (introduced in the original Kachanov's paper [18]) was discussed by Krajcinovic [33], Jansson and Stigh [34] and Hult [35].

It is rather simple to show that in the case of plane strain the non-vanishing components of the overall compliance tensor are

$$\bar{S}_{11} = \bar{S}_{22} = \frac{1-\nu_0^2}{E_0(1-\omega)} \tag{90}$$

and

$$\bar{S}_{12} = -\nu_0 \frac{1+\nu_0}{E_0}$$

Once the compliances are determined, the computation is quite straightforward. The nonlinearity of the response can be attributed solely to the gradual growth of cracks in all directions. The damage evolution equation, however, must be determined from macroscopic considerations (fitting a given set of experimental data).

For the sake of completeness, it should be pointed out that the elastic moduli for an isotropic damage case are determined in Budiansky and O'Connell [8] and Horii and Nemat-Nasser [20].

As pointed out by Budiansky and O'Connell [8], the presented self-consistent theory breaks down at a certain crack concentration. Inverting (88), the axial stiffness

$$\bar{C}_{22} = \bar{S}_{22}^{-1} = E_0(1-\omega)$$

Thus, the axial stiffness changes its sign and becomes negative when

$$ND^2 > 4/\pi \qquad \text{for two-dimensional cases}$$

The same problem occurs according to Budiansky and O'Connell [8] if

$$ND^3 > 9/16 \qquad \text{for three-dimensional cases}$$

since at least one of the overall elastic moduli becomes non-positive. This is actually a useful feature of the model since the strong interaction among adjacent cracks must be expected at those levels of the crack interaction. The situation is remedied (at considerable cost in computational effort) using the effective field method (Kunin [9], Ch. 7) predicting positive values for all elastic constants at all stress levels. However, if the restriction on the intersection of cracks is introduced, the overall moduli determined using the effective field theory can again vanish but, fortunately, at much higher stress levels. It must also be pointed out that the entire difference may be purely of academic interest in the case of energy barriers of finite strength.

The situation when some of the cracks are not open will be discussed elsewhere in this manuscript.

Numerical Results

In practical applications, it would be obviously desirable to use the analytical expressions of the Taylor model whenever possible providing that the attendant error is within the acceptable limits. The performed computations comparing the results obtained using the Taylor and self-consistent models are focused on determining the conditions under which the former become of limited utility. The results are computed using $ND^2/4$ as a parameter. Roughly speaking, this parameter is associated with the grain structure of the solid, number of aggregate-cement paste interfaces in a unit cell of concrete, etc. The relation between the geometry of the mesostructure of the solid and the parameter $ND^2/4$ will be discussed at some length in the next section related to replication of the experimental data for concrete.

The relative error of the lateral compliances (solid line) and the axial compliances (dashed line) S^S_{ij}/S^T_{ij} for three values of the parameter $ND^2/4 = 0.1$, 0.2 and 0.3 are shown in Fig. 3. Even for a rather large damage (the upper limit for $ND^2/4$ is $1/\pi$), the relative error appears to be rather small if $q/q_o < 3$ to 4. Since, for example, the rupture of a concrete specimen in uniaxial tension occurs at stress levels 2 to 2.5 times in excess of those at the outset of nonlinear behavior, the error associated with the Taylor model appears to be within the acceptable limits.

The normalized stress vs. the normalized strain computed by two models for the same three values of the parameter $ND^2/4$ is plotted in Figs. 4 to 6. The errors associated with the Taylor model are again within acceptable limits for moderate damage levels and physically meaningful load magnitudes.

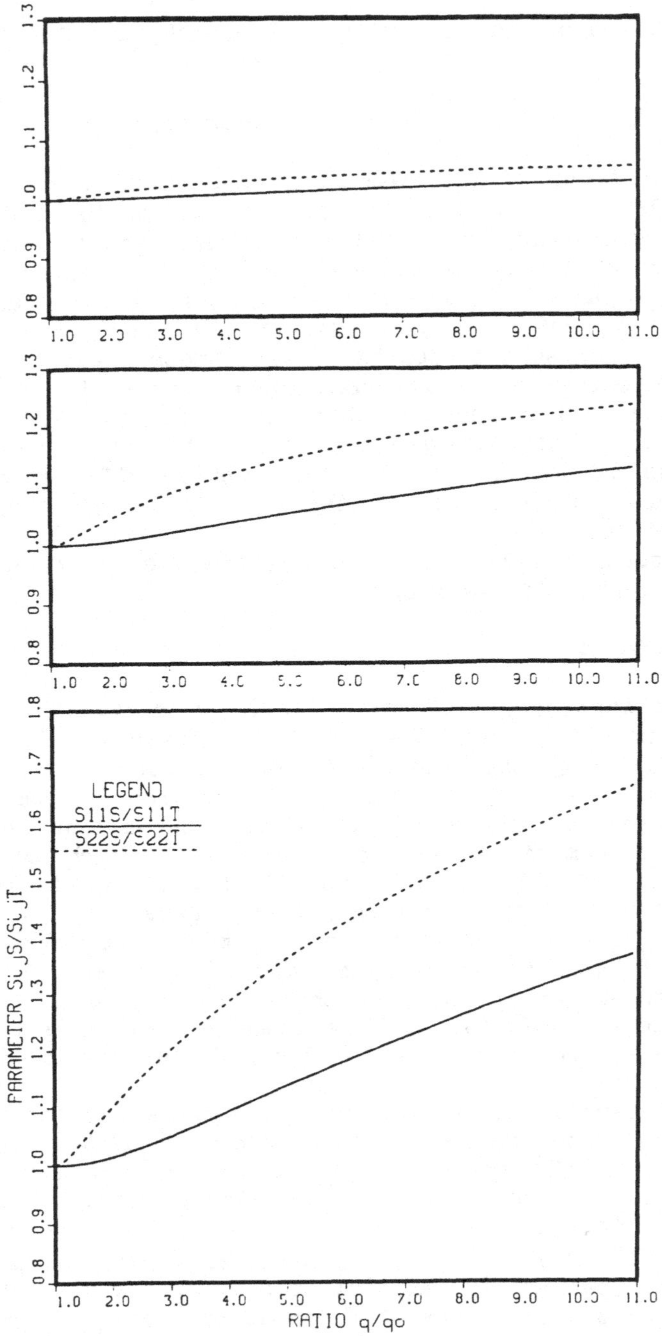

Fig. 3. The ratios of compliances computed using the self-consistent and Taylor models in axial (dashed) and lateral (solid line) directions vs. the nondimensional tensile load. Curves are plotted for $ND^2/4 = 0.1$, 0.2 and 0.3.

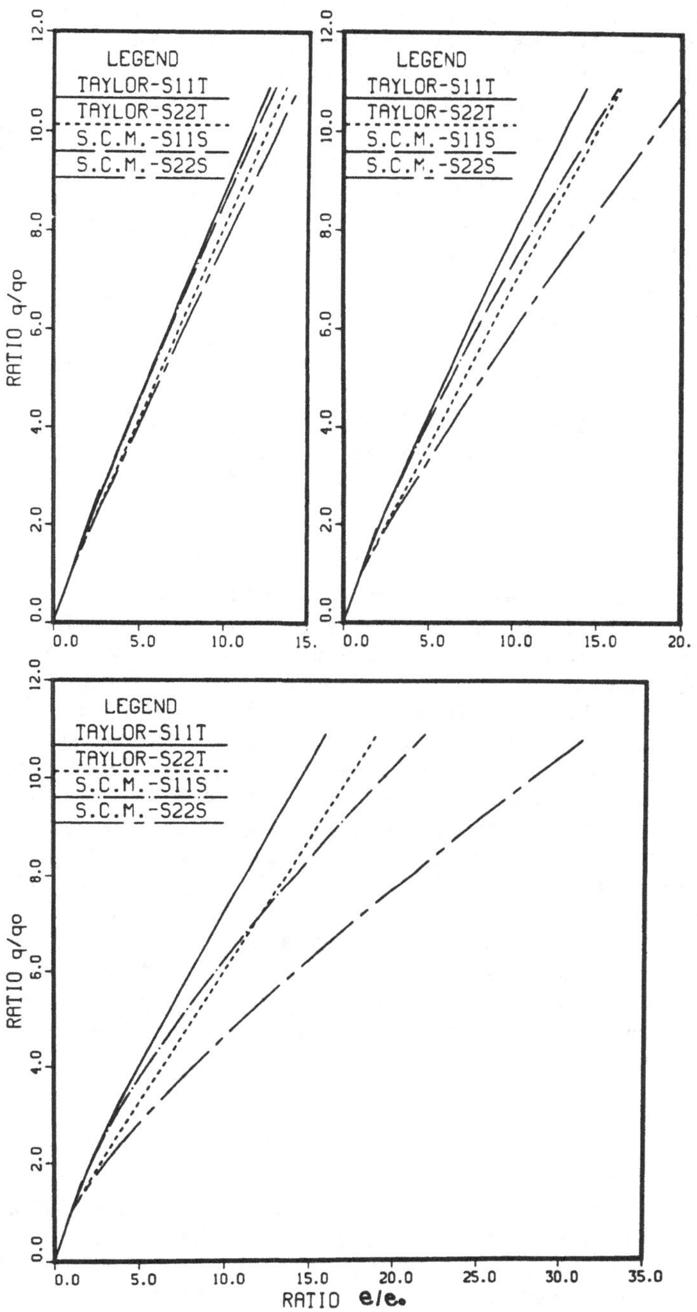

Fig. 4-6. The normalized stress in axial (dashed) and lateral (solid line) directions vs. normalized strain computed using Taylor and self-consistent models for $ND^2/4$ = 0.1, 0.2 and 0.3.

The plot of the parameter m (58) vs. the normalized axial tensile stress q/q_o, computed for both models, is shown in Fig. 7. It is important to note that the initially isotropic material (m=1) becomes anisotropic (m≠1) and eventually tends to isotropy again as the externally applied tension q approaches infinity. This must obviously be expected since all of the cracks will eventually acquire the length 2a = D contingent on the conditions set at the beginning. Consequently, as demonstrated in the Fig. 8, all compliances of the material are bounded from above and below by the inequalities

$$\begin{aligned} (1+\omega_a)\overset{o}{S}_{22} &< \bar{S}_{22} < \overset{o}{S}_{22}(1-\omega_D)^{-1} \\ (1+\omega_a)\overset{o}{S}_{11} &< \bar{S}_{11} < \overset{o}{S}_{11}(1-\omega_D)^{-1} \end{aligned} \tag{91}$$

where ω_a and ω_D are defined in (89) and (82), respectively.

Naturally, the upper bound is physically unattainable since the rupture will occur well before the onset of the instability of a crack in a plane parallel to the tensile axis.

Failure Criterion

The two models presented in the preceding section predict the existence of a lower bound for all components of the overall compliance tensor (much in the same way as hardening in the conventional plasticity theory). In other words, these models in the present form lack an inherent, built-in failure criterion. From a physical standpoint, this is a logical consequence of the assumption that the energy barriers are of infinite strength. In other words, it was assumed that no crack can exceed in length the distance D separating the adjacent energy barriers.

Theoretically, Rudnicki [36] identified two possible mechanisms which might lead to the overall instability of a rock specimen subjected to compressive stresses typical of crustal conditions. Firstly, a rapid increase of the strain increment in a weakened zone may trigger a runaway instability. In other words, the eigenstrains become unbounded in a given zone while the overall strains remain finite. The condition leading to the runaway instability can be written as

$$\det |S^*| = 0 \tag{92}$$

Secondly, the onset of rupture may occur as a result of localization, i.e., inception of zones of large deformation localized within narrow bands (Rudnicki and Rice [37]).

However, from a purely practical standpoint it seems more likely (see, for example, Tetelman and McEvily [38]) that the stress concentration factor of the most preferentially oriented already arrested crack of length 2a = D will at some load

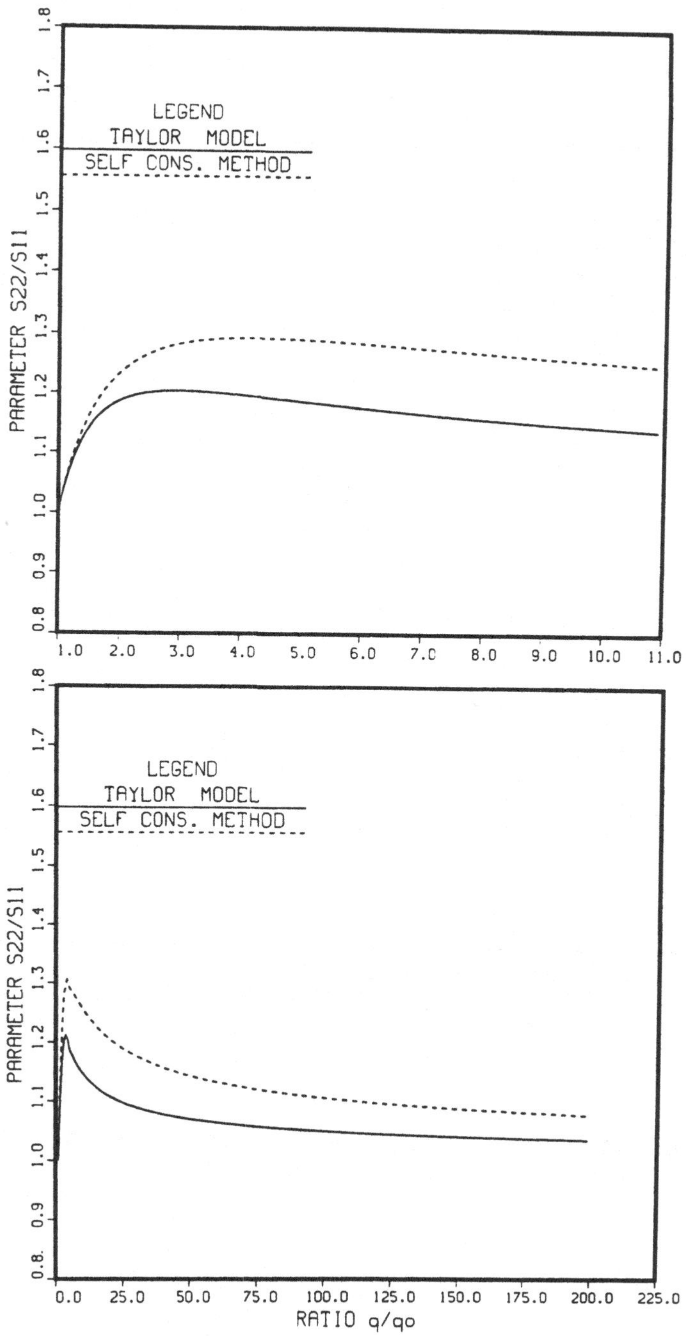

Fig. 7. The parameter m vs. the applied tensile traction determined using Taylor's (solid) and self-consistent (dashed line) methods for $ND^2/4 = 0.2$.

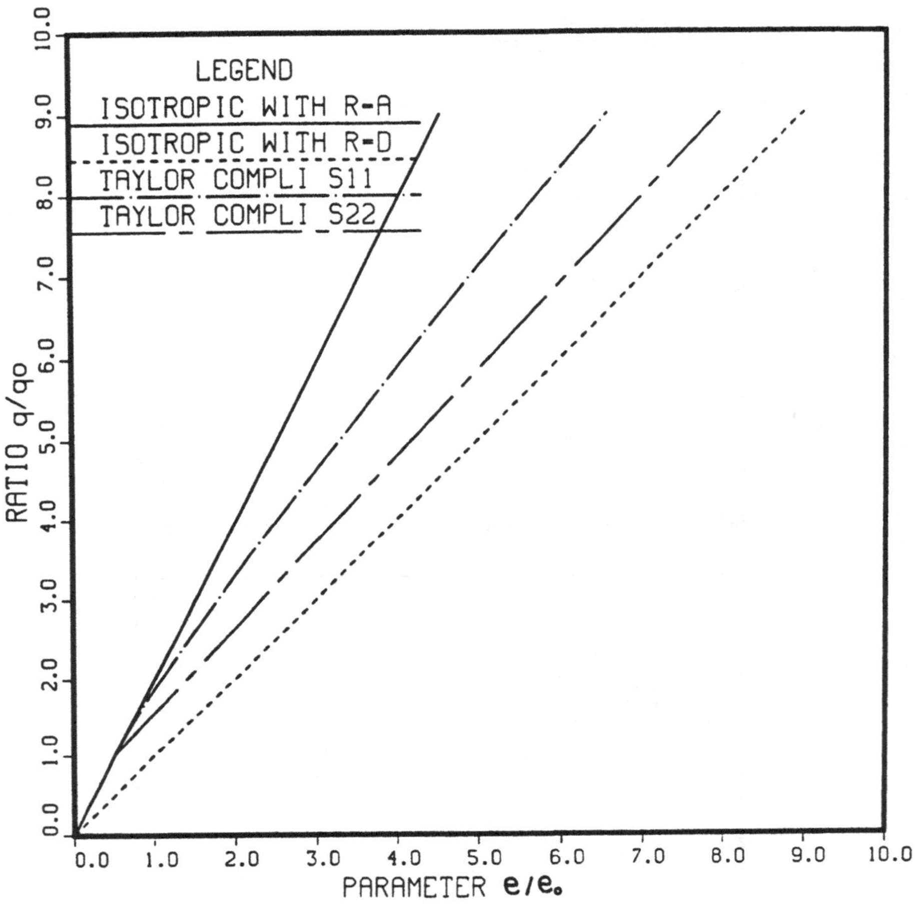

Fig. 8. The bounds on the solution using Taylor's and isotropic damage models.

$$q = q_u = \frac{K_{IC}^{B}}{\sqrt{\pi D/2}} \tag{93}$$

exceed the toughness of the energy barrier K_{IC}^{B}. When that happens, the crack will almost instantaneously double in size. Thus unless the next energy barrier is at least 2 times tougher, the final rupture is in all probability imminent.

Three-Dimensional Model

Assuming that the two-dimensional and three-dimensional models are not radically different in their behavior, it seems reasonable to expect that the Taylor model will, in all probability, lead to useful results over a large spectrum of loads and crack concentrations. To avoid purely computational problems, the considerations will be limited to uniaxial tension and compression of a prismatic concrete specimen assuming total absence of ductile effects and time-dependent deformation. The loading process will also be assumed to be monotonically increasing and quasi-static.

The geometry of the concrete on the mesoscale is characterized by the random shape, volume and disposition of aggregate. The major features of the random geometry (average volume and surface area of inclusions, mean path between the inclusions, etc.) can be estimated by random sectioning or some other method of quantitative stereology or some less rigorous empirical methods commonly used in practice. Thus it seems to be safe to conclude that the reasonably reliable estimates of the mesostructural geometry can be readily obtained with a desired degree of sophistication.

It is common knowledge that the microdefects are present in concrete even before the application of initial loads. These crack-like defects are attributable to bleeding, shrinkage, cement hydration heat, etc. While the distribution of other defects is relatively isotropic (if not always homogeneous), the cracks attributable to bleeding are concentrated on the underside of coarse aggregates which are more efficient in trapping the upward migrating excess moisture than the finer aggregates. Thus if the characteristic linear dimension (say diameter of the inscribed circle) of an aggregate facet is D and θ the angle subtended by the normal to the facet and the vertical axis, it appears reasonable to assume that the radii of the initial defects (located on the aggregate-cement paste interface) due to the bleeding can be defined by the expression

$$a_0 = \frac{1}{2} D(\rho\cos\theta)^2 \tag{94}$$

where D and θ are random variables while ρ is a scalar (with magnitude less than unity) which defines the fraction of the interface originally debonded by the trapped moisture.

Additionally, it is necessary to define the weakest link in the mesostructure of concrete to determine the pattern in which the cracks will grow. The key to understanding the crack evolution patterns is in the hierarchy of the critical stress intensity factors of various phases in concrete. According to experimental data [39,40], the hierarchy of the critical stress intensity factors in concrete is

$$K_{IC}^{a} \gg K_{IC}^{c} \simeq 2K_{IC}^{if} \simeq K_{IIC}^{if} \tag{95}$$

where the superscripts 'a', 'c' and 'if' stand for aggregate, cement paste and aggregate-paste interface. Consequently, a bleeding crack at the interface will at least initially propagate along the same interface. Thus knowing the stochastic distribution of aggregate facets (in terms of their size D and orientation θ), it is possible to deduce both the initial distribution of microcracks and the direction of their growth. Naturally, the angle θ is perfectly random while the distribution of the facet sizes can be determined from the sieve grading curve for aggregates and their shape coefficients.

Uniaxial Tension

Consider first the case of uniaxial tension applied in the direction of the vertical axis e_3, such that

$$\sigma_{33}^{o'} = q\cos^2\theta \tag{96}$$

where q is again the magnitude of the externally applied load.

The stress intensity factor for a penny-shaped crack embedded into an infinitely extended, isotropic and elastic medium is

$$K_I = 2\sigma_{33}^{o'}\sqrt{\frac{a_o}{\pi}} \tag{97}$$

Consequently, from the stability condition (37), in conjunction with expressions (94) to (97), it is possible to derive the relation between the applied load, facet size and orientation which must be satisfied in order for a crack to become unstable

$$q(\cos^3\theta)\sqrt{D} = \frac{1}{2\rho}K_{IC}^{if}\sqrt{2\pi} \tag{98}$$

From expression (98) it follows that cracking will commence along the largest $D = D_M$ facet embedded in a horizontal plane $\theta = 0$ at the force level $q = q_{min} = q_o$ given by

$$q_0 = \frac{K_{IC}^{if}}{2\rho}\sqrt{\frac{2\pi}{D_M}} \tag{99}$$

Once destabilized, the crack will increase its radius until it reaches the edge of the facet where it becomes arrested by the cement paste having superior toughness (95). In the process, the crack changes its radius from a_0 (given by (94)) to D/2. Again, if the distribution of facet sizes can be determined from the sieve grading curve, the distributions of the destabilized and subsequently arrested cracks (domain H_a) is known as well.

The synopsis describing the process of the gradual degradation of material attributable to the growth of microcracks in the considered case consists of several distinctly different phases:

(a) In its initial state, prior to the destabilization of the first crack, the material responds like a linear, homogeneous and slightly anisotropic elastic solid. Within the range $q < q_0$, no energy is being dissipated.

(b) As soon as $q > q_0$, the solid becomes anisotropic (or in this particular case transversely isotropic) and the response becomes nonlinear due to the dissipation of energy associated with the creation of new surfaces. The destabilized cracks (for which the condition (98) is satisfied) will increase their radius in a runaway fashion from a_0 to D/2. At the edge of the facet, the crack propagation will get arrested by the superior toughness of the cement paste (95). In complete analogy to expression (77) for the two-dimensional case, the compliance attributable to the the existence of cracks can be written as a sum of two terms; these terms reflect the displacement discontinuities across the surface of the previously destabilized cracks and the cracks retaining their initial size. Thus from (35), in conjunction with (42) and (43),

$$S* = \frac{1}{V}\left\{ N_a \int_{H_a} D^3 \underset{\sim}{b}(\phi,\theta)p(D)p(\phi)p(\theta)\ dH_a \right.$$

$$\left. + N_0 \int_{H_0} a_0^3 \underset{\sim}{b}(\phi,\theta)p(a_0)p(\phi)p(\theta)\ dH_0 \right\} \tag{100}$$

The regions H_a and H_0 are separated by the hyperplane (98) in the (D,θ,q) space shown in Fig. 2. Using (99), the equation of the hyperplane (98) can be rewritten in a more elegant form as

$$f = \left(\frac{D}{D_M}\right)^{1/2} (\cos^3\theta)\ \frac{q}{q_o} - 1 = 0 \tag{101}$$

The intersection of the surface $f = 0$ with the plane $\theta = 0$

$$D_1 = D_M(q_o/q)^2 \tag{102}$$

gives the smallest initial crack which will become unstable at the given load $q > q_o$. Also, the intersection of the surface $f = 0$ with the plane $D = D_M$

$$\theta_1 = \cos^{-1}\{(q_o/q)^{1/3}\} \tag{103}$$

defines the angle of the fan separating the two domains of integration (as a three-dimensional analogue of (74)).

(c) As the externally applied tension increases further, more and more cracks increase their radius to $D/2$, i.e., the number of cracks within the domain H increases at the expense of the number of cracks which are still dormant. At some magnitude of the externally applied tension $q = q_u$, the stress intensity factor at the tip of one of the previously arrested cracks will exceed the critical stress intensity factor of the cement paste

$$q_u = \frac{1}{2} K_{IC}^{c} \sqrt{\frac{2\pi}{D_M}} \quad ; \qquad \frac{q_u}{q_o} = \rho\ \frac{K_{IC}^{c}}{K_{IC}^{if}} = \rho\kappa \tag{104}$$

triggering the unstable propagation of the crack through the cement paste. At this point the final failure is, in all probability, imminent. In principle, the crack propagating through the paste can be arrested either as a consequence of the inertial effects or, more likely, by an aggregate in its path. However, for a typical concrete mix the mean free path [1,16,24] between the adjacent aggregates is quite large making even this scenario not a very probable one as well. Thus for all practical purposes, $q = q_u$ is the ultimate tensile load which can be supported by the specimen.

As indicated above, each element of the overall compliance tensor is a sum of three components

$$\bar{\underset{\sim}{S}} = \underset{\sim}{S}^{o} + \underset{\sim}{S}^{*}(H_o) + \underset{\sim}{S}^{*}(H_a) \tag{105}$$

The derivation is somewhat more complex than in the previously considered case due to the presence of two random variables D and θ. In the absence of precise data, it appears reasonable to assume that the probability density function of the facet sizes is uniform and band-limited

$$p(D) = (D_M - D_m)^{-1} \tag{106}$$

where D_M and D_m are the diameters of the largest and smallest facet, respectively.

Using the empirical formulas from Harr [41], the volume of coarse aggregates contained within a volume V of concrete is

$$\sum V_a = f_v V = n_a \alpha_v \int_{D_m}^{D_M} \left(\frac{D}{2}\right)^3 p(D)\, dD \tag{107}$$

where α_v is the volume shape factor which is approximately equal to $8x0.5 = 4$ (see [41]), f_v the volume fraction of coarse aggregate (taken to be 0.33 as in [42]), and n_a the total number of aggregates in the volume V. Substituting (106) into (107) and performing the integration, the total volume of concrete can be written as

$$V = \frac{n_a \alpha_v}{4 f_v} \left(\frac{D_M}{2}\right)^2 \left(\frac{1-\gamma^4}{1-\gamma}\right) \tag{108}$$

where

$$\gamma = D_m / D_M \tag{109}$$

The derivation of all non-vanishing components of the overall compliance tensor is now quite routine involving double integrals of simple trigonometric functions. For example, the explicit expression for the axial compliance attributable to cracks arrested at the facet edge is [1,16,24]

$$S^*_{3333}(H_a) = 2S^o \eta \int_0^{2\pi} d\phi \int_0^{\theta_1} 2(2\cos^2\theta - \cos^4\theta)\sin\theta\, d\theta \int_{D(\theta)}^{D_M} a^3\, da \tag{110}$$

where $D(\theta)$ is obtained solving (101) for D, while θ_1 is given by (103). The scalar S^o in front of the integral in (110) is

$$S^o = \frac{4 f_v}{\pi^2 \alpha_v} \frac{N}{n_a\ (1-\gamma^4)} \left(\frac{2}{D_M}\right)^4 \tag{111}$$

where N is the number of cracks in volume V. The fraction of the total number of cracks $\eta = N_a/N$ which have the radius $D/2$, can be readily determined [1,16,24] as

$$\eta = \frac{2}{\pi(1-\gamma)} \left\{ \theta_1 - \left(\frac{q_o}{q}\right)^2 \frac{\sin\theta_1}{15\cos^5\theta_1} (3+4\cos^2\theta_1 + 8\cos^4\theta_1) \right\} \tag{112}$$

Thus after some lengthy but elementary quadratures

$$S^*_{3333}(H_a) = 2\pi\eta S^o \left(\frac{D_M}{2}\right)^4 \left[-\frac{2}{3}\cos^3\theta + \frac{\nu}{5}\cos^5\theta - \frac{2}{21\lambda^3\cos^{21}\theta} + \frac{\nu}{19\lambda^8\cos^{19}\theta} \right]_0^{\theta_1} \tag{113}$$

where θ_1 is given by (103) while $\lambda = q/q_o$ is the normalized tension in the direction of the tensile axis.

The contribution of cracks which are still of original size is determined in an entirely similar fashion from

$$S^*_{3333}(H_o) = 2\pi\eta S^o D_M^4 \left\{ \int_0^{2\pi} d\phi \int_1^{\pi/2} d\theta \int_{D_m}^{D_M} \cos^6\theta b_{3333} D^3 \sin\theta \, dD + \int_0^{2\pi} d\phi \int_0^{\theta_1} d\theta \int_{D_m}^{D(\theta)} \cos^6\theta b_{3333} D^3 \sin\theta \, d\theta \right\} \tag{114}$$

i.e,

$$S^*_{3333}(H_o) = 2\pi S^o (1-\eta)\rho^6 \left(\frac{D_M}{2}\right)^4 \left\{ (1-\gamma^4)\left[-\frac{2}{9}\cos^9\theta + \frac{\nu}{11}\cos^{11}\theta \right]_{\theta_1}^{\pi/2} - \gamma^4 \left[-\frac{2}{9}\cos^9\theta + \frac{\nu}{11}\cos^{11}\theta \right]_0^{\theta_1} + \frac{1}{\lambda^8}\left[\frac{2}{15\cos^{15}\theta} - \frac{\nu}{13\cos^{13}\theta} \right]_0^{\theta_1} \right\}$$

The determination of all other elements of the overall compliance tensor follows the identical routine.

Numerical computations for an unnotched concrete specimen with a volume fraction of coarse aggregate of $f_v = 0.33$ and initial Poisson's ratio of 0.2 were reported in [1,15,24]. Since the available experimental data [42] did not specify the sieve grading curve, the parameter γ was taken to be 0.8. A typical value for the relative size of initial cracks was established viewing micrographs [43] as $\rho = 0.6$. The ratio of the critical stress intensity factors for the cement paste and interface was taken to be 2 [39,40].

Figure 9 demonstrates a surprisingly good fit between the experimental determined and computed stress-strain curves. The nonlinearity becomes noticeable at approximately 80 percent of the ultimate rupture stress which is in good agreement with experimental observations. At rupture, the strain attributable

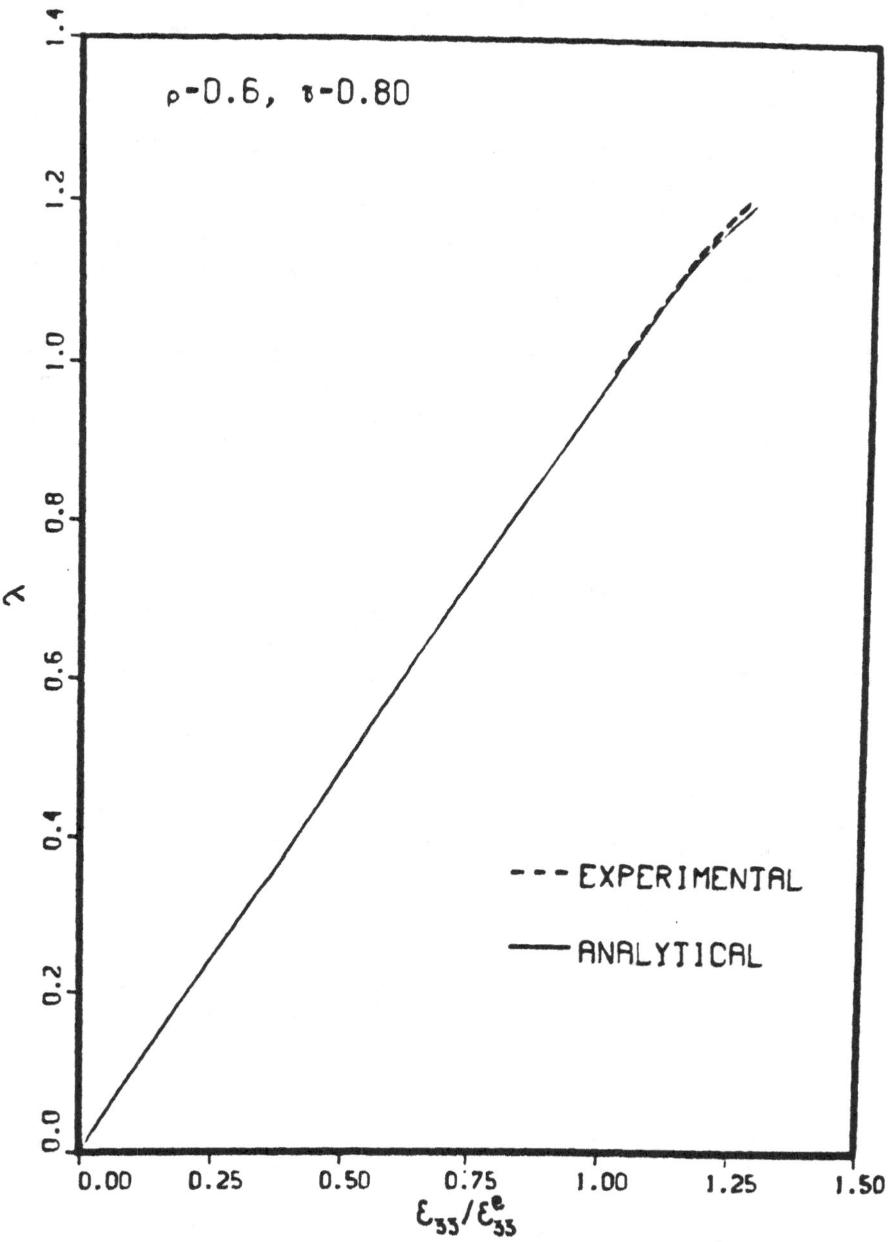

Fig. 9. The stress-strain curve for a specimen in uniaxial tension. Analytical model (solid line) and experiments (dashed line).

to the presence of cracks accounts only for 5 percent of the total strain.

The dependence of the material response on the size of initial defects is depicted in Fig. 10. As expected, larger defects result in a 'softer' material. Finally, the change in the apparent Poisson's ratio (lateral strain divided by the axial strain) is shown in Fig. 11.

Inherent to the proposed model is a tacit assumption that the concrete specimen is large, i.e., that the probability that a large facet is horizontal is close to unity. In case of small specimens, i.e., small number of initial defects, this is not necessarily true. The fact that the joint probability of the crack size and orientation is related to the volume of the specimen allows for establishment of a rational analytical model of the size effect. The computations reported in [2,16], based on the presented model, proved to be in excellent agreement with the experimental results. This feature of the proposed theory will be even more significant in analyses of loading conditions for which only a small fraction of the total volume is subjected to high stresses (three point bending, etc.).

Uniaxial Compression

A much more interesting problem (but more complicated as well) is the case of uniaxial compression of an unconfined concrete specimen. The mesostructural mechanisms of the propagation of the initially existing flaws (which are again located at the aggregate-cement paste interface) are radically different from those responsible for the nonlinear behavior in tension.

In absence of the tensile stress, the initial destabilization of existing cracks occurs in Mode II slip along the interfaces with the largest effective shear stress

$$\tau_{(u)} = q(\sin\theta\cos\theta - \mu\cos^2\theta) = qF(\theta) \tag{114}$$

where q is the externally applied compression (96) and μ the coefficient of sliding friction along the crack surface. The stability condition for the Mode II frictional sliding can then be written [16,39] as

$$q = \frac{2-\nu}{4\rho F(\theta)} K_{IIC}^{if}\sqrt{\frac{2\pi}{D}} \tag{115}$$

where K_{IIC}^{if} is the critical Mode II stress intensity factor for the aggregate cement interface (95). The initial loss of stability will occur along the interface at an angle θ_o rendering the expression (115) maximum

$$\theta_o = \tan^{-1}\theta\left[\mu \pm \sqrt{1+\mu^2}\right] \tag{116}$$

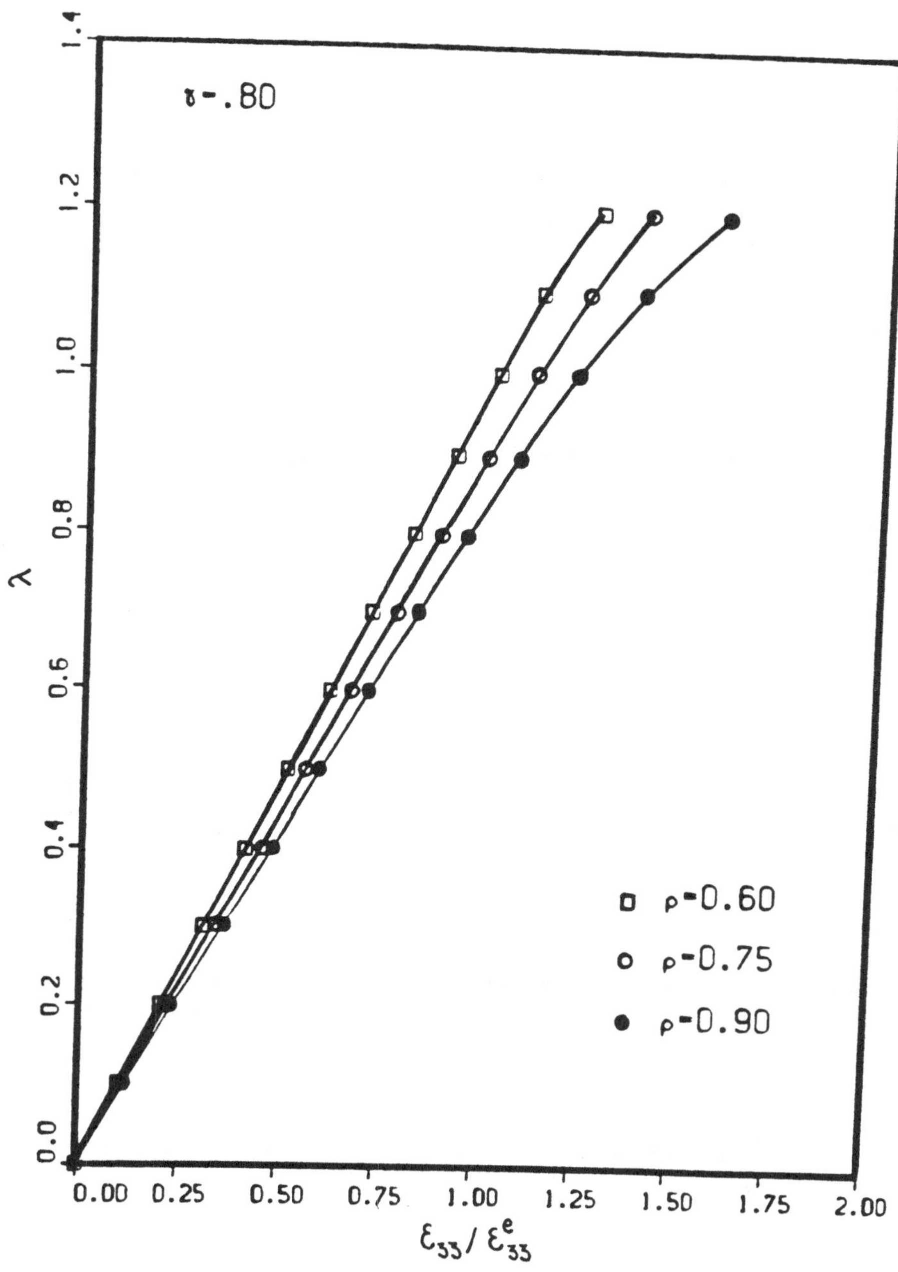

Fig. 10. Dependence of the response (in uniaxial tension) on the size of the initial defects.

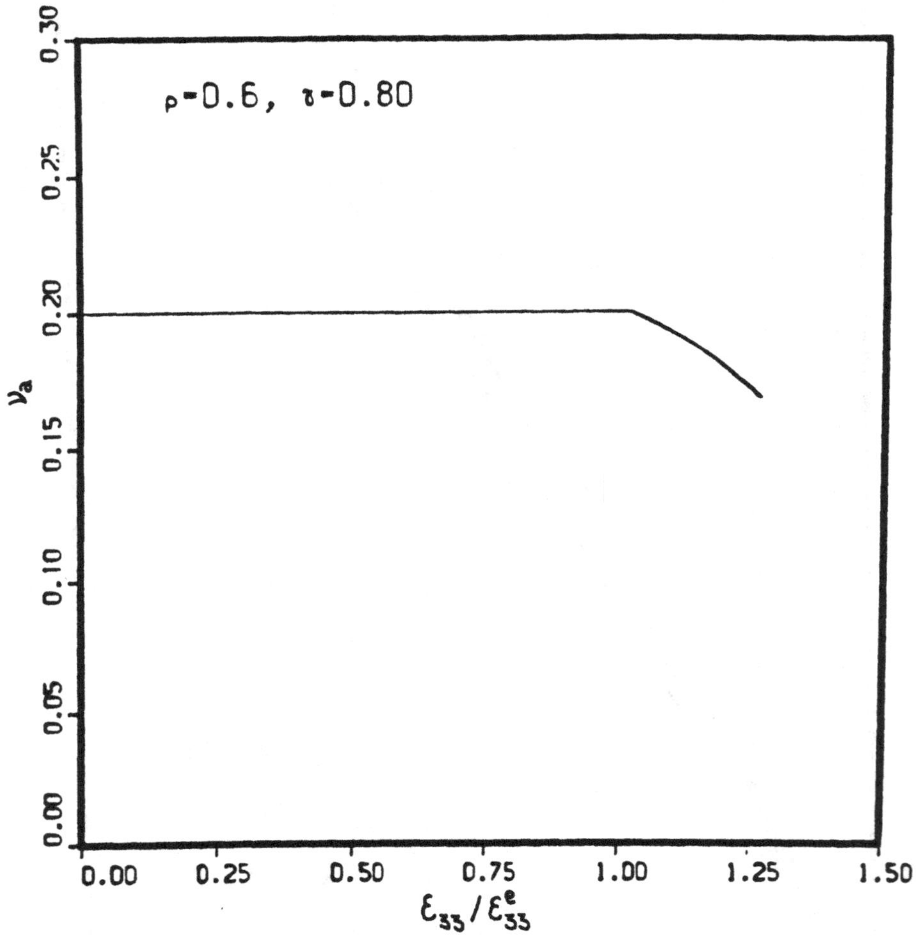

Fig. 11. The apparent Poisson's ratio in uniaxial tension.

The stress of the onset of the nonlinear behavior is thus

$$q_o = \frac{2-\nu}{4\rho F(\theta_o)} K_{IIC}^{if} \sqrt{\frac{2\pi}{D_M}} \tag{117}$$

As the external load is further incremented, more and more initial cracks will get destabilized in the frictional sliding mode only to be arrested at the edge of the aggregate by the superior toughness of the cement paste. At some value of the load, the stress intensity factor at the tip of one of these cracks will exceed the critical stress intensity factor of the cement paste (in Mode I) and the crack will kink in a direction which is roughly parallel to the compressive axis (see the results of the impressive set of experiments reported in [44,45], etc.).

Since the closed-form analytical expressions for the length and the crack opening displacements of a kinked crack are not available, a simplified procedure must be devised [16,23,44] in order to determine the compliances attributable to the presence of cracks. Every element of the compliance tensor is a sum of four components corresponding to the elasticity of the matrix, kinked cracks, cracks which underwent the slip only and the cracks which are still of the original length.

Omitting the laborious details available in [16], it is of some interest to highlight the analytical results and the ability of the model in replicating the experimental data reported in the literature. For a given set of parameters defining the mesostructural geometry, the stress-strain curves are computed and plotted in Fig. 12 along with the experimentally obtained ones. The onset of the nonlinear behavior becomes noticeable at 45 percent of the rupture stress followed by the onset of kinking at about 50 percent of the ultimate stress. The volumetric strain changes it slope at about 90 percent of the ultimate stress which is also in good agreement with experimentally observed data.

The influence of the sieve grading on the stress-strain curve is is apparent from Fig. 13. The apparent Poisson's ratio is plotted vs. the axial strain in Fig. 14. All of the computed results closely replicate the experimentally observed trends. A relative paucity of data related to the mesostructural geometry, with a possible exception of the volume fraction of coarse aggregate, makes some of the comparisons difficult. While all of the introduced parameters are geometrically identifiable, the simplifying assumptions introduced in derivation might eventually necessitate application of at least one corrective fudge parameter.

Since the growth of the kinked parts (wings) of the crack is stable in the presence of a very small confinement [44], it is also necessary to introduce a failure criterion based on the possibility of coalescence of neighboring cracks. In [16], such a criterion is proposed based entirely on the change of the

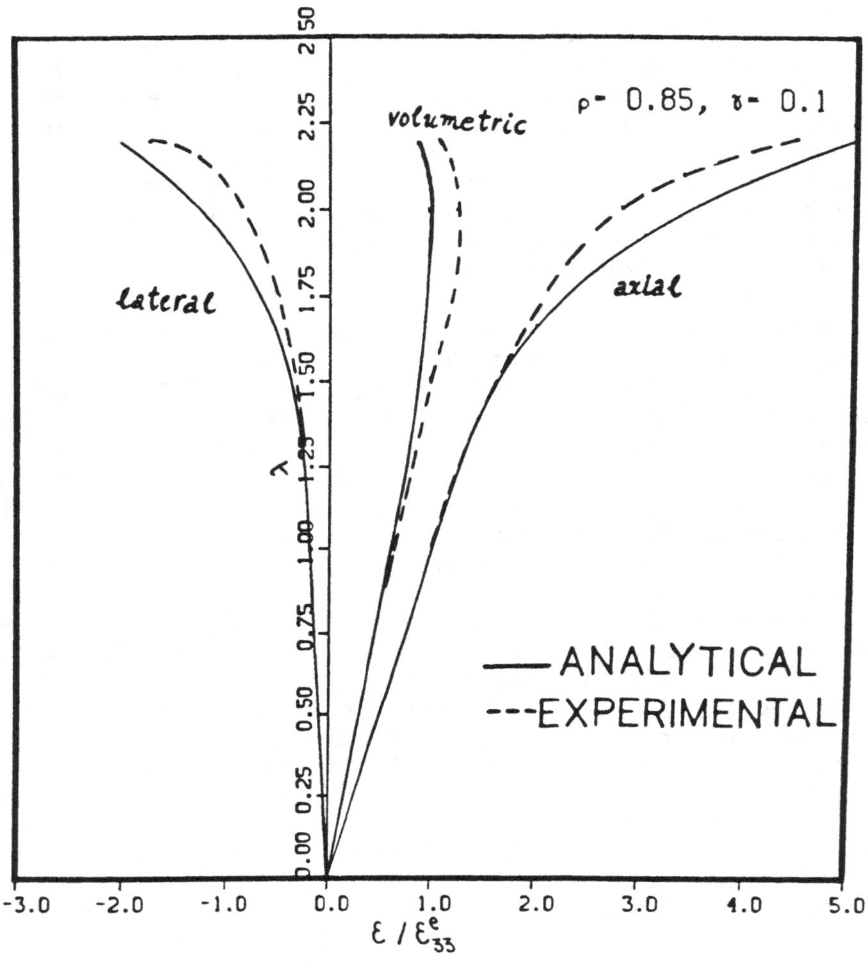

Fig. 12. The stress- strain curve for uniaxial compression of an unconfined specimen.

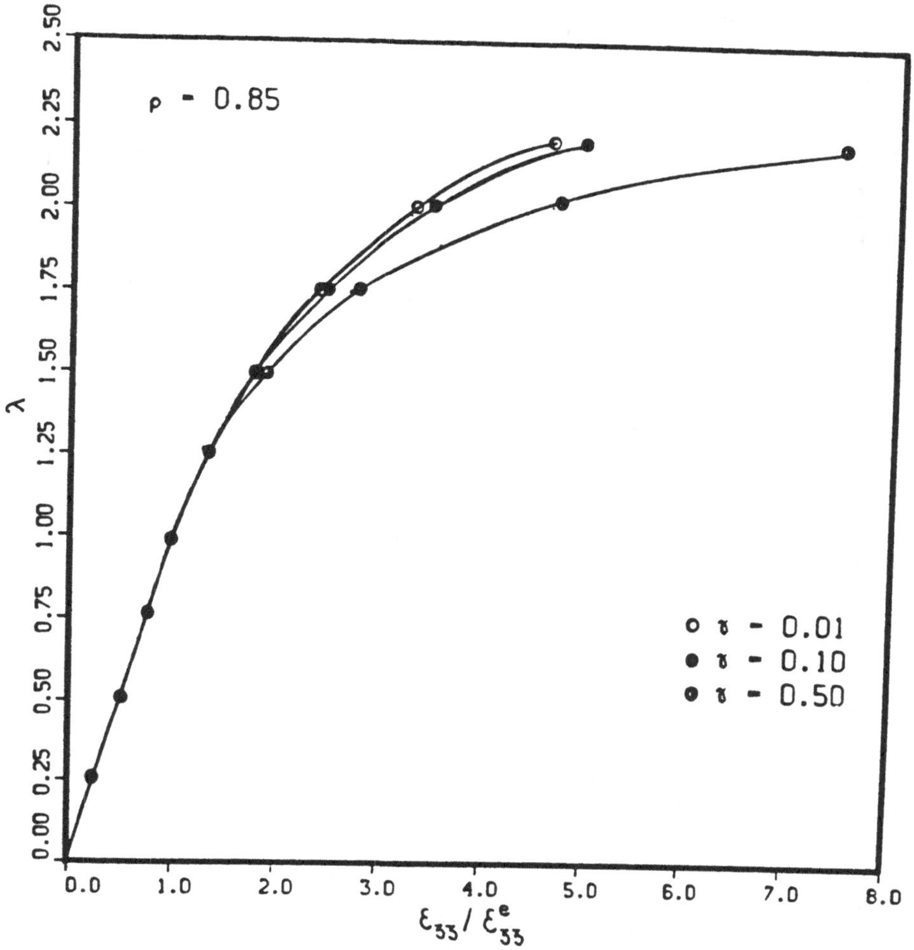

Fig. 13. The influence of the sieve grading (aggregate size distribution) on the stress-strain relationship in uniaxial compression.

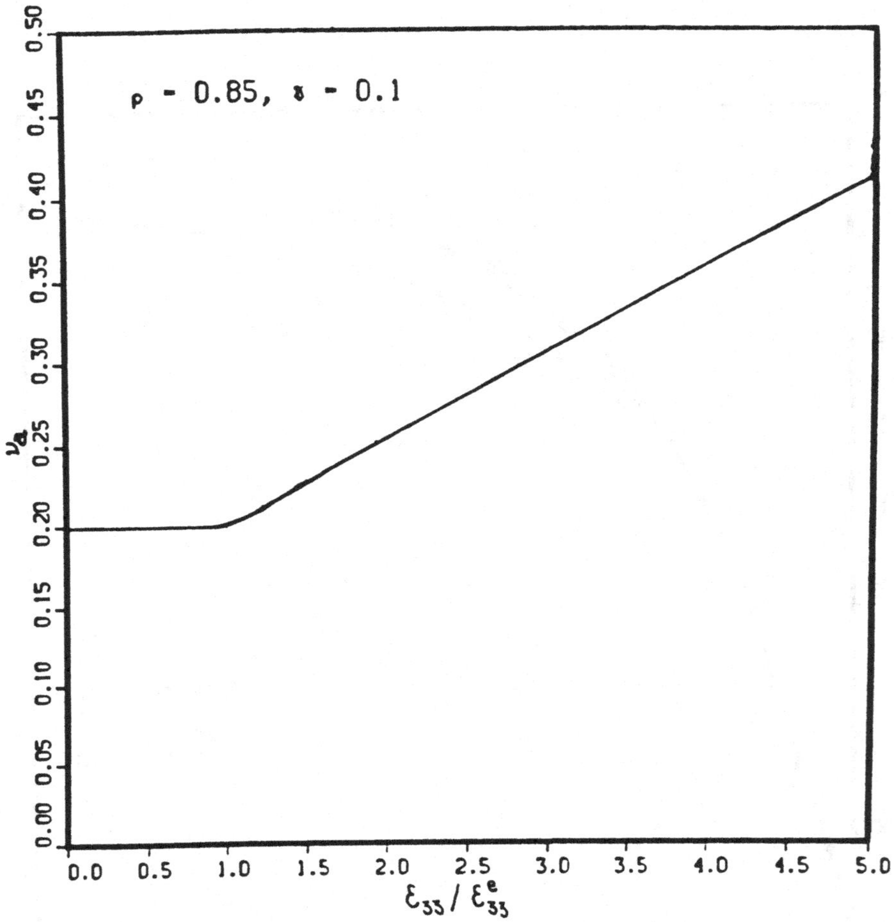

Fig. 14. The apparent Poisson's ratio in uniaxial compression of an unconfined prismatic specimen

average distance between the neighboring cracks. The basic assumption, supported by experimental observations, was that the failure in the analyzed case will occur when the length of the secondary cracks equals the distance between the aggregates.

SUMMARY AND CONCLUSIONS

The presented study summarizes a two-year effort on the formulation of a rational analytical model for the establishment of a constitutive theory for an elastic solid weakened by a large number of small crack-like defects distributed over a large portion of the volume. The primary objective was to formulate a model which will reflect the salient aspects of the physical phenomenon blurring simultaneously the less important and experimentally irreproducible details. The emphasis was also placed on simplicity in order to keep the theory competitive in practical engineering applications.

Perhaps the greatest attraction of the model is the total absence of the ubiquitous 'material' (fudge) parameters, which are all too often indiscriminately introduced in various models. In many, if not all of the cases, these additional 'material' parameters have but a tenuous relation with the physical reality creating difficulties in trying to relate them to measurable and observable attributes of the process. The fact that a particular set of these 'material' parameters fits a certain set of curves offers little assurance of their utility under different circumstances.

The basic premise on which the proposed theory is based is that the knowledge of the material parameters of the constituent phases and their mesostructural geometry suffice for an accurate analytical description of the phenomenon over a wide spectrum of circumstances. The degree of the accuracy is, naturally, inversely proportional to the complexity of the model. It is very reassuring that even the simplest of the models, admitting a closed-form analytical solution, leads to the results which are well within the scatter of the experimental measurements.

At this stage of the development, the analyses are restricted to the quasi-static and isothermal processes involving perfectly elastic materials. Consequently, the attention was focused on the brittle processes during which a lion share of the dissipated energy is spent on the formation of many new surfaces within the solid, i.e., nucleation and growth of microcracks. The further limitation to the 'cleavage 1' processes is a result of the dictates of economy in time and space.

The physical state of the material obviously depends on the distribution, concentration, size, type and orientation of the microdefects (internal variables of the thermodynamic problem). The change of state depends on the rate at which these attributes of the microdefect distribution change (fluxes). Thus the formulation of the analytical model is also comprised of two major tasks: derivations of the compliance tensor (defining the state) and the kinetic equations (defining the rate at which the state changes). In both cases the strategy was identical. The observed process is analyzed on the mesoscale and the transition to the macroscale is accomplished through an appropriate homogenization (averaging) scheme. All of the models discussed in this manuscript are based on the assumption that the mechanical state and the response of the observed solid are sufficiently well defined by volume averages of the all involved state and internal variables.

The present study focuses on two simple analytical models:

- Taylor's model assuming no interaction between the microcracks, and the
- self-consistent theory assuming weak interaction between the microcracks.

The first of the two models, applicable in case of dilute microcrack concentrations, admits closed-form, analytical solutions. The determination of the components of the overall compliance tensor according to the self-consistent model requires application of a predictor-corrector iterative scheme in conjunction with a solution of a system of algebraic equations.

In view of a strong temptation to minimize the numerical effort, it is indeed fortunate that the numerical results obtained through the application of these two models are reasonably close over a large part of the stress-strain curve. This seems to diminish, if not obviate, the need for the formulation of more sophisticated and, computationally, much more demanding models involving higher statistical momenta of the stress and strain field. The fact that the discussed models offer no clues regarding the fluctuations and dispersions of state variables seem to have a limited significance on the result. It is quite possible, though, that a better estimate of the material behavior in the close neighborhood of the apex of the stress-strain curve might indeed be contingent on the introduction of the higher statistical momenta of the stress and strain distribution. The same would apply to the analytical determination of the onset of localization which, by its very nature, involves close proximity of the cracks.

In sharp contrast to traditional routes taken in continuum mechanics, the kinetic equations defining the rate at which the microcrack distribution changes are in this study derived rather than a priori postulated on the basis of macroscopic observations. According to the experimental observations provided by the acoustic emission tests, the seemingly gradual degradation of the solid is a macroscopic reflection of a sequence of mesostructural events consisting of spasmodic changes of length of a large number of microcracks. In other words, as the level of the externally supplied energy is raised some of the existing microcracks will become unstable and start propagating until arrested by an energy barrier in its path. The order in which the cracks get destabilized depends entirely on their size and orientation. Since the number of cracks is typically very large, their distribution over all orientations and sizes is almost continuous. Consequently, the macroscopic behavior must be rather smooth.

The major point of this study is to emphasize that the kinetics of the microcrack growth is fully determined by the stochastic description of the mesostructural geometry (i.e., distribution and size of the defects and energy barriers). Thus knowing the sieve grading of the concrete mix or the grain size distribution for polycrystalline ceramics, it becomes possible to derive the kinetic equations using only the basic principles of the fracture mechanics. Naturally, the result will be just the expected value taken over the entire ensemble of realizations.

Both of the models formulated in this study are, in their final form, reassuringly simple and free of any apparent computational problems and pitfalls. The clarity of the mathematical structure and its unambiguous relation to the physics of the phenomenon eliminates the need for endless arguments common to most phenomenological theories. The absence of additional fudge parameters should prove to be a relief for the embattled experimentalist.

At its present state of the development, the model is practically in its early childhood. However, there seems to be no reason why the essential structure of the proposed model cannot be used, modified and refined for application over a much wider spectrum of problems. For example, even the extensions to the 'cleavage 2' and 'cleavage 3' processes have not yet been accomplished. The extensions to the solid weakened by a combination of voids and cracks-like defects seems to be rather straightforward. Since the spheroidal voids (cavities) distributed in a rather random fashion over large volumes of material have an isotropic effect on the compliance they will simply alter the elastic moduli (i.e., the tensor $\underset{\sim}{S}^{o}$) as demonstrated by Davison, et al. [46].

The analyses of the brittle-ductile processes involving interaction of slips and microcracks have yet to be attempted. The same applies to time-dependent processes (such as creep and shrinkage of concrete, stress corrosion of rocks, creep of metals, etc.), strain-rate effects, healing of damage common to some polymers and biomaterials (remodeling of bones), and many other problems as well. In many cases the necessity for the introduction of non-mechanical sources of energy will result in complex analytical models but the price might be worth paying.

In summary, despite the limited scope of the present study, the proposed model has all the earmarks of a theory which can be used to form a general framework for the development of the CDM theories. The presented theory establishes the basic structure of the governing equations and allows for establishment of a sequence of approximate models dealing with the problem at a varying level of sophistication. Each of the models represent a trade-off between the desired accuracy and the magnitude of the needed computational effort. Fortunately, the performed computations indicate that even the simplest of the models seems to be quite adequate for all practical purposes.

ACKNOWLEDGEMENT

The authors gratefully acknowledge the financial support rendered by the Air Force Office of Scientific Research, Directorate of the Aerospace Sciences, Civil Engineering Program Grant to the University of Illinois at Chicago. The authors would also like to acknowledge their appreciation to Dr. David A. Fanella whose help in many phases of this work greatly facilitated their effort.

REFERENCES

1. Krajcinovic, D.: Continuum damage mechanics, to appear in Mechanics Today.
2. Lemaitre, J. and J.L. Chaboche: Aspect phenomenologique de la rupture par endommagement, J. de Mec. Applique, 2 (1978), 317-365.
3. Stroh, A.N.: A theory of fracture of metals, Advances in Physics, 6 (1957), 418-465.
4. Wittmann, F.H.: Structure of concrete with respect to crack formation, in Fracture Mechanics of Concrete (Ed. F.H. Wittmann), Elsevier, Amsterdam, 1983, 43-74.
5. Chudnovski, A.: Crack Layer Theory, Case Western Reserve Univ., NASA Contractor Report 174634, 1984.
6. Hoagland, R.H., G.T. Hahn and A.R. Rosenfield: Influence of microstructure on fracture propagation in rock, Rock Mechanics, 5 (1973), 77-106.
7. Meyers, M.A.: Discussion of 'Pressure-shear impact and the dynamic viscoplastic response of metals' by R. W. Klopp, et al., Mech. of Mat., 4 (1985), 387-393.
8. Budiansky, B. and R.J. O'Connell: Elastic moduli of a cracked solid, Int. J. Solids Struct., 12 (1976), 81-97.
9. Kunin, I.A.: Elastic Media with Microstructure II, Springer Verlag, Berlin 1983.
10. Mura, T.: Micromechanics of Defects in Solids, M. Nijhoff Publ., The Hague 1982.
11. Delameter, W. and G. Herrmann: Weakening of elastic solids by doubly periodic arrays of cracks, in Topics in Applied Mechanics (Eds., J.L. Zeman and F. Ziegler), Springer Verlag, Berlin 1974, 156-173.
12. Margolin, L.G.: Elastic moduli of a cracked body, Int. J. of Fracture, 22 (1983), 65-79.
13. Eshelby, J.D.: The determination of the elastic field of an ellipsoidal inclusion and related problems, Proc. Royal Soc., A241 (1957), 376-396.
14. Hoenig, A.: The behavior of a flat elliptical crack in an anisotropic body, Int. J. Solids Struct., 14 (1978), 925-934.

15. Wu, C.H.: Tension-compression test of a concrete specimen via damage theory, in Damage Mechanics and Continuum Modeling (Eds., N. Stubbs, D. Krajcinovic), ASCE Publ., New York, 1985, 1-12.
16. Fanella, D.A.: A Micromechanical Continuous Damage Model for Plain Concrete, Ph. D. Thesis, Univ. of Illinois at Chicago, 1986.
17. Kanninen, M.F. and C.H. Popelar: Advanced Fracture Mechanics, Oxford Univ. Press, New York 1985.
18. Kachanov, L.M.: On the creep rupture time, Izv. AN SSSR, Otd. Tekhn. Nauk, 8 (1958), 26-31.
19. Krajcinovic, D.: Continuous damage mechanics, Applied Mech. Rev., 37 (1984), 1-6.
20. Horii, H. and S. Nemat-Nasser: Overall moduli of solids with microcracks: load induced anisotropy, J. Mech. Phys. Solids, 31 (1983), 155-171.
21. Kanaun, S.K.: A random crack field in an elastic continuum, Isled. po Uprug. i Plast., 10 (1974), 66-83.
22. Kachanov, M.: Elastic solids with many cracks: a simple method of analysis, to appear in Int. J. Solids Struct.
23. Kachanov, M.: A microcrack model of rock inelasticity - Part I, Mech. of Materials, 1 (1982), 19-27.
24. Krajcinovic, D. and D. Fanella: A micromechanical model for concrete, to appear in Eng. Fracture Mech.
25. Ashby, M.F.: Micromechanics of fracture in static and cyclic failure, in Fracture Mechanics (Ed: R.A. Smith), Pergamon Press, Oxford 1979, 1-27.
26. Broek, D.: Elementary Engineering Fracture Mechanics, Sijthoff and Noordhoff Publ., The Netherlands, 1978.
27. Holcomb, D.J.: Using acoustic emissions to determine in-situ stress: problems and promise, in Geomechanics - AMD Vol. 57 (Ed: S. Nemat-Nasser), ASME 1983.
28. Holcomb, D.J. and L.S. Costin: Detecting damage surfaces in brittle materials using acoustic emissions, to appear in J. Appl. Mech.
29. Sih, G.C., P.C. Paris and G.R. Irwin: On cracks in rectilinearly anisotropic bodies, Int. J. Fract. Mech., 1 (1965), 189-203.
30. Lekhnitski, S.G.: Theory of Elasticity of an Anisotropic Body, Mir Publ., Moscow 1981.
31. Hill, R.: The essential structure of constitutive laws for metal composites and polycrystals, J. Mech. Phys. Solids, 15 (1967), 79-95.
32. Lemaitre, J. and J.L. Chaboche: Mecanique des Materiaux Solides, Dunod, Paris 1985.
33. Krajcinovic, D.: Continuum damage mechanics revisited: basic concepts and definitions, J. Appl. Mech., 52 (1985), 829-834.
34. Jansson, S. and U. Stigh: Influence of cavity shape on damage parameter, J. Appl. Mech., 52 (1985), 609-614.
35. Hult, J.: Effect of voids on creep rate and strength, in Damage Mechanics and Continuum Modeling (Eds., N. Stubbs and D. Krajcinovic), ASCE Publ., New York 1985, 13-24.

36. Rudnicki, J.W.: The inception of faulting in a rock mass with a weakened zone, J. Geophys. Res., 82 (1977), 844-854.
37. Rudnicki, J.W. and J.R. Rice: Conditions for the localization of deformation in pressure-sensitive dilatant materials, J. Mech. Phys. Solids, 23 (1975), 371-394.
38. Tetelman, A.S. and A.J. McEvily, Jr.: Fracture of Structural Materials, J. Wiley and Sons, New York 1967.
39. Zaitsev, Y.: Deformation and Strength Models for Concrete Based on Fracture Mechanics, Stroiizdat, Moscow 1982.
40. Mindess, S. and J. Young: Concrete, Prentice-Hall Inc., Englewood Cliffs N.J. 1981.
41. Harr, M.E.: Mechanics of Particulate Media, McGraw Hill Co., New York 1977.
42. Gopalaratnam, V.S. and S.P. Shah: Softening response of plain concrete in direct tension, J. Am. Concrete Inst., 82 (1985), 310-323.
43. Moavenzadeh, F. and T.W. Bremner: Fracture of Portland cement concrete, in Structure, Solid Mechanics and Engineering Design (Ed., M. Te'eni), Wiley-Interscience Publ., New York, 1971, 997-1008.
44. Nemat-Nasser, S. and H. Horii: Compression induced nonplanar crack extension with application to splitting, exfoliation and rockburst, J. Geophys. Res., 87 (1982), 6805-6821.
45. Horii, H. and S. Nemat-Nasser: Compression induced microcrack growth in brittle solids: axial splitting and shear failure, J. Geophys. Res., 90 (1985), 3105-3125.
46. Davison, L., A.L. Stevens and M.E. Kipp: Theory of spall damage accumulation in ductile metals, J. Mech. Phys. Solids, 25 (1977), 11-28.

MICROMECHANICAL BASIS OF PHENOMENOLOGICAL MODELS

Dusan Krajcinovic
University of Illinois at Chicago

ABSTRACT

The objective of this short study is to show that the majority of the existing models have a common base which can be readily derived from the micromechanical considerations. Most of the models known from the literature are, in fact, just the truncations of the general model. As such these models are applicable only in special cases (isotropic, orthotropic, etc.) of microcrack distribution.

INTRODUCTION

It is often argued that the ultimate task of engineering research is to provide not so much a better insight into the examined phenomenon but to supply a rational predictive tool applicable in design. Assuming this argument to be valid it is probably fair to state that a design oriented phenomenological model, with all of its limitations, has the best chance to be accepted by the general engineering community. Indeed, leaving aside the conceptual rigors even the purely bookkeeping complexities associated with tracking down the multitude of defects (such as slips or microcracks) influencing the state of stress and strain in a material point are bound to tax the ingenuity and patience of the analyst and the capacity of the available computing device. Thus it is often concluded that 'a direct calculation from microscopic models ... entails substantial complexity (and) ... is unlikely to displace the phenomenological and less-rigorously based structure-parameter models' (Rice [1]).

The literature on the purely phenomenological CDM models is not only very extensive but is also growing at an ever increasing rate. Thus, even with a limitation to CDM models in a narrow sense (defined as those introducing a special damage variable and containing the 'damage law') a comprehensive review of all interesting phenomenological models goes far beyond the objective of this study. In view of the time and space limitations it appears appropriate to focus on the general philosophy in developing of phenomenological models ans select at least some of the existing models as illustration. The selection of these models is based not necessarily on some perceived merit but more on their compatibility with the general tenor of the text.

The opinion that a phenomenological model does not have to have a lot in common with the underlying physical phenomena is surprisingly persistent. A proliferation of plasticity models, in a variety of modifications and incarnations, custom made to fit a particular experimental curve or a set of data is but one testimony to the state of affairs. The essential difference between ductile and brittle mode of response only seldom seem to concern the casual analyst.

To focus the scope of this section the discussion will be restricted on CDM models in the strict definition of this word and the crack-like (high aspect ratio) defects.

GENERAL FRAMEWORK OF THE THEORY

In general sense, the objective of structural analyses is to determine the state of the material locally as defined by a set of basic mechanical and thermal variables such as the stress tensor, heat flux vector and the specific Helmholtz free energy density (Coleman and Gurtin [2]). In case of a reversible thermodynamic process (deformation of a perfectly elastic solid, change of the fluid volume, etc.) the system traverses a continuous sequence of equilibrated states. In absence of the irreversible changes of the microstructure the state of the solid is fully determined by a single kinematic variable (displacement gradient) which is, consequently, a parameter of state (Kestin [3]).

During an irreversible thermodynamic process a system passes through a continuous sequence of near equilibrated states. In contrast to the reversible process the microstructure of the material changes (typically either in a viscous mode such as slip or twinning or in a brittle mode such as microcracking). Consequently, the elastic strain and stress (as its conjugate force) are insufficient to describe the state of the material locally. The changes in the microstructure must be defined by an additional set of internal or hidden variables (Kestin and Rice [4]). The ensuing formalism allowing use of the concepts and methods of the reversible thermodynamics is known as the thermodynamics with internal variables. The assumptions that:

- every irreversible thermodynamic process may be approximated by a sequence of constrained equilibrium states corresponding to the instantaneous values of internal variables (Kestin and Rice [4], Rice [5], etc.), and
- that the mechanical response of the material depends only on the current arrangement of the mesostructure, and not on the entire history of the process,

are inherent to the entire class of these models. In effect the described stratagem replaces the dependence on the entire history 'by a dependence on what it has produced, namely, the current pattern of structural arrangement, on the microscale, of material elements' (Rice [1]).

Moreover, for the theory of this type to be useful (Kestin and Bataille [6]) the approximation of the irreversible thermodynamic process by a sequence of constrained states must be achieve through the introduction of a finite set of internal variables. The contrasting requirements of simplicity (reflected in minimizing the number of internal variables) and accuracy and rigor in modelling the complex geometry of the microstructural rearrangement of polycrystalline, amorphous and composite solids demand careful considerations leading to the selection of internal variables. Although the internal variables can, at least in principle, be selected in an arbitrary fashion (without any regard to the underlying physical phenomenon) it seems highly desirable (if not actually mandatory) to identify 'one among the multitude of possible parametrizations that is based on elements with a clear physical meaning ' (Kestin and Bataille [6]).

Casting certain unavoidable idiosyncrasies aside the most frequently followed recipe for the formulation of a phenomenological CDM theory develops along the following lines. Restricting the discussion, for simplicity, to isothermal processes, during an arbitrary irreversible mechanical deformation the increment of the total strain energy U stored within a value V of the solid must not exceed the increment of the total mechanical work δW imparted to the solid by external forces where (:) stands for the contraction with respect to two indices.

$$V\underset{\sim}{\sigma} : \delta\underset{\sim}{\varepsilon} - \delta U \geq 0 \tag{1}$$

According to the principle of omnipresence the strain energy density is a function of all internal and state (kinematic) variables. Thus, the entropy production (Clausius - Duhem) inequality can be rewritten in a more familiar form as

$$\left(V\underset{\sim}{\sigma} - \frac{\partial U}{\partial \underset{\sim}{\varepsilon}}\right) : \delta\underset{\sim}{\varepsilon} - \sum_{i} \frac{\partial U}{\partial \omega_i} \delta\omega_i \geq 0 \tag{2}$$

where ω_i is an internal variable which in some, physically acceptable, manner serves as a measure defining the influence of the microcracks on the mechanical state of the material.

Since the energy is preserved in all purely elastic processes for which the fluxes vanish

$$\underset{\sim}{\sigma} = \frac{1}{V} \frac{\partial U}{\partial \underset{\sim}{\varepsilon}} \tag{3}$$

Thus, the entropy production inequality (2) can finally be cast in a form of a scalar product

$$\sum_i R_i \, \delta\omega_i \geq 0 \tag{4}$$

between fluxes $\delta\omega_i$ and affinities (or conjugate thermodynamic forces

$$R_i = - \frac{\partial U}{\partial \omega_i} \tag{5}$$

The ultimate goal of most models is to derive the constitutive equation relating the increments of stresses and strains. The incremental form of (3) is obviously

$$V\delta\underset{\sim}{\sigma} = \frac{\partial^2 U}{\partial \underset{\sim}{\varepsilon} \, \partial \underset{\sim}{\varepsilon}} : \partial \underset{\sim}{\varepsilon} - \sum_i \frac{\partial^2 U}{\partial \varepsilon \, \partial \omega_i} \, \delta\omega_i \tag{6}$$

where the summation extends over all active microcrack and systems (defined as those for which the fluxes have non-zero values). The final step in the model formulation necessitates prescription of kinetic laws relating the fluxes and the increments in stress or elastic strain. Assuming that it is possible to devise such a law from experimental observations and theoretical speculations it follows that

$$\delta\omega_i = \underset{\sim}{A} : \delta\underset{\sim}{\varepsilon} \tag{7}$$

where $\underset{\sim}{A}$ is a tensor reflecting the initial properties of the material and the recorded history. In conjunction with (7) the expression (8) can be finally cast into the desired form

$$\delta\underset{\sim}{\sigma} = (\underset{\sim}{K}^o + \underset{\sim}{K}^*) : \delta\underset{\sim}{\varepsilon} \tag{8}$$

where the fourth order tensors within the parentheses on the right-hand side of (9) reflect the properties of the virgin material and the subsequent changes of its mesostructure in some appropriately averaged sense. The second term in the parentheses, labeled by the asterisk is non-zero only during loading (i.e. for non-zero values of fluxes). As a consequence of coupling between the elastic strain $\underset{\sim}{\varepsilon}$ and the 'damage' variable ω_i in the expression for the strain energy density the stiffness tensor $\underset{\sim}{K}^o$ in (8) depends on the already accumulated damage in form of microcracks. Hence the unloading segment of the stress-strain curve will not, in general, be parallel to the initial segment of the loading curve for the virgin state of the material.

The relatively straightforward manipulation leading to the incremental form of the stress-strain law (8) while enticing in its simplicity implies the necessity for making at least three choices in a somewhat arbitrary fashion. These choices very often reflect little but the preference of the author based on the expedience rather than rigor. In either case the specific choices to a large extent account to a somewhat bewildering array of different and sometimes contrasting CDM models. Specifically, the three contentious aspects of the competing CDM theories, based on the thermodynamics of internal variables, are associated with the selection of:

(a) a 'proper' mathematical representation for the 'damage' variable ω,

(b) the particular objective form for the strain energy density U reflecting the load induced anisotropy and the relative order of magnitude of involves variables, and

(c) the appropriate form of the kinetic laws (7) defining the defect interaction, conditions for their stable and unstable growth and possible the condition defining the incipient global loss of the integrity of the solid locally.

Due to the intrinsic importance of these three aspects of the theory it is important to examine them separately in a rather detailed fashion. In the process of the formulation of a CDM model, it must be kept in mind that the thermodynamic principles alone place 'only the most trivial constraints on the form of macroscopic constitutive equations' and that their deduction 'solely from the macroscopic considerations is a rather perilous procedure' (Hart [7]).

DAMAGE VARIABLE

The selection of the 'damage' variable is perhaps the most noticable, if not the most important, distinction separating various CDM models (which were actually often even labeled as scalar, vectorial or tensorial with respect to the representation of the 'damage' variable). There is even some uncertainty whether the damage itself, defined as the impairment of the stress transmitting capacity as a result of the presence of microcracks, is actually an appropriate choice for the damage variable (Krajcinovic [8]).

However, one other side of the story appears to be even more important. It has become a belief, bordering on being considered a law of nature, that a 'proper' mathematical representation for the damage must be selected up front, as the very overture to the proposed model. None holds this belief more firmly, and misguidedly, as those in search of clever little artifices demonstrating perceived advantages of a particular representation.

In the present discussion a different route will be taken. The representation of the damage will be derived, rather than selected, on the basis of simple micromechanical considerations.

STRAIN ENERGY DENSITY

In a purely mathematical sense the determination of the strain energy density function as an objective scalar valued function of two or more tensors of any rank is a relatively straightforward task involving the theory of invariants (see, for example, Spencer [9], etc.). Once the mathematical representation for the 'damage' variable ω and the level of approximation (i.e., highest orders of involved variables) is decided upon it is possible to determine the minimum integrity basis for the strain energy density function which contains all desired polynomials (coupled and uncoupled) and reflects all symmetries characteristic of the material. While this approach has been widely used in the past (Vakulenko and Kachanov [10], Davison and Stevens [11], Krajcinovic and Fonseka [12], Krajcinovic [1] and [13], Litewka and Sawczuk [14], Betten [15], etc.) it is preferable to adopt a somewhat different approach more in tune with the physics of the phenomenon (Dienes and Margolin [16], Ilankamban [17] and Krajcinovic [18]).

In deriving the expression for the strain energy density of a specimen weakened by an ensemble of crack-like defects it will be assumed that the model must reflect the underlying physics of the phenomenon in some approximate sense. In deriving this expression it is necessary to invoke the notion of the unit cell. The minimum prescription for the size of the unit cell was discussed by Hill [19].

To derive the expression for the strain energy density in a material point P(x) consider, for simplicity, that the unit cell surrounding the point P(x) contains a single penny-shaped crack with radius a and normal $n(\phi,\theta)$ defined by Euler angles ϕ and θ. According to the equations (26) and (32) of the preceding Chapter (authored by D. Krajcinovic and D. Sumarac) the macro stresses and macro strains are related by

$$\bar{\varepsilon}_{ij} = \frac{1+\nu_0}{E_0}\bar{\sigma}_{ij} - \frac{\nu_0}{E_0}\delta_{ij}\bar{\sigma}_{mm} + \frac{a^3}{V}\frac{8(1-\nu_0^2)}{3E_0(2-\nu_0)}(\delta_{im}n_j n_n + \delta_{in}n_j n_m + \delta_{jm}n_n n_i + \delta_{jn}n_i n_m - 2\nu_0 n_i n_j n_m n_n)\bar{\sigma}_{mn} \qquad (9)$$

where E_o and ν_o are the elastic modulus and the Poisson's ratio of the elastic matrix and V the volume of the unit cell. The nondimensional parameter (a^3/V) is an appropriate measure of damage introduced originally by Budiansky and O'Connell [20].

The constitutive relation (9) implies a sequence of simplifying assumptions listed below for the sake of completeness.

(a) The strains are assumed to be infinitesimal.
(b) The matrix is assumed to be homogeneous, isotropic and perfectly elastic.
(c) The response of the solid is assumed to be perfectly brittle (no energy being dissipated in plastic slip).
(d) The crack is circular in plan form.
(e) The expected values of state and internal variables describe the process in a sufficiently accurate manner.

The last of the simplifying assumption implies that the exact position of the crack within the unit cell is inconsequential. However, all of these simplifying assumptions are introduced purely for the computational convenience.

The expression for the complementary energy can now after some elementary manipulations be written [16, 18, 21] as

$$W = \frac{1}{2}\bar{\sigma}_{ij}\bar{\varepsilon}_{ij} = \frac{1}{4\mu}\bar{\sigma}_{ij}\bar{\sigma}_{ij} - \frac{\lambda}{4\mu(2\mu+3\lambda)}\bar{\sigma}_{kk}\bar{\sigma}_{mm} + \frac{8(1-\nu_o^2)}{3E_o(2-\nu_o)}\frac{a^3}{V}\left[2\bar{\sigma}_{ki}\bar{\sigma}_{kj}n_i n_j - \nu_o(\bar{\sigma}_{ij}n_i n_j)(\bar{\sigma}_{mn}n_m n_n)\right] \tag{10}$$

where μ and λ the Lame parameters of the virgin solid.

Assuming that $(a^3/V) \ll 1$, i.e., that the crack is small compared to the unit cell the corresponding expression for the strain energy density can be obtained as a Fenchel transform of (10) in form

$$U = \mu\bar{\varepsilon}_{ij}\bar{\varepsilon}_{ij} + \frac{\lambda}{2}\bar{\varepsilon}_{kk}\bar{\varepsilon}_{mm} - \frac{32}{3}\frac{(\lambda+2\mu)\mu}{3\lambda+4\mu}\left[\bar{\varepsilon}_{ik}\bar{\varepsilon}_{kj}n_i n_j - \frac{\nu_o}{2}(\bar{\varepsilon}_{ij}n_i n_j)(\bar{\varepsilon}_{mn}n_m n_n) + \frac{\lambda(2-\nu_o)}{2\mu}\bar{\varepsilon}_{kk}\bar{\varepsilon}_{ij}n_i n_j + \frac{1-\nu_o}{8}\left(\frac{\lambda}{\mu}\right)^2\bar{\varepsilon}_{kk}\bar{\varepsilon}_{ii}\right]\frac{a^3}{V} \tag{11}$$

However the objective of this study is to derive the expression for the strain energy of a solid weakened by an ensemble of many cracks (more than one per unit cell). This task is accomplished rather easily within the context of the Taylor's model (see the preceding Chapter) which leads to accurate results for dilute crack concentrations. According to the

Taylor's model the interaction between the cracks is negligible. Consequently, every crack can be analyzed as a single crack embedded into the infinitely extended undamaged (virgin) medium. Neglecting the crack interaction the strain energy density can be obtained simply through superposition of the contributions (11) for each active crack within the unit cell, i.e.

$$U = \mu\bar{\varepsilon}_{ij}\bar{\varepsilon}_{ij} + \frac{\lambda}{2}\bar{\varepsilon}_{kk}\bar{\varepsilon}_{mm} - \frac{32}{3}\frac{(\lambda+2\mu)\mu}{3\lambda+4\mu}\sum_{I=1}^{N}\left\{\omega_I\left[\bar{\varepsilon}_{ik}\bar{\varepsilon}_{kj}n_i^I n_j^I - \right.\right.$$

$$- \frac{\nu_o}{2}(\bar{\varepsilon}_{ij}n_i^I n_j^I)(\bar{\varepsilon}_{mn}n_m^I n_n^I) + \frac{\lambda(2-\nu_o)}{2\mu}\bar{\varepsilon}_{kk}\bar{\varepsilon}_{ij}n_i^I n_j^I + \tag{12}$$

$$\left.\left. + \frac{1-\nu_o}{8}\left(\frac{\lambda}{\mu}\right)^2 \bar{\varepsilon}_{kk}\bar{\varepsilon}_{ii}\right]\right\}$$

where $\omega_I = a^3/V$. The sum is taken over all N active cracks.

This expression, derived in [18] using a somewhat different but related argument, is exact within the limitations of the Taylor's model and the simplifying assumptions listed above. In the case when the crack concentration is not dilute the constants of the effective solid into which the crack is embedded must reflect the crack induced anisotropy in the sense of the effective fields theory.

The expression derived in [18] was based on the representation of the microcrack field in a given point by a set of doublets $[\tilde{\omega}, \underset{\sim}{n}]$ (or axial vectors) [8, 17] where $\tilde{\omega}$ is the relative void area density (i.e., the measure of damage in the sense of Kachanov [22]) in the plane with normal $\underset{\sim}{n}$. The only difference is that the two damage measures are not identical, i.e., taking the unit cell to be a sphere of radius R it is easy to prove that the two damage measures are related as

$$\omega_I = \frac{3}{4}\tilde{\omega}^{3/2} \tag{13}$$

where the tilde refers to the damage measure in the sense of the original Kachanov's (in fact Rabotnov's [23]) model. The two damage measures become linearly proportional through the introduction of weighting functions [8].

Since the orientation of cracks is a random variable it is reasonable to introduce the probability density distribution of the relative void area $\hat{\omega}(n)$ for the planes of varying orientation through the material point P(x). Introducing notation

$$\omega^{o} = \frac{1}{V} \int \hat{\omega}(\underset{\sim}{n})dV$$

$$\omega_{ij} = \frac{1}{V} \int \hat{\omega}(\underset{\sim}{n})n_i n_j dV \tag{14}$$

$$\omega_{ijmn} = \frac{1}{V} \int \hat{\omega}(\underset{\sim}{n})n_i n_j n_m n_n dV$$

Thus, in the sense of probabilities over the entire ensemble of realizations the expression (12) for the strain energy density can be rewritten as

$$U = \lambda\delta_{ij}\, \bar{\varepsilon}_{ij}\bar{\varepsilon}_{kk} + 2\mu\bar{\varepsilon}_{ij}\bar{\varepsilon}_{ij} - \frac{32}{3}\, \frac{(\lambda+2\mu)\mu}{3\lambda+4} \left\{ \frac{1-\nu_o}{8} \left(\frac{\lambda}{\mu}\right)^2 \omega^{o}\bar{\varepsilon}_{kk}\bar{\varepsilon}_{ii} + \right.$$

$$\left. + \omega_{ij}\left[\bar{\varepsilon}_{ik}\bar{\varepsilon}_{kj} + \frac{\lambda(2-\nu_o)}{2}\, \bar{\varepsilon}_{kk}\bar{\varepsilon}_{ij}\right] - \frac{\nu_o}{2}\, \omega_{ijmn}\bar{\varepsilon}_{ij}\bar{\varepsilon}_{mn} \right\} \tag{15}$$

Similar expressions can be readily derived for the affinity (5) and the stiffness tensor (8).

In the case when the influence of the crack interaction ceases to be negligible the derived expressions are not any more exact. The compliance (or stiffness) is not any more a linear function of the damage variable. However, a common strategy in circumventing this problem consists in introducing fudge constants multiplying each term in the expressions for the strain energy density (12) or (15), in order to extend the validity of the model for larger concentrations of microcracks.

KINETIC EQUATIONS

The formulation of kinetic laws (7) is, perhaps, the single most arbitrary link in the process of the formulation of phenomenological theories. Guided by the plasticity and slip theories it is often convenient to assume [8, 13, 17, 18, 25] that a potential exists in the space of affinities and that the fluxes can be determined from the properties of the convexity and normality,

$$\partial\omega_i = c\, \frac{\partial F}{\partial R_i} \tag{16}$$

where c is a nonnegative scalar ensuring that the flux is directed along the outward normal to the flow potential F.

Since the thermodynamic force R_i (5) is obviously related to the stress intensity factors [8, 13, 25] in Mode I and II it is, at least in principle, possible to select the flow potential in a form resembling the crack stability criterion in the mixed mode response.

SUMMARY AND CONCLUSIONS

This objective of this short precis of the micromechanical basis for the derivation of phenomenological theories is to demonstrate several important aspects of the problem. The most important conclusion is to show that very little arbitrariness is involved in deriving a wide spectrum of models. The fundamental assumption was that the external fields of every crack weakly depend on the actual crack pattern rendering the effective field approximation applicable [26]. In Taylor's approximation the strain energy density is derived to be a linear function of the Budiansky, O'Connell [20] damage variable. For larger volume crack densities this relation ceases to be linear.

This short study also demonstrates the direct relation between the micromechanical model and the model based on the representation of damage in form of a set of doublets (or axial vectors) [8]. In the limit, introducing density functions instead of the discrete values of the damage in many planes the same model becomes identical to the Leckie and Onat [24] model containing as special cases:

- the scalar (isotropic damage) model [27],
- the second order tensor (orthotropic damage) model [28, 29], and
- the fourth order tensor (anisotropic damage) model [30].

Consequently, the differences between various models are not as pronounced as their appearance would suggest. They are simply different approximations of the general model [8, 24] based on the micromechanics of the process.

ACKNOWLEDGEMENT

The author gratefully acknowledges the financial support in form of a research grant from the U.S. Department of Energy, Office of Basic Energy Sciences, Engineering Research Program (monitored by Dr. O. Manley) to the University of Illinois at Chicago.

REFERENCES

1. Rice, J. R.: Continuum mechanics and thermodynamics of plasticity in relation to microscale deformation mechanisms, in: Constitutive Equations in Plasticity (Ed. A. S. Argon), The MIT Press, Cambridge, Mass. 1975, 23-79.

2. Coleman, B.C. and M. Gurtin: Thermodynamics with internal state variables, J. Chem. Phys., 47 (1967), 597-613. 3. Kestin, J.: On the application of the principles of thermodynamics to strained solid materials, in: Proc. IUTAM Symp., Vienna (Ed. H. Parkus), Springer-verlag, Berlin 1966, 177-212.
4. Kestin, J. and J. R. Rice: Paradoxes in application of thermodynamics to strained solids, in: A Critical Review of Thermodynamics (Ed. E. B. Stuart, et al.), Mono Book Corp., Baltimore Md., 1970, 275-313.
5. Rice, J.R.: On the structure of stress-strain relations for time dependent plastic deformations in metals, J. Appl. Mech., 37 (1970), 728-737.
6. Kestin, J. and J. Bataille: Irreversible thermodynamics of continua and internal variables, in: Continuum Models of Discrete Systems (Ed. H.H.E. Leipholz), Univ. of Waterloo Press, Waterloo, Canada 1977, 39-67.
7. Hart, E. W.: A micromechanical basis for constitutive equations with internal state variables, J. Eng. Mat. and Techn., 106 (1984), 322-325.
8. Krajcinovic, D.: Continuum damage mechanics revisited: basic concepts and definitions, J. Appl. Mech., 52 (1985), 829-834.
9. Spencer, A.J.M.: Theory of invariants, in: Continuum Physics, Vol. 1 - Mathematics (Ed. A. C. Eringen), Academic Press, New York 1971, 239-353.
10. Vakulenko, A. A. and M. Kachanov: Continuum theory of cracked media, Mekh. Tverdogo Tela, 4 (1971), 159-166.
11. Davison, L. and A. L. Stevens: Thermomechanic constitution of spalling elastic bodies, J. Appl. Phys., 44 (1973), 668-674.
12. Krajcinovic, D. and G. U. Fonseka: The continuous damage mechanics of brittle materials - Part I, J. Appl. Mech., 48 (1981), 809-815.
13. Krajcinovic, D.: Constitutive theory of damaging materials, J. Appl. Mech., 50 (1983), 355-360.
14. Litewka, A. and A. Sawczuk: A yield criterion for perforated sheets, Ing. Archiv, 50 (1981), 393-400.
15. Betten, J.: Damage tensors in continuum mechanics, J. de Mec. Theor. at Appl., 2 (1983), 13-22.
16. Dienes, J. K. and L. K. Margolin: A computational approach to rock fragmentation, Los Alamos Scientific Laboratory, Report LA-UR-79-3015, 1979.
17. Ilankamban, R.: Continuum damage mechanics for progressively degrading brittle solids with application to geomaterials, Ph.D. thesis, University of Illinois at Chicago, 1985.
18. Krajcinovic, D.: Continuum damage mechanics, to appear in Mechanics Today.
19. Hill, R.: The essential structure of constitutive laws for metal composites and polycrystals, J. Mech. Phys. Solids, 15, 79-95.

20. Budiansky, B. and R. J. O'Connell: Elastic moduli of a cracked solid, Int. J. Solids Struct., 12 (1976), 81-97.
21. Wu, C. H.: Tension-compression test of a concrete specimen via damage theory, in: Damage Mechanics and Continuum Modeling (Ed. N. Stubbs and D. Krajcinovic), ASCE Publ., New York (1985), 1-12/
22. Kachanov, L.M.: On the creep rupture time, Izv. AN SSSR, Otd. Tekhn. Nauk, 8 (1958), 26-31.
23. Rabotnov, Yu. N.: On the equation of state for creep, in: Progress in Applied Mechanics, Prager Anniversary Volume, MacMillan Corp., New York (1963), 307-315.
24. Leckie, F. A. and E. T. Onat: Tensorial nature of damage measuring internal variables, in: IUTAM Symp. on Physical Non-Linearities in Structural Analysis (Eds. J. Hult and J. Lemaitre), Springer-Verlag, Berlin 1981, 140-155.
25. Chaboche, J. L.: Description thermodynamique et phenomenologique de la viscoplasticite cyclique, Office National d'Etudes et de Recherches Aerospatiales, Publ. 1978-3, 1978.
26. Kunin, I.A.: Elastic Media With Microstructure II, Three-Dimensional Models, Springer Verlag, Berlin 1983.
27. Lemaitre, J. and J. L. Chaboche: Aspect phenomenologique de la rupture par endommagement, J. de Mec. Applique, 2 (1978), 317-365.
28. Kachanov, M.: A continuum model of medium with cracks, J. Eng. Mech. Div., ASCE, 106 (1980), 1039-1051.
29. Murakami, S. and N. Ohno: A continuum theory of creep and creep damage, in: Creep of Structures, IUTAM Symp. (Ed. A.R.S. Ponter), Springer Verlag, Berlin 1981, 422-444.
30. Chaboche, J. L.: Le concept de contrainte effective applique a l'elasticite et la viscoplasticite en presence d'un endommagement anisotrope, Report Colloq. Euromech 115 (1979), Grenoble Fr.

BOUNDING METHODS AND APPLICATIONS

F.A. Leckie
Department of Theoretical and Applied Mechanics
University of Illinois, Urbana, Illinois

Abstract

Using the constitutive equations which describe the rate of growth of damage, bounding theorems are developed which can be used to determine the failure life of components. An example of the technique is illustrated by determining the rupture life of a shell component subjected to cyclic loading.

1. Experimental Determination of Constitutive Laws

A large body of experimental data has been collected on the fracture properties of a number of materials under both uniaxial and multiaxial states of stress. In general these experiments have been conducted at constant stress, or, when the stress has been varied, this has been done in a proportional manner. An exception to this are the experiments of Hayhurst et al. [1] on thin walled tubes of copper, an aluminum alloy and a nimonic. In these experiments the axial load was kept constant while the torque experienced by the tube was cycled between two prescribed limits.

First we consider the case of constant load and develop constitutive equations that can deal specifically with this situation. Then we consider the situation of non-proportional loading. In particular we concentrate on developing equations for copper and an aluminim alloy, which have been tested extensively by Leckie and Hayhurst [2] over the temperature range of 150-300°C.

1.1. General features of material behaviour at constant stress

When tested in uniaxial tension it is often found that the time to failure, t_f, and the uniaxial stress σ are related by an equation

$$t_f = A\sigma^{-\nu} \tag{1.1}$$

of the form over a range of stress, where A and are material constants. In most materials it is found that ν is less than the creep exponent n. For the copper tested by Leckie and Hayhurst [2] it was found that $\nu = 5.6$ and $n = 5.9$. For the aluminum alloy, however, it was found that ν was greater than n, with $\nu = 10$ and $n = 9$. We explain this observation later through consideration of the strain softening mechanism of section 6.

Any constitutive equations we develop for the material behaviour, as well as reflecting the stress dependence on eqn. (1.1), must also reflect the shape of the uniaxial creep curve. Fig. shows two typical creep curves for aluminum and copper. The important characteristic of these curves is the value of the quantity λ (the creep damage tolerance [3]):

$$\lambda = \frac{\varepsilon_f}{\dot{\varepsilon}_{ss}\, t_v} \tag{2.2}$$

which is a measure of the materials ability to redistribute stress in a structural situation. $\lambda \approx 10$ for aluminum and $\lambda \approx 4.0$ for copper.

The results of multiaxial stress state tests are conveniently plotted as isochronous surfaces in stress space, Fig. 1. These surfaces connect points which give the same time to failure. Fig. 1 shows the isochronous surfaces found experimentlaly for copper and the aluminum alloy [2]. Failure in copper is a function of the maximum princpal stress, while aluminum fails according to an effective stress criterion.

1.2 Non-proportional cyclic loading

In the previous sub-sections we concentrated primarily on proportional, or constant, loading situations where we could describe the material behaviour in terms of a single damage parameter. In some structural situations components experience non-proportional cyclic loading. Examples of this can be found in a number of cómponents of the Liquid Metal Cooled Fast Breeder Nuclear Reactor which experience cyclic thermal loading as the reactor is shut down and started up. It is therefore important to understand how a material behaves under these types of loading conditions.

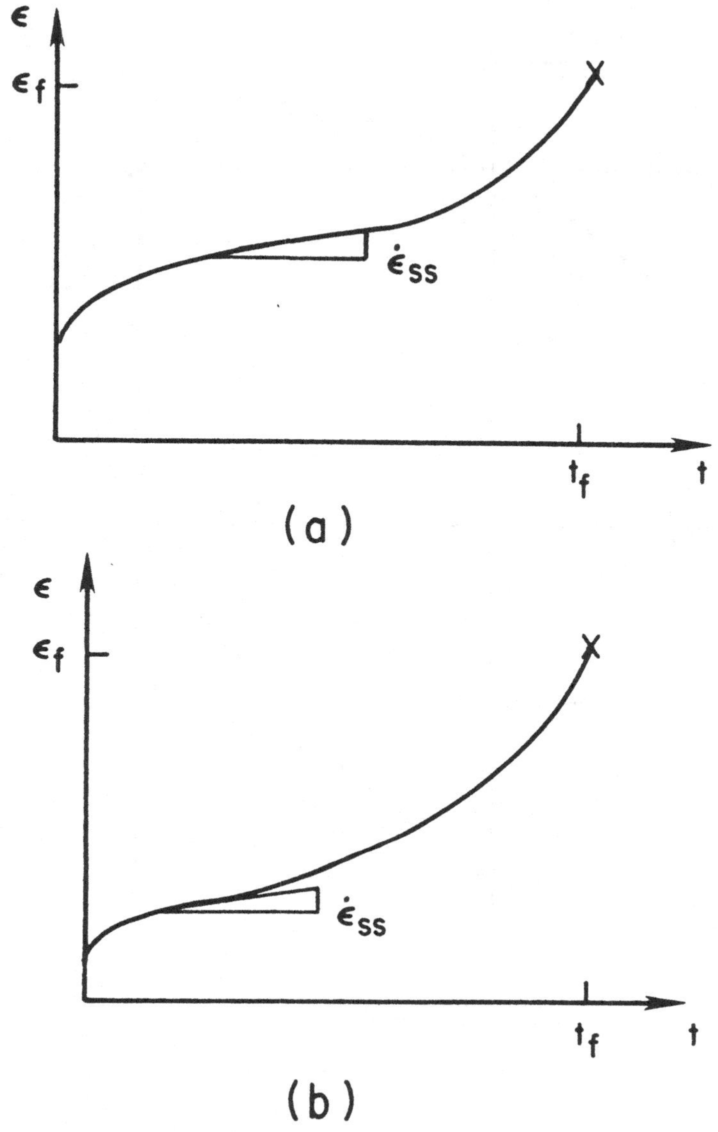

Fig. 1. Uniaxial creep curves for (a) copper and (b) an aluminum alloy.

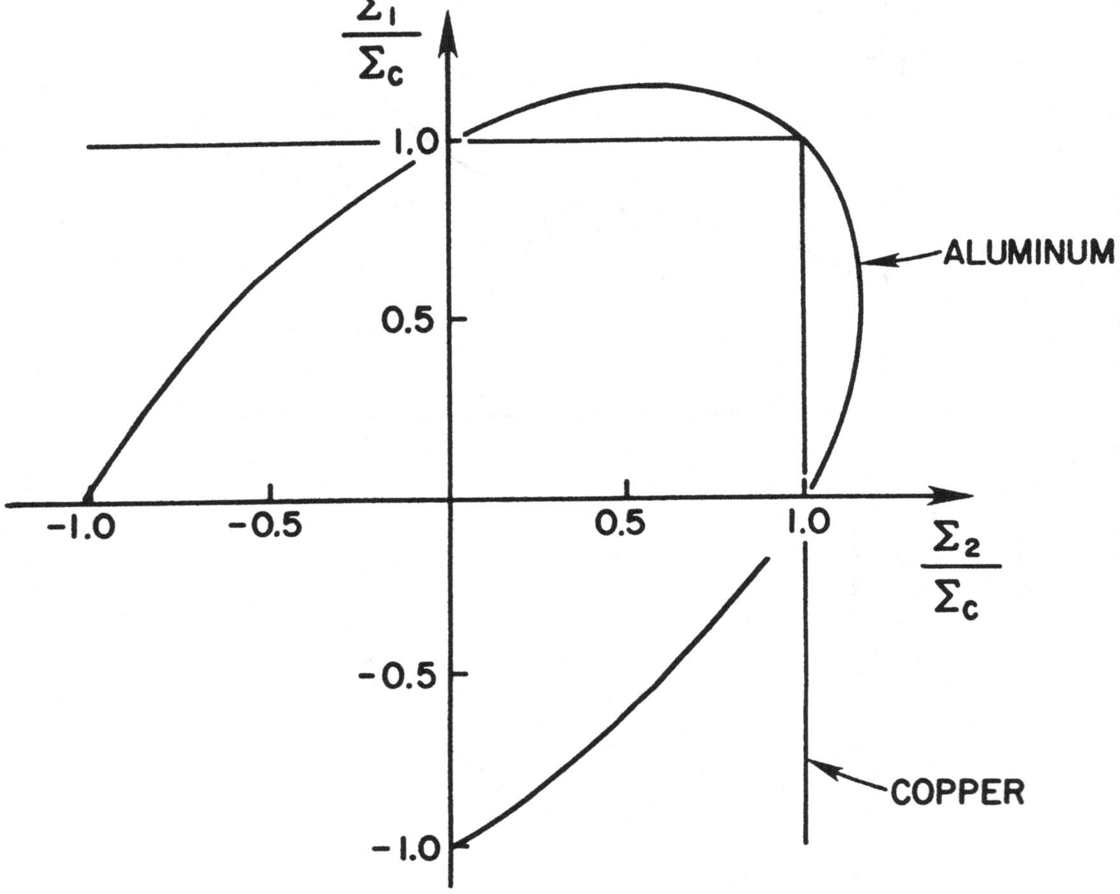

Fig. 2. Isochronous surfaces in plane stress space for copper and aluminum.

Trampczynski et al. [1,4] have performed non-proportional cyclic loading experiments on thin walled tubes of copper and aluminum. In these experiments the axial stress was maintained constant while the shear stress was cycled between ± τ. The stress levels were chosen such that the direction of maximum principal stress rotated through 32°.

In the tests on the aluminum alloy it was found that the magnitude of the components of strain-rate were the same before and after the reversal. It was also found that the time to failure was the same as when the stress-state was held constant for the entire life. These results suggest that the damage in the aluminum can be treated as a scalar quantity. This conclusion is again consistent with the predictions of the strain softening mechanism.

In tests on copper it was found that the magnitude of the strain-rates decreased after a single reversal of stress and the life was increased by a factor of 2 over that of a constant test. Metallographic examination of the failed specimen revealed that one set of damage grew at one end of the cycle and another set at the other extreme. The fact that the time to failure in the cyclic test is twice that in the static test indicates that there is no interaction between these two sets of damage. Two state variables are now needed in the constitutive law.

2. Life Bounds for Creeping Materials

When the damage in the material is in the form of dislocation loops, or even in some situations where it is in the form of voids, the damage rate equation can be written in the form

$$\dot{\omega} = f(\omega)\ \chi(\sigma_{ij}) \tag{2.1}$$

where $f(\omega)$ is a function of the damage ω and $\chi(\sigma)$ is a function of stress. For damage growth rates in the form of eqn. (2.1) it is possible to extend the results of Ponter [ ] to obtain upper bounds on the life of a component for situations of constant and cyclic loading. First we obtain bounds when the load remains constant and then in section 2.2 we examine cyclic loading.

2.1 Bounds for Constant Applied Load

A convenient measure of damage in these situations is

$$\psi = \int_{\omega_i}^{\omega} \frac{d\omega}{f(\omega)}$$

where ω_i is the initial damage

then $\dot{\psi} = \dfrac{\dot{\omega}}{f(\omega)}$ (2.2)

and eqn. (2.1) becomes

$$\dot{\psi} = \chi(\sigma_{ij}) \tag{2.3}$$

Now consider a structure subjected to a constant load P, Fig. 3 , then as an element of material deforms and becomes damaged the stress it experiences changes. If $f(\omega)$ is a monotonically increasing function of ω then when the structure fails after a time t_f.

$$\int_0^{t_f} \dot{\psi}\, dt < \psi_c \tag{2.4}$$

where ψ_c is the value of ψ when an element of material fails. Integrating eqn. (2.3) from $t = 0$ to t_f and integrating over the volume gives

$$\int_V \int_0^{t_f} \chi(\sigma_{ij})\, dt\, dV < \psi_c V \tag{2.5}$$

A further bound on the l.h.s. of eqn. (2.5) can be obtained by considering the same structure composed of a model material. Here we consider two such model materials.

2.1.1. Model creeping material

Consider a material which creeps according to the relationship

$$\dot{\varepsilon}^p_{ij} = \frac{\partial\chi(\sigma_{ij})}{\partial\sigma_{ij}} \tag{2.6}$$

where, we further assume that $\chi(\sigma_{ij})$ is a convex function of stress. Then if σ^1_{ij} and σ^2_{ij} are two arbitrary stress states the convexity condition is

$$(\sigma^1_{ij} - \sigma^2_{ij}) \frac{\partial\chi(\sigma^1_{ij})}{\partial\sigma^1_{ij}} + \chi(\sigma^2_{ij}) - \chi(\sigma^1_{ij}) \geq 0 \tag{2.7}$$

If we identify σ^2_{ij} with the actual stress field in the material, σ_{ij}, and σ^1_{ij} with the solution for the model material, σ^s_{ij} integrating eqn. (2.7) between $t = 0$ and $t = t_f$ and then over the volume gives

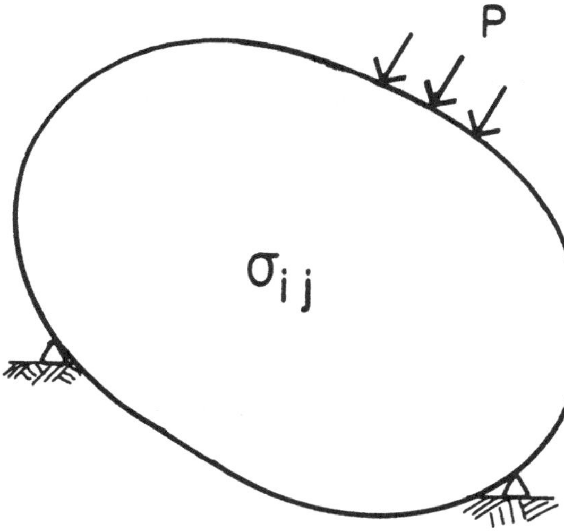

Fig. 3 Structure subjected to a constant load P fails after a time t_f.

$$\int_V \int_0^{t_f} \chi(\sigma_{ij})\, dt\, dV \geq \int_V \int_0^{t_f} \chi(\sigma^s_{ij})\, dt\, dV = \int_V \chi(\sigma^s_{ij})\, dV\, t_f \quad (2.8)$$

since both σ_{ij} and σ^s_{ij} are in equilibrium with the same load P. Substituting eqn. (2) into eqn. (2) gives

$$t_f \leq \frac{\psi_c V}{\int_V \chi(\sigma^s_{ij})\, dV} \quad (2.9)$$

2.1.2 Perfectly plastic material model

Now consider a perfectly plastic material with a yield surface given by an equation of the form

$$\chi(\sigma_{ij}) - \chi_c = 0 \quad (2.10)$$

where χ_c is a material constant equivalent to the yield stress. The value of χ_c is chosen such that a structure composed of the model material collapses at the applied load P. The inelastic strain-rate is given by the associated flow rule

$$\dot{\varepsilon}^p_{ij} = \dot{\mu} \frac{\partial \chi(\sigma_{ij})}{\partial \sigma_{ij}} \quad (2.11)$$

where $\dot{\mu}$ is a plastic multiplier.

Multiplying both sides of eqn. (2.3) by $\dot{\mu}$, integrating over the life and the volume we obtain, after noting eqn. (2):

$$\int_V \int_0^{t_f} \dot{\mu}\, \chi(\sigma_{ij})\, dt\, dV < \psi_c \int_V \dot{\mu}\, dV \tag{2.12}$$

Similarly eqn. (2.7) can be multiplied through by $\dot{\mu}$, integrated between $t = 0$ and $t = t_f$ and, after setting σ^1_{ij} to the stress field at the limit load for the perfectly plastic material and σ^2_{ij} to the stress field in the real material, integrated over the volume to give

$$\int_V \int_o^{t_f} (\sigma^1_{ij} - \sigma^2_{ij})\, \dot{\varepsilon}^{pl}_{ij}\, dt\, dv + \int_V \int_o^{t_f} \dot{\mu}\, \chi(\sigma^2_{ij})\, dt\, dV$$

$$- \int_V \int_o^{t_f} \dot{\mu}\, \chi(\sigma^1_{ij})\, dt\, dV > 0 \tag{2.13}$$

where $\dot{\varepsilon}^{pl}_{ij}$ is the strain-rate under a stress σ^1_{ij}.

The stress fields σ^1_{ij} and σ^2_{ij} are in equilibrium with the same load P_i and, for the model material, plastic flow can only occur when $\chi(\sigma^1_{ij}) = \chi_c$. Equation (2) then becomes

$$\int_V \int_o^{t_f} \dot{\mu}\, \chi(\sigma^2_{ij})\, dt\, dV > \chi_c\, t_f \int_V \dot{\mu}\, dV \tag{2.14}$$

Substituting this into eqn. (2.12) we obtain

$$t_f < \frac{\psi_c}{\chi_c} \tag{2.15}$$

If it is not convenient to obtain the exact limit load for the model material the above bound applies when an upper bound to the unit load is obtained. Then, if P is an upper bound to the limit load for a given yield function $\bar{\chi}_c$, and is the exact solution for χ_c:

$$\bar{\chi}_c < \chi_c$$

and eqn. (2.15) becomes

$$t_f \leqslant \frac{\psi_f}{\chi_c} \tag{2.16}$$

2.2 Cyclic Loading Bounds

For situations where the cycle times are large compared to a characteristic time for stress redistribution the results of the last section can be readily extended.

Consider the global damage rate

$$\dot{\Psi} = \int_V \dot{\psi}\, dV = \int_V \chi(\sigma_{ij})\, dV \tag{2.17}$$

then for a given constant stress state $\chi(s_{ij})\, dV$ we can define a time

$$\bar{t} = \frac{\psi_c V}{\int_V \chi(\sigma_{ij}) dV} \tag{2.18}$$

and eqn. (2.17) becomes

$$\dot{\Psi} = \frac{\psi_c V}{\bar{t}} \tag{2.19}$$

In the present context it can be readily seen that eqns. (2.9) and (2.16) are bounds on $\bar{t}$ for a constant equilibrium stress field σ_{ij}. First consider the bound of eqn. (2.9), then eqn. (2.19) becomes

$$\frac{d\Psi}{dt} \geqslant \int_V \chi(\sigma^s_{ij})\, dV \tag{2.20}$$

where σ^s_{ij} is in equilibrium with the same applied loads as σ_{ij}. If the applied load changes then eqn. (2.20) can be integrated to give

$$\int_o^{\dot{\Psi}_f} d\Psi \int_o^{t_f} \int_V \chi(\sigma^s_{ij})\, dV\, dt \tag{2.21}$$

where t_f is the time to failure and Ψ_f is the value of Ψ at failure, which is less that $\psi_c V$. Therefore eqn. (2.21) becomes

$$\psi_c V \geq \int_o^{t_f} \int_V \chi(\sigma_{ij}^s)\, dV\, dt . \tag{2.22}$$

If the loading is applied in a cyclic manner with a cycle time of t_c, eqn. (2.22) can be expressed as

$$\psi_c V \geq \bar{t}_f \int_o^1 \int_V \chi(\sigma_{ij}^s)\, dV\, d\tau \tag{2.23}$$

where τ is a dimensionless measure of time: $\tau = t/tc$; and $\bar{t}_f$ is the beginning of the cycle at which failure occurs. Rearranging eqn. (9.23) gives

$$\bar{t}_f < \frac{\psi_c V}{\int_o^1 \int_V \chi(\chi_{ij}^s)\, dV\, d\tau} \tag{2.24}$$

If the number of cycles to failure is large then $\bar{t}_f$ will be close to t_f.

Similarly, if eqn. (2.16) is used as the bound on $\bar{t}$, we find

$$\bar{t}_f < \frac{\psi_c}{\int_o^1 \chi_c\, d\tau} \tag{2.25}$$

where χ_c is a function of τ, and is the value of the yield function that will give plastic collapse at the instantaneous load. Again the bound is retained if χ_c is replaced by $\bar{\chi}_c$, where $\bar{\chi}_c$ corresponds to the yield function which results from an upper bound limit load calculation.

A more interesting, and, perhaps more important, situation is when the load is cycled fast compared to the characteristic relaxation time for the structure. The above bounds, eqs. (2.24) and (2.25), also apply to this situation, but in certain situations it is possible to obtain better bounds in terms of rapid cycle solutions. As before rapid cycle solutions can be obtained in terms of the deformation solution for a model creeping material or a model plastic material.

Under rapid cycle loading conditions the only variation of stress during a cycle is that due to the elastic response of the material, so that

$$\sigma_{ij}(T) = \hat{\sigma}_{ij}(t) + \rho_{ij} \tag{2.26}$$

where $\hat{\sigma}_{ij}(t)$ is the elastic stress field at a given instant in time and ρ_{ij} is a residual stress field. We assume here that the elastic constants are unaffected by the presence of the damage so that $\sigma_{ij}(t)$ is the same for each cycle. This assumption breaks down when the first element of material fails so that it cannot support any stress. The following are therefore strictly bounds for the time to initiate failure in the components.

2.2.1 Model creeping material

A bound on the life of a component can again be obtained by considering the result of eqn. (2.5). Let σ_{ij}^{2} of eqn. (2.7) be the actual stress field in the component, σ_{ij}, under conditions of rapid cycling and σ_{ij}^{1} the distribution obtained from the solution for the model creeping material σ_{ij}^{rc}. Integrating eqn. (2.7) over a cycle and then over the volume gives, after noting eqns. (2.6) and (2.26),

$$\int_V (\rho_{ij}^{rc} - \rho_{ij}) \Delta\varepsilon_{ij}^{p} \, dV + \int_V \int_0^{t_c} \chi(\sigma_{ij}) \, dt \, dV$$

$$\geq \int_V \int_0^{t_c} \chi(\sigma_{ij}^{rc}) \, dt \, dV \qquad (2.27)$$

where $\Delta\varepsilon_{ij}^{p} = \int_0^{t_c} \dot{\varepsilon}_{ij}^{p} \, dt$ is the compatible inelastic strain accumulated during a cycle of duration t_c. The first term on the l.h.s. of eqn. (2.27) is then identically zero. Substituting eqn. (2.27) into eqn. (2.5) then give

$$t_f^{i} = \frac{\phi_c V}{\int_V \int_0^1 \chi(\sigma_{ij}^{rc}) \, d\tau \, dV}$$

where, as before, $\tau - t/t_c$.

2.2.2 Model plastic material

Under conditions of rapid cycling shakedown boundary solutions can be used to facilitate the construction of bounds on the time to initiate failure in a component. Ponter [5] considers a general cyclic loading history. Here, however, we limit our attention to the class of problems where the load is cycled between two prescribed limits, Fig. 16. The results can easily be generalized to include the situations considered by Ponter [5].
Again we make use of the inequality of eqn. (2.3) and the convexity condition of eqn. (2.7). As before we identify σ_{ij}^{2} with the actual stress field σ_{ij}^{1} with the shakedown solution for an elastic perfectly plastic material of yield strength

$$\chi(\sigma_{ij}) = \chi_c \tag{2.29}$$

where the magnitude of χ_c is chosen such that the structure composed of the model plastic material just shakes down. If we now write the associated flow rule in incremental form,

$$d\varepsilon^p_{ij} = \mu \frac{\partial\chi(\sigma_{ij})}{\partial\sigma_{ij}}$$

and apply eqn. (2.7) at each extreme of the cycle, we obtain

$$\mu_1\ \chi(\sigma^1_{ij}) - \mu_1\ \chi(\sigma^{1s}_{ij}) - \mu_1 \frac{\partial\chi(\sigma^{1s}_{ij})}{\partial\sigma^{1s}_{ij}} (\sigma^1_{ij} - \sigma^{1s}_{ij}) \geq 0$$

$$\mu_2\ \chi(\sigma^2_{ij}) - \mu_2\ \chi(\sigma^{2s}_{ij}) - \mu_2 \frac{\partial\chi(\sigma^{2s}_{ij})}{\partial\sigma^{2s}_{ij}} (\sigma^2_{ij} - \sigma^{2s}_{ij}) \geq 0 \tag{2.30}$$

where the first eqns. (2.30) applied when $0 \leq \tau \leq \lambda$ and the second when $\lambda < \tau \leq 1$. Here we will assume that $\lambda \leq 1 - \lambda$. Combining eqns. (2.30) and noting that plastic straining can only occur when eqn. (2.29) is satisfied leads to the result

$$\frac{\mu_1}{\lambda} \lambda\ \chi(\sigma^1_{ij}) + \frac{\mu_2}{(1-\lambda)} (1-\lambda)\ \chi(\sigma^2_{ij}) - (\mu_1 + \mu_2)\ \chi_c - d\varepsilon^p_{ij}\ \pi_{ij} \geq 0 \tag{2.31}$$

where $\rho_{ij} = \sigma^1_{ij} - \sigma^1_{ij} = \sigma^2_{ij} - \sigma^{2s}_{ij}$ is a residual stress field and $d\varepsilon^p_{ij}$ is the increment of plastic strain experienced by an element of material during a cycle. The inequality of eqn. (9.31) is still retained if μ_1/λ and $\mu_2/(1 - \lambda)$ is replaced by $\bar{\mu}$, where μ is the maximum of μ_1/λ and $\mu_2/(1 - \lambda)$. Integrating eqn. (2.31) over the volume then gives

$$\int_V \bar{\mu}\ \lambda\chi(\sigma^1_{ij}) + (1 - \lambda)\ \chi(\sigma^2_{ij})\}dV \geq \chi_c \int_V (\mu_1 + \mu_2)dV \tag{2.32}$$

Multiplying both sides of eqn. (9.3) by $\bar{\mu}$ and integrating over the time to initiate failure and the volume gives

$$t^i_f \int_V \bar{\mu}\ \{\lambda\chi(\sigma^1_{ij}) + (1 - \lambda)\ \chi(\sigma^2_{ij})\}dV = \int_V \int_o^{t^i_f} \dot{\psi}\ \bar{\mu}\ dtdV \tag{2.33}$$

Combining eqns. (2.32) and (2.33) and noting the inequality of eqn. (2.4) gives

$$t_f^i < \frac{\psi_c \int_V \overline{\mu}\, dV}{\chi_c \int (\mu_1 + \mu_2) dV} < \frac{\psi_c}{\chi_c \lambda} \tag{2.34}$$

The above bound still holds if χ_c is replaced by $\overline{\chi}_c$, the yield function resulting from a kinematic bound for the shakedown solution. As discussed by Ponter [5] this bound can drastically overestimate the time to failure of a component. The eqn. (2.28), (2.24) and (2.25) can then give more accurate estimates of the life.

3. Creep Rupture of Shell Structures Subjected to Cyclic Loading

3.1. Introduction

When metals are subjected to stress at temperatures in excess of $T_m/3$, where T_m is the melting temperatures in °K, the metal suffers time-dependent creep deformations. In addition, internal damage increases with time so that the metal ultimately ruptures. Consequently, when designing shell structures which operate at such elevated temperatures, consideration must be made to ensure that creep deformations do not exceed operational requirements during the life of the component. Common allowable deformations are 1% average and 5% maximum strain. In addition, the rupture conditions are that no part of the component may separate nor that local leakage can occur.

By establishing suitable constitutive equations which give the strain rates and the rate of internal damage of the material, it is possible in principle to establish by numeric means the strain, stress and damage history at all points in the shell. Such procedures tend to be very complex and it is difficult on the basis of the calculations to draw conclusions of the type which can help to reach a deeper understanding of the shell behavior.

In a companion study [6] it was shown how bounds could be obtained on the creep deformations which occur when shells are subjected to cyclic loading. It was shown how the role of the shakedown concept, so strongly associated with plasticity theory can be extended by replacing the yield stress by the uniaxial stress σ_o which causes the steady state strain rate $\dot{\varepsilon}_o$, which in turn is a design parameter given by the relationship

$$\dot{\varepsilon}_o t_L = \varepsilon_D$$

where t_L is the life of the component and ε_D is the allowable design strain.

This paper is concerned with the determination of the creep rupture life of shell components subjected to cyclic loading. It shall be shown that the shakedown concept can again be extended to give estimates of the creep rupture life. The role of the yield surface in plasticity is replaced by the so-called Isochronous Surface which is a surface in stress space for which the rupture time is constant. It proves convenient to let this time correspond to the design life of the component. For the purpose

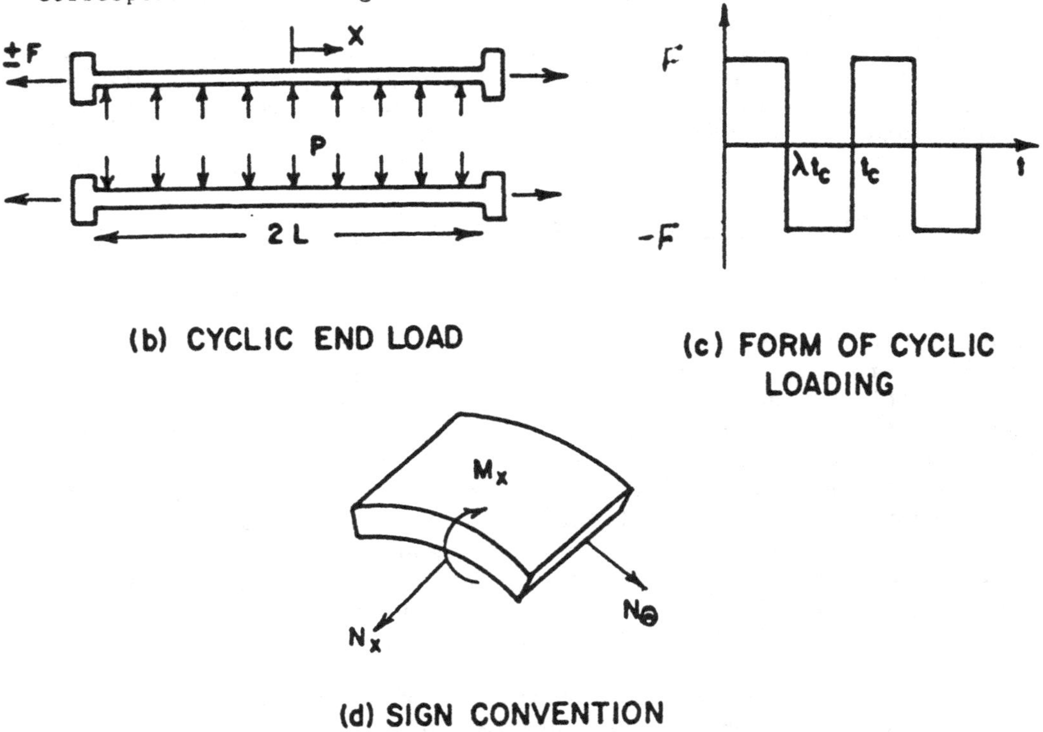

Fig. 4. Cylindrical shell of subjected to cylic load.

of illustration the creep rupture behavior of a cylindrical shell shall be studied. The loading case to be considered is that of a constant internal pressure p and cyclic end load ±F (Fig. 4).

3.2. Material Behavior

The high temperature strain/time response of a metal subjected to constant stress has the form shown in Fig. 5. After the initial time independent response, the strain rate decreases with time during the so-called transient period I when the hardening processes which occur within the material exceed the effects of thermal softening. In the region II, referred to as the steady state, the hardening and thermal rate are equal and opposite so that the strain rate is constant. In the tertiary region III, the effects of internal damage become evident so that the strain rate increases until rupture eventually occurs. In a previous paper [6] constitutive equations were developed which describe the deformations in portion I and II of the creep curve. For rupture life predictions, equations are necessary which describe the internal damage occurring within the metal and the increasing strain rate characteristic of the tertiary behavior. It is these equations which shall be used in the present investigation.

2.1 Creep Rupture Constitutive Equations

The constitutive equations which describe the tertiary portion of the creep curve are given by,

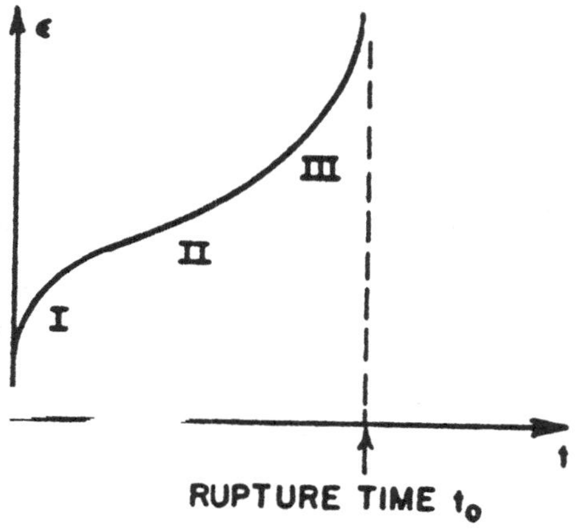

Fig. 5. Uniaxial creep curve showing the three stages of creep.

$$\frac{\dot{\varepsilon}_{ij}}{\dot{\varepsilon}_o} = \phi^n \left(\frac{\sigma_{ij}}{\sigma_o}\right) \frac{\partial \phi}{\partial \sigma_{ij}} g_1(\omega)$$

$$\dot{\omega} = A\, \Delta^{\nu}\left(\frac{\sigma_{ij}}{\sigma_o} \right) g_2(\omega) \qquad (2.1a\text{-}b)$$

where ω is a measure of the internal damage. The physical nature of the damage depends upon the material and its operating conditions. It is possible to identify ω rather precisely with specific forms of damage and special forms of $g_1(\omega)$ and $g_2(\omega)$ in eqns. (2.1) can indeed be used to describe the range of mechanisms which have been identified [3] [7]. In the case when grain-boundary diffusion is responsible for the growth of voids it is found that $\omega = (r/\ell)^{1/3}$ where r is the radius of the void and 2ℓ is the spacing between voids. In other situations, especially those associated with precipitate hardened materials, the damage is in fact a softening mechanism associated with dislocation loops left behind on the precipitate particles by dislocations which account for creep deformation. The presence of the dislocation loops left increases the climb rate [7]. In those circumstances it is the density of the dislocation loops which provides the measure of the damage. The function of stress $\Delta(\sigma_{ij}/\sigma_0)$ describes the so-called Isochronous Surface which is the locus of multiaxial stress states for which the rupture time is constant (Fig. 6). The constant A is selected to give the rupture time t_0 observed for an applied stress σ_0. For the uniaxial loading with $\sigma = \sigma_0$, Eq. (2.4b) becomes

$$\dot{\omega} = A\, g_2(\omega)$$

which on integration gives the rupture time t_0

$$t_0 = \int_0^1 \frac{1}{A} \frac{d\omega}{g_2(\omega)}$$

Two specific forms of Δ are discussed which represent the extremes of material behavior. For some materials strengthened by precipitate hardening $\Delta(\frac{\sigma_{ij}}{\sigma_0}) = \frac{\bar{\sigma}}{\sigma_0}$ where $\bar{\sigma}$ is the effective stress, and Eq. 2.1b then has the form

$$\dot{\omega} = A\left(\frac{\bar{\sigma}}{\sigma_0}\right)^{\nu} g_2(\omega)\ . \qquad (2.4c)$$

A material within this class shall be referred to as a $\bar{\sigma}$ material. For another class of materials the growth of damage is dictated by the maximum stress σ_I when diffusion of vacancies along the grain boundary is the dominant mechanism. Then $\Delta(\frac{\sigma_{ij}}{\sigma_0}) = \frac{\sigma_I}{\sigma_0}$ and the damage growth equation becomes

$$\dot{\omega} = A\left(\frac{\sigma_I}{\sigma_0}\right)^{\nu} g_2(\omega)\ . \qquad (2.4d)$$

Materials satisfying this relationhip are referred to as σ_I materials. These extreme forms of the isochronous surface for plane stress conditions are shown in Fig. 4, and both forms shall be used in this study. Nonproportional loading tests [4] on a precipitate hardened aluminum which is a $\bar{\sigma}$ material indicate that the damage is isotropic and that the rate of damage is dependent only on the magnitude of $\bar{\sigma}$ and not on its direction. Similar tests on copper indicate however that damage grows on planes independent of each other. Consequently if the stress field is rotated, the life of the material is dictated by the rupture of the plane which suffers longest exposure to maximum stress, and the failure time is independent of the damage in other directions. Consequently the concept of independent damage directions becomes useful and this property shall be used in the next section. The function $\phi(\sigma_{ij}/\sigma_o)$ is a homogeneous function of degree one and in practice is found [2] to be proportional to the effective stress $\bar{\sigma}$.

3.3. Generalized Forces and Moments

The generalized forces to be considered in this problem are the hoop stress resultant N_θ the moment resultant M_x, and the axial stress resultant N_x.

From the constitutive equations (2.1) it is possible to determine the expression for the steady state creep energy dissipation rate as

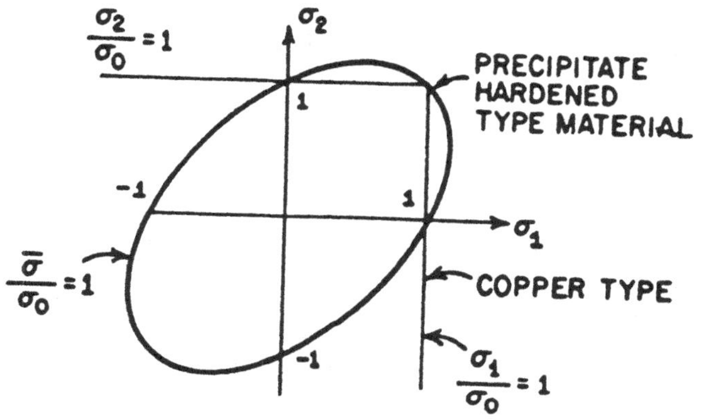

Fig. 6. Isochronous surfaces for $\bar{\sigma}$ and σ_I materials in plane stress space.

$$\dot{D}(\sigma_{ij}) = \sigma_o \, \dot{\varepsilon}_o \, \phi^{n+1} \left(\frac{\sigma_{ij}}{\sigma_o}\right) . \tag{3.1}$$

To simplify the shell analysis expressions for the isochronous surfaces are expressed in terms of the stress resultants N_θ, M_x and N_x. This can be achieved by making the usual shell assumptions about the variations in displacement through thickness and integrating the resulting stress fields. This can prove to be a lengthy exercise in itself and instead of following this route, use shall be made of the results of Hodge [8].

The constant energy dissipation rate $h\sigma_o \dot{\varepsilon}_o$ per unit length is given by the relations,

$$n_\theta = \pm 1$$

$$n_\theta - n_x = \pm 1$$

$$n_x + m_x = \pm 1$$

$$n_\theta - m_x = \pm 1 \qquad (3.2)$$

$$n_\theta - n_x - \frac{m_x}{2} = \pm 1$$

$$n_\theta - n_x + \frac{m_x}{2} \pm 1$$

where M_o and N_o are the normalizing factors

$$M_o = \frac{\sigma_o t^2}{4} \quad \text{and} \quad N_o = \sigma_o t \,. \qquad (3.3)$$

These expressions are recognized as those used in Hodge's expressions with σ_o replacing the yield stress σ_y. In forming this surface the same through thickness stress fields as exist in plasticity are assumed to exist also in creep. This procedure gives an upper bound on energy dissipaton rates [4], which is nevertheless close to the exact value. It also is convenient for shell rupture calculations to express the Isochronous Surface in terms of stress resultants. For materials whose rate of creep damage is governed by the effective stress $\bar{\sigma}$, damage occurs irrespective of the sign of the stress. Experiments by Hayhurst [10] on precipitate hardened aluminum beams in bending verify the validity of the assertion. He also demonstrates that when the applied moment is

$$M_o = \frac{\sigma_o t^2}{4} \qquad (3.3)$$

that the rupture time of the beam is t_o. From this behavior it can be deduced that the isochronous surface for rupture time t_o has the same form as the constant energy dissipation rate surface. The resulting isochronous surface for a $\bar{\sigma}$ material and for the given shell is then defined by the inequalities (3.2).

When materials suffer the type of creep damage which is dependent on the value of the maximum stress, it is implied that the stress is tensile. Damage does not grow when the stress is compressive. This means in the case of a beam in bending, for example, that damage only occurs in those portions where stress is tensile. Hayhurst [10] has shown in this case that the moment which gives a rupture time t_o is given by

$$M_o = \frac{\sigma_o t^2}{4.2^{1/\nu}} . \tag{3.4}$$

Since the beam is symmetric a moment of opposite sign shall give the same rupture time except that it is the opposite face that shall suffer damage.

Finally it was noted from previous experiments on non-proportional loading [4] that faces in different directions do not interact. Hence since M_x and N_x, do not interact with N_θ, the isochronous surface is defined by the equations (3.5). Since no damage is caused by compressive stress the surface extends infinitely far to the left so that the so-called Non-interacting Isochronous Surface is

$$n_\theta = 1$$

$$n_x + m_x = 1 \tag{3.5}$$

$$n_x - m_x = 1$$

3.4. Upper Bound on Rupture Time

An upper bound can be obtained on the rupture time in terms of the shakedown load associated with a yield surface defined by $\Lambda(\sigma_{ij}/\sigma_o) = 1$ which coincides with the isochronous surface for the material. We follow Ponter [5] in deriving a bound on the time for initiation of rupture in a structure subjected to cyclic loading of the type shown in Fig. 1, where one extreme of loading is maintained for $0 \leq \tau < \lambda$ and the other extreme for $\lambda \leq \tau \leq 1$, where $0 \leq \lambda \leq 0.5$ and $\tau = t_c$, t_c being the cycle time.

From eqn. (2.1b) we note that prior to failure for a material element which experiences a stress σ_{ij} that

$$\int_o^t \Lambda^\nu\left(\frac{\sigma_{ij}}{\sigma_o}\right) dt = \int_o^{\bar{\omega}} \frac{1}{A} \frac{d\omega}{g_2(\omega)} \leq t_o \tag{4.1}$$

where $\bar{\omega} \leq 1$. For conditions of rapid cycling the stress at each end of the cycle remains approximately constant and the convexity condition for Λ^ν at any instant is given by

$$\Delta^{\nu}\left(\frac{\sigma_{ij}}{\sigma_o}\right) - \Delta^{\nu}\left(\frac{\sigma^s_{ij}}{\sigma_o}\right) - \frac{\partial\Delta^{\nu}}{\partial\sigma_{ij}{}^s}\left(\sigma_{ij} - \sigma^s_{ij}\right) \geqslant 0 \qquad (4.2)$$

We identify σ_{ij} with the actual solution and σ^s_{ij} with the shakedown solution for a perfectly plastic material of yield strength σ_o, where

$$\Delta^{\nu}\left(\frac{\sigma_{ij}}{\sigma_o}\right) = 1 \;; \quad d\varepsilon^p_{ij} = \mu\,\frac{\partial\Delta^{\nu}}{\partial\sigma_{ij}}, \quad \mu \geqslant 0 \qquad (4.3)$$

represent the yield surface and associated flow rule. Applying eqn. (4.2) at each extreme of the cycle:

$$\mu_1\Delta^{\nu}\left(\frac{\sigma^1_{ij}}{\sigma_o}\right) - \mu_1\Delta^{\nu}\left(\frac{\sigma^{1s}_{ij}}{\sigma_o}\right) - \mu_1\frac{\partial\Delta^{\nu}}{\partial\sigma^{1s}_{ij}}\left(\sigma^1_{ij} - \sigma^{1s}_{ij}\right) \geqslant 0$$
$$\mu_2\Delta^{\nu}\left(\frac{\sigma^2_{ij}}{\sigma_o}\right) - \mu_2\Delta^{\nu}\left(\frac{\sigma^{2s}_{ij}}{\sigma_o}\right) - \mu_2\frac{\partial\Delta^{\nu}}{\partial\sigma^{2s}_{ij}}\left(\sigma^2_{ij} - \sigma^{2s}_{ij}\right) \geqslant 0 \qquad (4.4)$$

where the first of eqns. (4.3) applies when $0 \leqslant \tau < \lambda$ and the second when $\lambda \leqslant \tau \leqslant 1$. Combining eqns. (4.4) and noting that plastic straining can only occur when $\Delta^{\nu}\left(\frac{\sigma^s_{ij}}{\sigma_o}\right) = 1$ for the perfectly plastic material we obtain

$$\frac{\mu_1}{\lambda}\,\lambda\Delta^{\nu}\left(\frac{\sigma^1_{ij}}{\sigma_o}\right) + \frac{\mu_2}{(1-\lambda)}(1-\lambda)\,\Delta^{\nu}\left(\frac{\sigma^2_{ij}}{\sigma_o}\right) - (\mu_1 + \mu_2) - d\varepsilon^p_{ij}\rho_{ij} \geqslant 0 \qquad (4.5)$$

where $\rho_{ij} = \sigma^1_{ij} - \sigma^{1s}_{ij} = \sigma^2_{ij} - \sigma^{2s}_{ij}$ is a residual stress field and $d\varepsilon^p_{ij}$ is the plastic strain experienced by the element of material during the cycle at shakedown. The inequality of eqn. (4.5) is still retained if μ_1/λ and $\mu_2/(1-\lambda)$ are replaced by $\bar{\mu}$, where $\bar{\mu}$ is the maximum of μ_1/λ and $\mu_2/(1-\lambda)$. Integrating eqn. (4.5) over the volume then gives

$$\int_V \bar{\mu}\left\{\lambda\Delta^{\nu}\left(\frac{\sigma^1_{ij}}{\sigma_o}\right) + (1-\lambda)\Delta^{\nu}\left(\frac{\sigma^2_{ij}}{\sigma_o}\right)\right\}dV \geqslant \int_V (\mu_1 + \mu_2)\,dV \qquad (4.6)$$

Integrating eqn. (4.6) from $t = 0$ to $t = t_i$, the initiation time for rupture, and making use of the inequality of eqn. (4.1) yields

$$t_i < t_o \frac{\int_V \bar{\mu}\, dV}{\int_V (\mu_1+\mu_2) dV} < \frac{t_o}{\lambda} \qquad (4.7)$$

where t_o is the time to failure in a uniaxial test under a stress σ_o.

As discussed by Ponter [5] this bound can drastically overestimate the time to rupture if λ is small. In the present paper we limit our attention to situations where $\lambda = 1/2$ and eqn. (4.7) becomes

$$t_i < 2t_o \qquad (4.8)$$

The bounding technique therefore reduces the problem to the determination of the shakedown load for the shell structure when the yield stress is σ_o and the yield condition is $\Delta(\sigma_{ij}/\sigma_o) = 1$.

3.5. Shakedown Solutions for Shell Problem

For the present class of problems for which a structure is subjected to a constant load and a cyclic load the shakedown boundary is best obtained using the method due to Gokhfeld and Cherniavsky [14]. The method involves the construction of a modified yield surface which is used in a limit load calculation for the structure subjected only to the constant load. For the shell problem considered here it proves advantageous to obtain a modified yield condition directly in terms of N_θ, N_x and M_x.

The problem to be considered is that of the cylindrical shell subjected to a constant internal pressure and a cyclic end load ± F (Fig. 1). Since the solution for this problem was solved previously in connection with creep deformation [6], it means that the current problem is solved for materials whose Isochronous Surface is defined by $\Delta = \bar{\sigma}$ defined by the equalities (3.2). In this section therefore the shakedown boundary shall be determined for the case when the shell consists of a σ_1 material whose Isochronous Surface is defined by the Limited Interaction Surface defined by the equalities (3.5).

First the Limit Load is obtained for the constant internal pressure p (Fig. 4) and F = 0.

The equilibrium equation is,

$$\frac{\partial^2 M_x}{\partial x^2} + \frac{N_\theta}{R} - p = 0 \qquad (5.1)$$

With

$$m = \frac{M_x}{M_o}\ ,\ n = \frac{N_\theta}{N_o}\ ,\ \bar{p} = \frac{pR}{N_o} \quad \text{and} \quad \bar{x} = x/\sqrt{RM_o/N_o}$$

Eqn. (5.1) then becomes

$$\frac{\partial^2 \bar{m}}{\partial \bar{x}^2} + 4(\bar{n} - \bar{p}) = 0 \tag{5.2}$$

The yield condition for non-interaction surfaces (Eq. 3.5) when $n_x = 0$ has the form shown in Fig. 7

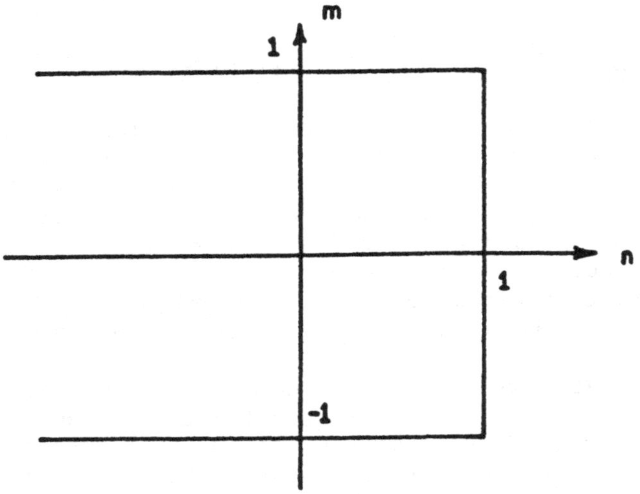

Fig. 7. Limited Interaction Isochronous Surface

Assume a solution exists for which $n = 1$ together with the boundary conditions,

$$\frac{\partial m}{\partial \bar{x}} = 0 \quad \text{and} \quad m = -1 \quad \text{at} \quad \bar{x} = 0 \tag{5.3(a)}$$

$$m = 1 \quad \text{at} \quad \bar{x} = \pm L \tag{5.3(b)}$$

Eqn. (5.2) then becomes

$$\frac{\partial^2 m}{\partial \bar{x}^2} = 4(\bar{p} - 1) \tag{5.4}$$

The solution is

$$m = 2(\bar{p} - 1)\bar{x}^2 + A\bar{x} + B$$

Substituting in the boundary conditions of eqn. 5.3 gives

$$A = 0 \ , \quad B = -1$$

so that

$$m = 2(\bar{p} - 1)\bar{x}^2 - 1$$

Using the boundary condition of eqns. (5.3b) then gives

$$\bar{p} = (1 + \frac{1}{\bar{L}^2}) \tag{5.5}$$

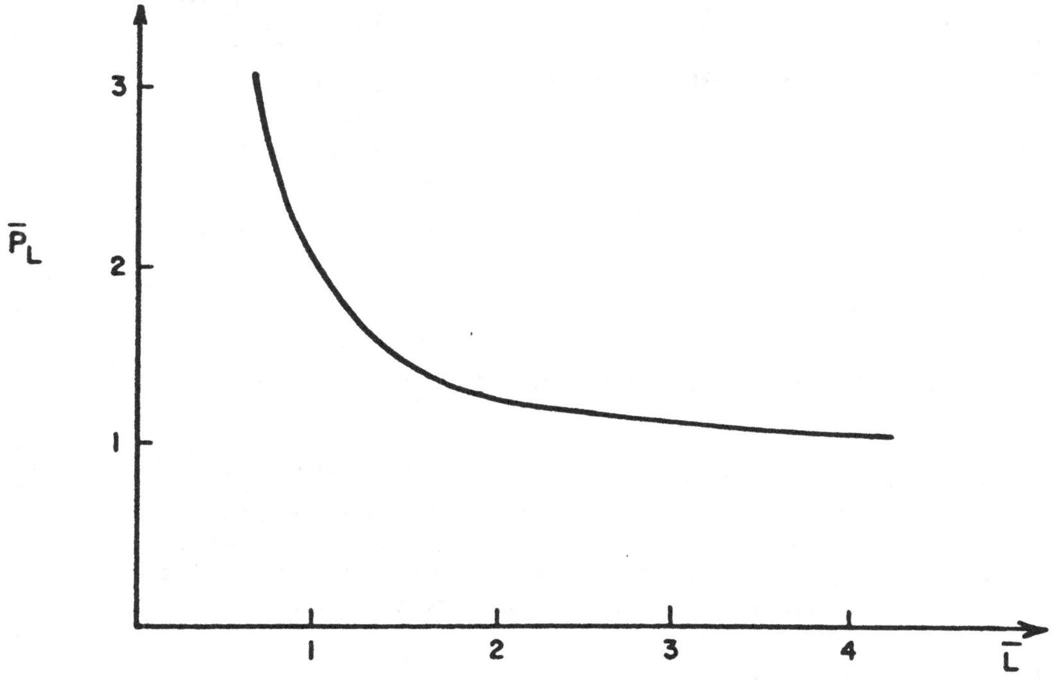

Fig. 8. Variation of Limit Load with Length

where $\bar{L} = L/\sqrt{RM_o/N_o}$. Eqn. 5.5 is plotted in Fig. 8 for a range of values of $\bar{L}$.

To determine the shakedown solution, the yield surface in m, n, and n_x is required. The portion of this surface for $m > 0$ and its section through an n = constant plane are shown in Fig. 9. The elastic solution for cyclic loading is very simple and given by

$$N_x = \pm \frac{F}{2\pi R}$$

The limit load for F acting alone is $F_L = N_o 2\pi R$

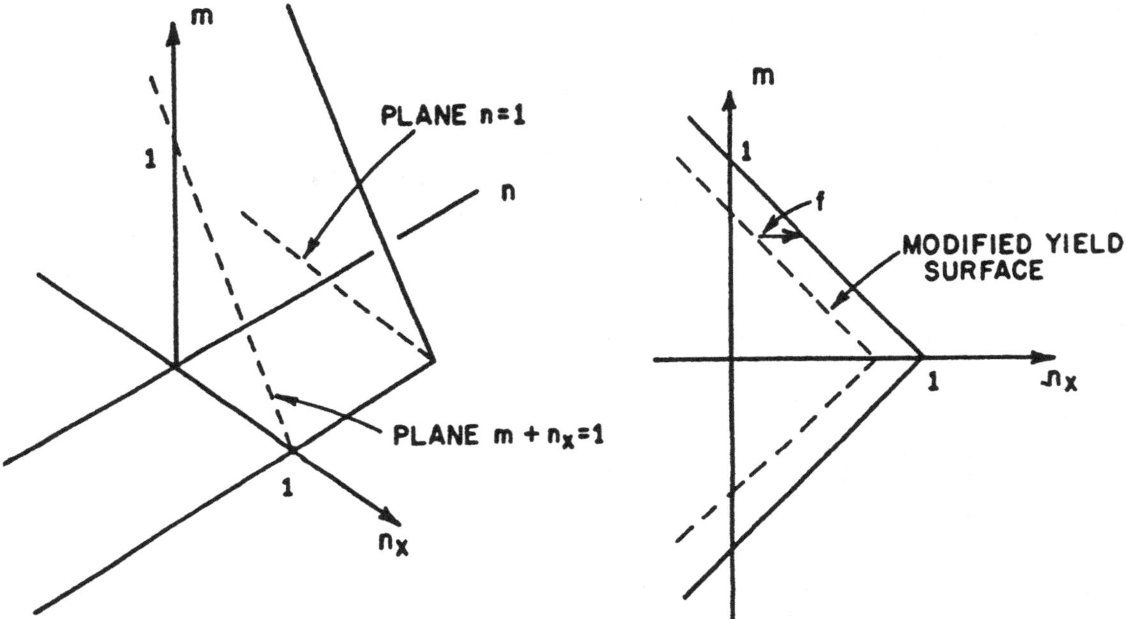

Fig. 9. Isochronous Surface and a Section at n = constant

Defining $n_x = \frac{N_x}{N_o} = \pm \frac{F}{F_L} = \pm f$ the yield surface modified by the elastic solution is shown in Fig. 6 for constant n , and in Fig. 10 for $n_x = 0$ The limit load for the pressure loading for the modified yield surface is determined using the governing equation (5.4) is solved subject to the boundary conditions

$$\frac{\partial m}{\partial \bar{x}} = 0 \quad \text{and} \quad m = -(1 - f) \quad \text{at} \quad \bar{x} = 0$$

$$m = (1 - f) \quad \text{at} \quad \bar{x} = \pm \bar{L}$$

The resulting solution is

$$\bar{P} = 1 + \frac{1 - f}{\bar{L}^2} \quad \text{or} \quad \frac{P}{P_L} = \frac{\bar{L}^2 + 1 - f}{\bar{L}^2 + 1} \tag{5.6}$$

When $f = 1$ the yield surface disappears and $\bar{p} = 0$. The shakedown boundary formed by eqns. (5.6) and (5.7) is plotted in Fig. 11 for $\bar{L} = 1$ and $\bar{L} = 10$.

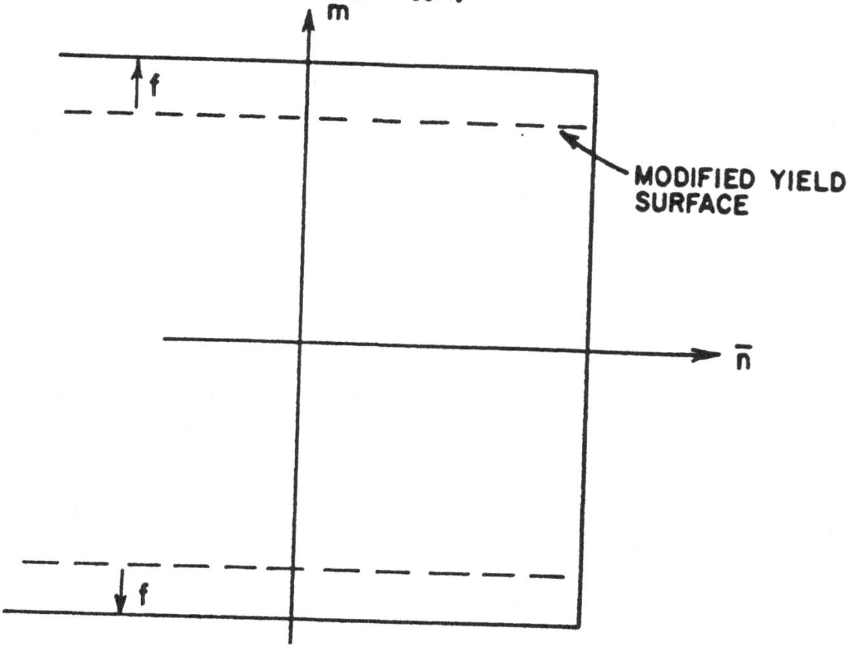

Fig. 10. Modified Surface for $n_x = 0$

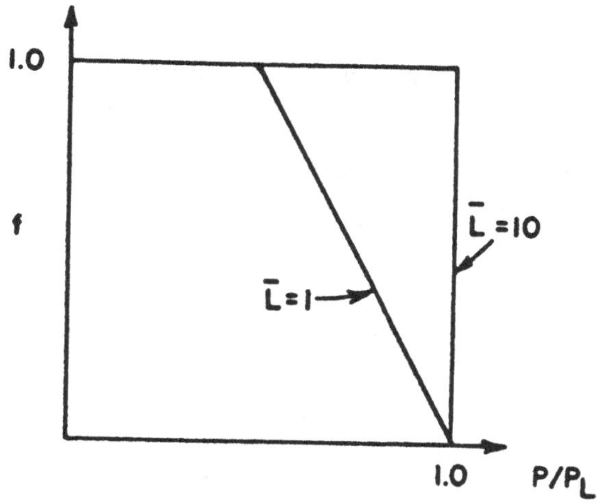

Fig. 11. Shakedown Boundaries

Acknowledgment

The authors acknowledge the support of the National Science Foundation through Grant No. MSM 82-10620.

References

1. Hayhurst, D. R., Trampczynski, W. A., and Leckie, F. A., "Creep Rupture under Non-Proportional Loading," Acta Met., 28, 1981, 1171.

2. F. A. Leckie and D. L. Hayhurst, "Creep Rupture of Structures," Proc. Royal Soc., A340, 1974, 324.

3. Ashby, M. F. and Dyson, B. F., "Creep Damage Mechanics and Micromechanisms," National Physical Laboratory, Report DMA/1A/77 1984.

4. Trampczynski, W. A., Hayhurst, D. R. and Leckie, F. A., "Creep Rupture of Copper and Aluminum under Non-proportional Loading," J. Mech. Phys. Solids, 29, 1981, 353.

5. Ponter, A.R.S., "Upper Bounds on the Creep Rupture Life of Structures Subjected to Variable Load and Temperature," Int. J. Mech. Sci., 19, 1977, 79

6. A.C.F. Cocks and F. A. Leckie, "Deformation Bounds for Cyclically Loaded Shell Structures Operating under Creep Conditions." Report No. 1986, Department of Theoretical and Applied Mechanics, University of Illinois, Urbana, IL.

7. Cocks, A.C.F. and Leckie, F. A., "Creep Constitutive Equations for Damaged Materials," Theoretical and Applied Mechanics, Department Report 480, 1986.

8. P. G. Hodge, "Limit Analysis of Rotationally Symmetric Plates and Shells," Prentice-Hall, Englewood, 1963.

9. Ponter, A.R.S. and Leckie, F. A., "The Application of Energy Theorems to Bodies which Creep in the Plastic Range," J. Appl. Mech. 37, 1970, 753.

10. Hayhurst, D. R., "Estimates of the Creep Rupture Lives of Structures Subjected to Cyclic Loading," Int. J. Mech. Sci., Vol. 18, 1976, 75-83.

11. Gokhfeld, D. A. and Cherniavsky, O. R., "Limit Analysis of Structures at Thermal Cycling," Sijthoff and Noordhoof, Amsterdam, 1980.

CONTINUOUS DAMAGE OF BRITTLE SOLIDS

Jerzy Najar, University of Munich
Department of Mechanics A

ABSTRACT

Some elastic solids fail not through macro – fracture or plastic flow, but rather through growth of micro – defects. Modelling of such processes occurs through theories of continuous damage mechanics. Thermodynamic aspects of one such theory are discussed in this paper, focussing on limitations of admissible processes, energy dissipation in a loading – unloading cycle and the number of cycles leading to failure.

0. INTRODUCTION.

It has been experimentally established that some elastic (brittle) materials fail, if subjected to applied stress or strain, not through macro - fracture or plastic flow, but rather as a result of evolution of small defects or flaws, such as microvoids and microcracks. The process which results in this type of failure is called damage development. This observation led in recent years to the establishment of various theories, collectively termed continuous damage mechanics, cf.[1], [2].

It is the purpose of the present contribution to discuss the thermodynamic aspects of one such theory by postulating a simple form of the free energy function and invoking the Gibbs relation, as well as the 2nd law of thermodynamics. At first the strain and (as yet unspecified) dissipation are taken as state variables and a linearity in strain leads to a more special form of the free energy, permitting to consider both loading and unloading processes. Dissipation is then specialized to damage development by considering the energy release rate of a non - interacting distribution of growing voids.

The consequences of such a damage model are highlighted, in particular the limitations on admissible processes and the amount of energy dissipated per cycle. A simple formula is advanced, which permits to calculate the number of loading - unloading cycles at a given strain amplitude needed to achieve full damage, i.e. to bring the material to failure. This number depends on initial damage, on the strain amplitude and on the specific damage law relating the damage development to the strain energy.

Finally, particular forms of the damage law and damage energy development are discussed. Application of extremal principles is proposed, leading to the choice of brittle damage models with least dissipation and maximal entropy increase within a loading process.

I. DISSIPATION.

I.1. Two – parameter model of isothermal dissipation.

Let the *specific free energy* function W be dependent on the *strain* ε, the *dissipation state variable* D and the *temperature* T

(I.1.1) $$W = W(\varepsilon, D, T)$$

the *stress* σ and the *entropy* s being correspondingly defined as

(I.1.2) $$\sigma = W_{,\varepsilon} \qquad s = -W_{,T}$$

and, in general, non – linearly depending on ε, D and T. Here, as well as in the sequel, a comma denotes partial differentiation with respect to the indicated variable.

Introduce the *dissipative stress* θ as

(I.1.3) $$\theta = -W_{,D}$$

and consider an arbitrary *isothermal* loading process $\varepsilon(\tau)$, $D(\tau)$ in *time* τ. The increment of the specific free energy dW over a time increment $d\tau$ consists of two components

(I.1.4) $$dW = \delta W_\sigma - \delta W_\theta$$

where the first one

(I.1.5) $$\delta W_\sigma = \sigma\, d\varepsilon$$

represents the increment of the *specific external work* W_σ performed in *stretching* ($d\varepsilon > 0$) and corresponding to the time increment $d\tau$, while the second one

(I.1.6) $$\delta W_\theta = \theta\, dD$$

represents the increment of the *specific work of dissipation* W_θ at growing dissipation variable, $dD>0$. The symbol δ denotes here the dependence on the process. It is called a *loading process*, when both $d\varepsilon>0$ and $dD\geqslant0$ take place.

The relationships (I.1.4 – 5) yield, after integrating over the process $\varepsilon(\tau)$, $D(\tau)$, the following equation

$$W_\sigma = W + W_\theta \tag{I.1.7}$$

It represents the balance between the external work performed in loading and the sum of the dissipated and the free energy, recoverable at any instance of the process. Given initial conditions, the free energy W depends as a state function only on the final values of the state variables ε and D, while the amounts of work W_σ and W_θ are path – dependent, i.e. are determined by the process.

The specific free energy W is connected with the *specific internal energy* U through the relation

$$W = U - Ts \tag{I.1.8}$$

The increment dU of the internal energy is equal to the sum of the external work increment δW_σ and the *heat influx* δQ (1st law of thermodynamics), thus

$$dU = \sigma\, d\varepsilon + \delta Q \tag{I.1.9}$$

Applying the relationships (I.1.4 – 9) one obtains the relation

$$T\, ds = \delta W_\theta + \delta Q \tag{I.1.10}$$

which shows that the entropy increment ds depends both on the dissipation work increment as well as on the heat influx.

The 2nd law of thermodynamics for non – stationary processes, in the form $Tds \geqslant \delta Q$, yields now

$$\delta W_\theta \geqslant 0 \tag{I.1.11}$$

for any $\varepsilon(\tau)$, $D(\tau)$ of the class considered.

The non-negative increment of the dissipation work needs to be split now into two non-negative parts. The first one corresponds to the work $\delta W_{\theta q}$, which shall be converted further into heat, e.g. through inducing crystal lattice vibrations or internal friction. The second one is the work $\delta W_{\theta s}$ causing no heat production but rather responsible for internal structural changes in the material, like opening of new internal free surfaces, i.e. internal damage. This does not preclude, that internal structural changes cannot be accompanied by heat production, although it implies that some part of the dissipation work is due to the structural changes alone. Thus

(I.1.12) $$\delta W_{\theta} = \delta W_{\theta q} + \delta W_{\theta s}$$

Our assumption of an isothermal process can hold only, when the work $\delta W_{\theta q}$ converted into *heat increment is fully extracted* from the material and no other heat flux is present. We have to assume thus

(I.1.13) $$\delta Q = - \delta W_{\theta q} \leqslant 0$$

The entropy increment in (I.1.10) is equal now to

(I.1.14) $$ds = \delta W_{\theta s} / T \geqslant 0$$

and is determined exclusively by the part of the dissipation work, which is non-convertible into heat and depends only on the isothermal structural changes in the material, e.g. on the internal damage. The assumption T = const. results now in the path-independence of the dissipation losses $\delta W_{\theta s} = dW_{\theta s}$. Integrating over the process we obtain

(I.1.15) $$W_{\theta s} = T\,\Delta s$$

where Δs denotes the increment of the entropy during the process.

Having in mind the definition (I.1.6), we obtain from (I.1.11) that the dissipative stress θ must be non-negative, when the dissipation variable grows. Similarly as in (I.1.12), it may happen that the dissipative stress θ consists of two non-negative components

(I.1.16) $$\theta = \theta_q + \theta_s$$

responsible for the heat - convertible dissipation $\delta W_{\theta q}$ and for the structural losses $dW_{\theta s}$, respectively. At growing dissipation variable one obtains thus the following expression

$$T\, ds = \theta_s\, dD \geqslant 0 \qquad \text{(I.1.17)}$$

relating the increment of the entropy to the increment of the dissipation state variable. The dissipative stress θ_s should present an integrable multiplier for the dissipation variable D.

I.2. Linearity in strain.

Consider a particular case of (I.1.1 – 2) in the form

(I.2.1) $$\sigma(\varepsilon, D) = E_d \, \varepsilon$$

i.e. a linear dependence of stress σ on strain ε at constant dissipation variable D. The *secant modulus* E_d depends on the dissipation variable D only

(I.2.2) $$E_d = E_d(D)$$

and shall be assumed in the form of a monotonic decreasing function of D, with an initial value E_0 at the start of the loading process under consideration. Decreasing secant modulae in loading processes have been observed in most solids under tension, cf. [1 – 2]. We shall assume also that the beginning of the process corresponds to $\varepsilon = 0$ and $D = D_0$, i.e. the strain measure is defined for the current loading cycle, at some initial value D_0 of the dissipation parameter D.

For an arbitrary loading process $\varepsilon(\tau)$, $D(\tau)$ these assumptions correspond to a material, whose carrying capacity at loading decays after having reached a maximum, usually defined as the tensile strength of the material. The corresponding $\sigma(\varepsilon)$ – curve gradually deviates from the linear elastic dependence, exhibiting a stress maximum, Fig. I.2.1.

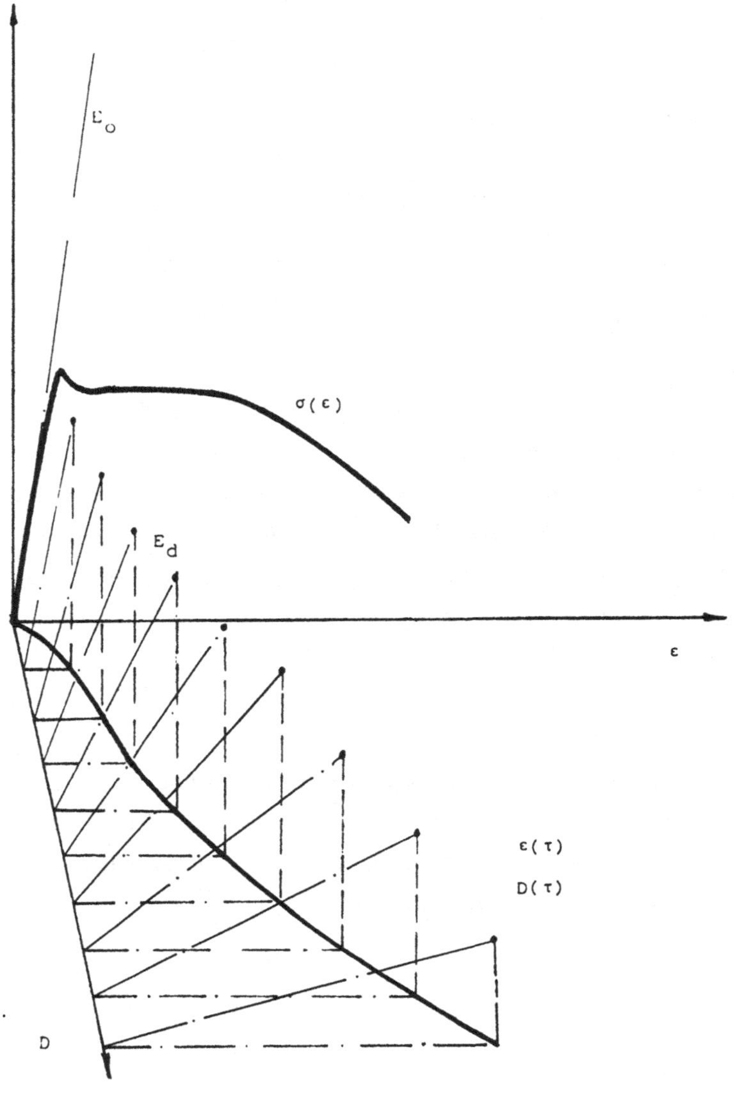

Fig. I.2.1 Stress – strain curve at linearity in strain for a general strain – dissipation path.

Assumptions (I.2.1 – 2) lead to the following relationships for the free energy

(I.2.3) $$W = \frac{1}{2} E_d \varepsilon^2 - \Delta\Phi(D)$$

comp. (I.1.2), for the dissipative stress

(I.2.4) $$\theta = -[\frac{1}{2} dE_d/dD \, \varepsilon^2 - d\Phi/dD]$$

comp. (I.1.3), and for the dissipative work increment

(I.2.5) $$\delta W_\theta = -\frac{1}{2} \varepsilon^2 \, dE_d + d\Phi$$

comp. (I.1.6) and (I.2.4). The quantity $\Delta\Phi(D)$ denotes in these relations the increment

(I.2.6) $$\Delta\Phi(D) = \Phi(D) - \Phi(D_o)$$

of an arbitrary function $\Phi(D)$ of the dissipation variable D, arising from the integration of the eqn. (I.1.2) within the range (D_o, D) upon substitution of the assumption (I.2.1).

The thermodynamic meaning of the function $\Phi(D)$ can be seen from the analogy between the formulae (I.2.5) and (I.1.12). Similarly as in the latter one, the dissipative work increment δW_θ can be split into two components

(I.2.7) $$\delta W_\theta = \delta W_{\theta\varepsilon} + \delta W_{\theta d}$$

corresponding to the addends in (I.2.5).

The first component, denoted $\delta W_{\theta\varepsilon}$, depends both on strain and dissipation variable. Its integral over the process is therefore path – dependent, similarly as the integral of the heat – convertible component $\delta W_{\theta q}$ in (I.1.12).

The second component, denoted $\delta W_{\theta d}$, is equal to the increment of the function $\Phi(D)$, i.e. does not depend on the strain variable. Its integral over the process depends thus only on the values D_o and D, which means the path – independence of this part of losses

(I.2.8) $d\Phi = \delta W_{\theta d}$

similar to that of the component of losses $\delta W_{\theta s}$ in (I.1.12) responsible for structural changes within the material, comp. (I.1.14).

Could the dissipation variable D be identified, for a class of materials or processes, with a *structural state variable*, e.g. damage, the direct consequence of the striking analogy between the relations (I.1.12) and (I.2.5) would be

(I.2.9) $\delta W_{\theta q} = \delta W_{\theta \varepsilon}$

(I.2.10) $dW_{\theta s} = dW_{\theta d}$

Since the histories $\varepsilon(\tau)$ and $D(\tau)$ are, in general, not interdependent, the 2nd law of the thermodynamics implies now $d\Phi > 0$ at loading, yielding thus

(I.2.11) $\Delta\Phi(D) \geqslant 0$

for any $D > D_o$. The condition (I.1.11) can be satisfied at loading only for $dE_d < 0$, in agreement with our assumption of a monotonically decreasing function $E_d(D)$, comp. (I.2.2).

Consider now an *unloading process*, defined by $\varepsilon,_{\tau} < 0$, beginning at a reversal point (r), corresponding to an arbitrary strain ε and an arbitrary value D of the dissipation parameter. Comparing the relation (I.2.3) with the formula (I.1.7) one obtains, that the dissipated energy W_θ exceeds the area $W_\sigma - E_d \varepsilon^2/2$ of the segment included between the $\sigma(\varepsilon)$-curve and the secant line connecting the origin (0) of the (σ, ε)-coordinates with the reversal point (r) exactly by the amount $\Delta\Phi(D)$, s. Fig. I.2.2.

This implies that the unloading curve in the (σ, ε)-plane must run, at least in some part, below the secant line (0r) and, in the case of *linear unloading*, results in a *residual strain* $\varepsilon_r > 0$.

Let us consider a process, in which *no dissipation at unloading* takes place. Then the *recoverable energy* is equal to the free energy W at the reversal point (r). Presenting the recoverable energy in the form

(I.2.12) $$W = \frac{1}{2}\sigma(\varepsilon)(\varepsilon - \varepsilon_r)$$

and comparing it with the expression (I.2.3) for the free energy W one is lead to

(I.2.13) $$\varepsilon\ \varepsilon_r = 2\ \Delta\Phi(D)/E_d(D)$$

Introduce now the *unloading modulus* E_u at linear unloading. Its value can be found from the continuity of stress at the reversal of the process, which yields

(I.2.14) $$E_u(\varepsilon - \varepsilon_r) = E_d\varepsilon$$

In view of the relationship (I.2.13) one obtains the following expression for the unloading modulus

(I.2.15) $$E_u = E_d\ /\ [1 - \Delta\Phi\ /\ (\frac{1}{2}E_d\varepsilon^2)]$$

Obviously, for a monotonically growing function $\Phi(D)$ at $D > D_o$, i.e. for $\Delta\Phi(D) > 0$, the residual strain ε_r is greater than zero and, correspondingly, the unloading modulus E_u is greater than the secant modulus E_d at the reversal point of the process.

Note, that although non-linear unloading could result in vanishing residual strain ε_r also at $\Delta\Phi > 0$, s. Fig. I.2.2, it would necessarily lead to higher values of the unloading (tangent) modulus E_u as compared to the corresponding linear case.

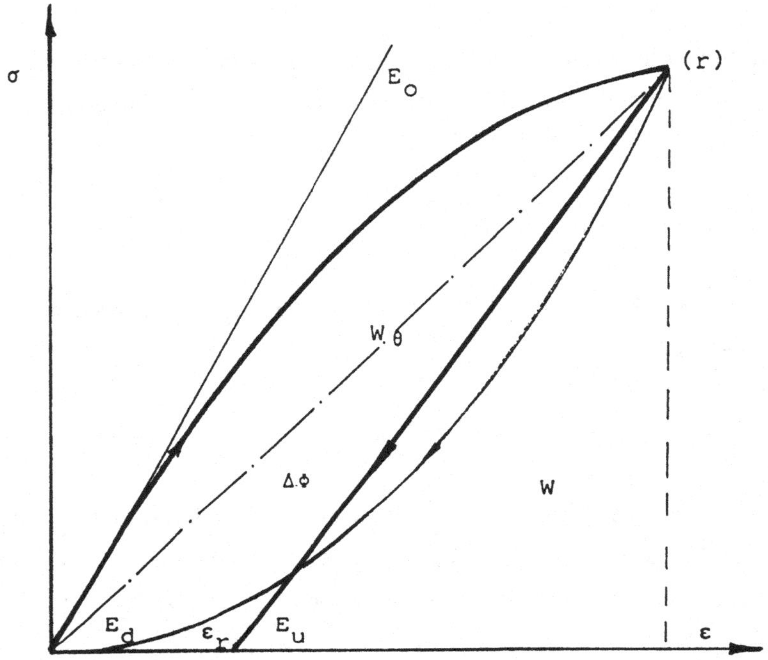

Fig. I.2.2 Loading and unloading at dissipation losses.

II. DAMAGE.

II.1. Microflaws in an elastic matrix.

Consider an elastic body of volume V subjected to uniform tensile strain ε. Let the material in the *perfect* (undamaged) state (I) be characterized by the Young's modulus E. The *specific strain energy* of the perfect body is equal to

$$W_{perf} = \frac{1}{2} E \varepsilon^2 \tag{II.1.1}$$

Compare the state (I) with the *damaged* state (II), in which the body contains a uniform distribution of n geometrically identical *flaws* of a characteristic *dimension* a at the tensile strain ε.

Consider the *change* of the internal energy of the whole body with the change of the characteristic flaw dimension a at a *fixed* level of strain ε and at a constant number n of *non-interacting* flaws. This change can be conveniently calculated on the basis of one of the so-called conservation (or balance) laws of elastostatics, applying the M-integral, s. [3],[19].

In two dimensions it is defined as

$$M = \oint (U^* n_k - T_i u_{i,k}) X_k \, dl \tag{II.1.2}$$

where U^* is the specific strain energy of the elastic matrix material, n_k the outward normal to the (closed) contour of integration, T_i the traction vector and $u_{i,k}$ the displacement gradient. The X_k are material coordinates, l denotes the length of the contour of integration.

It can be shown, that the M-integral is path-independent and vanishes, if the contour does not enclose any inhomogeneities. By contrast, if the contour does enclose a flaw, e.g. a circular void of radius a, then the M-integral will not vanish, but will represent the change of the specific strain energy for self-similar expansion of the void, i.e.

$$M / V = - a \, U^*_{,a} \tag{II.1.3}$$

while volume V in the two-dimensional problem is represented by the area encircled by the contour of integration in (II.1.2).

In evaluating the M-integral it is most convenient to take the rim of the circular void as the contour of integration, because there $T_i = 0$ and $U^* = \sigma_{\varphi\varphi}{}^2/2E$. Further, $n_k X_k = a$ and $dl = a\, d\varphi$, where φ denotes the angle of the polar coordinates, located in the center of the void. Thus, for a single circular two-dimensional flaw we obtain

(II.1.4) $$M = \frac{a^2}{2E} \int_0^{2\pi} \sigma_{\varphi\varphi}{}^2 \, d\varphi$$

For uniaxial tension $\sigma_o = E\varepsilon$ at strain ε, with

(II.1.5) $$\sigma_{\varphi\varphi} = \sigma_o(1 - 2\cos 2\varphi)$$

the result is

(II.1.6) $$M = 3\pi\, a^2\, E\, \varepsilon^2$$

For n identical circular flaws one can write the result as

(II.1.7) $$M = 6\pi\, a^2\, n\, W_{perf}$$

For non-interacting spherical flaws of radius a the corresponding formula is

(II.1.8) $$M = 6\pi\, a^3\, n\, f(\nu)\, W_{perf}$$

where $f(\nu) = (1 - \nu)(9 + 5\nu)/(7 - 5\nu)$, while for plane cracks of length 2a normal to the direction of strain one obtains

(II.1.9) $$M = 4\pi\, a^2\, n\, W_{perf}$$

Compute now the rate of change of the specific free energy W of the material in the state (II) with respect to the size a of the flaw, at a given tensile strain ε. Having in mind that the specific strain energy U^* of the elastic matrix material represents simultaneously the specific internal energy U of the material in the state (II), comp. Ch. I.1, we can apply the formula (II.1.3) to obtain the following relationship

(II.1.10) $U_{,a} = - M/aV$

Substituting the values of the M from (II.1.7 - 9) and integrating at constant strain ε over the length a, with the initial condition

(II.1.11) $U = W_{perf}$ at $a = 0$

which corresponds to the state (I), we obtain from the relation (I.1.8)

(II.1.12) $W = (1 - D^*) W_{perf} - T s$

Here D^* denotes a parameter of the *damage* introduced into the elastic matrix by the presence of the flaws. In the case of cylindrical circular holes we find from (II.1.7)

(II.1.13) $D^* = 3\pi a^2 n_v$

where $n_v = n/V$ corresponds to the specific density of the flaws. Similarly, for spherical flaws we obtain from (II.1.8)

(II.1.14) $D^* = 2\pi a^3 n_v f(\nu)$

and for plane cracks the formula (II.1.9) yields

(II.1.15) $D^* = 2\pi a^2 n_v$

The damage parameter D^* is, at least explicitly, independent of the strain ε and clearly *governs the deviation* of the state (II) of the body from the perfectly elastic state (I). It describes structural changes within the material and can be considered as a state variable, determining the specific free energy W and other functions of state. We can therefore *identify* the dissipation state variable D with the damage variable D^* for the *class of brittle damaging materials*, which shall be the subject of our further investigation, dropping in the sequel the upper star.

Under these conditions we may recall the remarks leading to the relations (I.2.9 – 10). Substituting the second one into the equation (I.1.14) and performing the integration over the range (D_o, D), we obtain from the formula

(I.2.8) the relation

$$T\,\Delta s = \Delta\Phi(D) \tag{II.1.16}$$

where Δs denotes the entropy increment in the process of damage development, while $\Delta\Phi$ is defined by (I.2.6). Substitution into the relation (II.1.12) leads to the following form of the specific free energy

$$W = \frac{1}{2}E\varepsilon^2\,(1 - D) - \Delta\Phi(D) \tag{II.1.17}$$

We observe immediately that formula (II.1.17) repeats the relation (I.2.3), yielding linearity of stress σ with respect to strain, s. (I.2.1), while the secant modulus $E_d(D)$ decreases linearly with D according to

$$E_d = E\,(1 - D) \tag{II.1.18}$$

from the initial value $E_o = E\,(1 - D_o)$, obeying thus the conditions formulated in Ch. I.2. Note, that in view of the independence of the variables ε and D, the initial value D_o of the damage parameter D can be chosen independently from the reference strain $\varepsilon = 0$.

Leaving to the following chapter a more detailed analysis of the model, let us give now a simple interpretation of the damage parameter D. From the relation (II.1.17) follows, that it can be presented in the form

$$D = 1 - (W + \Delta\Phi)/W_{perf} \tag{II.1.19}$$

comp. (II.1.1). Comparing formula (I.2.1) with (I.2.3) one comes to the following expression for the *strain energy* $W_\varepsilon = \sigma\varepsilon/2$ of the damaged material

$$W_\varepsilon = W + \Delta\Phi \tag{II.1.20}$$

while for the perfect (undamaged) material the strain energy is equal W_{perf}.

Define the *strain energy deviation* ΔW_ε of the damaged state (II) from the undamaged state (I) as

$$\Delta W_\varepsilon = W_{perf} - W_\varepsilon \tag{II.1.21}$$

The damage parameter D can be represented now as the ratio of the strain energy deviation ΔW_{ε} to the value W_{perf} of the strain energy in the perfect (undamaged) state

(II.1.22) $$D = \Delta W_{\varepsilon} / W_{perf}$$

s. Fig. II.1.1.

The above formula represents a new *energy-based measure* of the damage D. In the cases considered in this Chapter it is equivalent to the definitions relying on microstructural considerations, s. (II.1.9 - 11). From the phenomenological point of view it has, however, a more general range of applicability, being independent of the particulars of the distribution and the geometry of flaws, s. Ch. II.2. Additionally, it can serve as a guideline in choosing the constitutive relations for damage developement, s. Ch.III.4.

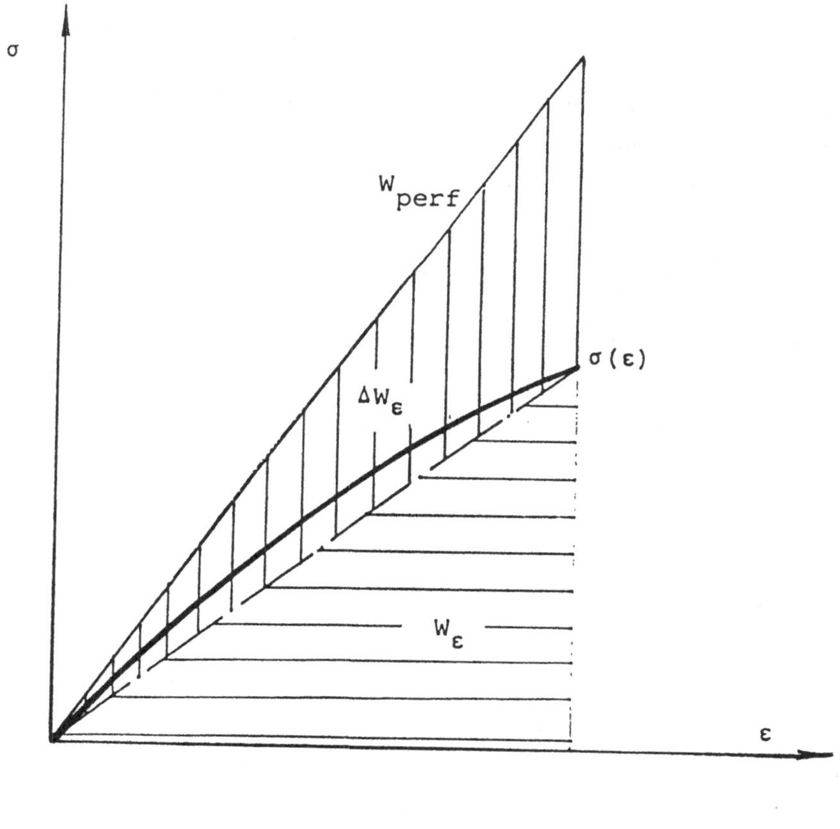

Fig. II.1.1 Energy-based definition of damage.

II.2. Analysis of the damage model.

The results obtained in Ch. II.1. for a simple micro-structural model of damage in an elastic matrix indicate that the particular case of the theory of dissipative thermomechanical processes, as presented in Ch. I.2, can be applied to continuous description of material damage of a general character, provided a suitable identification of the dissipation variable with some internal structural damage parameter is possible. Accordingly, the physical meaning of the expressions derived in Ch. II.1 should be properly interpreted.

Let us begin with the increment $\Delta\Phi$ of the function $\Phi(D)$, present in the derived expression (II.1.17) of the specific free energy $W(\varepsilon, D)$. The incremental definition (I.2.8) indicates that $\Delta\Phi$ is equal to the strain- and path-independent part $W_{\theta d}$ of the dissipation losses W_{θ}, comp. (I.2.7). It represents thus the amount of the energy, directly spent on the development of the internal structural damage, s. (I.2.10). The amount of losses converted into heat, s. (I.2.9), is irrelevant here, depending on the process as a whole.

Therefore, it should be proper to name $\Phi(D)$ the *damage energy function*, while the amount $\Delta\Phi$ be the *damage development energy*, corresponding to the damage increment from D_o to D, comp. (I.2.6). The definition (I.2.8) implies immediately

(II.2.1) $$d\Phi/dD \geqslant 0 \qquad \text{for any } D > 0$$

comp. (I.2.11). Without loss of generality we can take also

(II.2.2) $$\Phi(D) > 0 \qquad \text{for any } D > 0$$

The dissipative stress θ, s. (I.1.3), can be computed from (II.1.17) as

(II.2.3) $$\theta = \frac{1}{2}E\varepsilon^2 + d\Phi/dD$$

comp. (I.2.4), and is positive for any $\varepsilon > 0$. Using the same argument, which led us to accept the equations (I.2.9-10), we may observe that the two components of the sum in the expression (II.2.3) present exactly the form (I.1.16) of the dissipative stress. It is interesting to notice that the component θ_q, responsible for heat-convertible losses, is quantitatively equal to the specific strain energy W_{perf} of the perfect material

(II.2.4) $$\theta_q = \frac{1}{2}E\varepsilon^2$$

comp. (II.1.1), and depends solely on the level of strain ε. The second component, which obviously performs the work of structural damage, since

(II.2.5) $$\theta_s = d\Phi/dD$$

depends only on the damage variable D. At the same time it clearly obeys the demand of integrability with respect to D, comp. (I.1.17).

Accordingly, the energy W_θ dissipated in the process $\varepsilon(\tau)$, $D(\tau)$ consists of two parts. The *heat-convertible losses* $W_{\theta q}$ can be found from the expression

(II.2.6) $$W_{\theta q} = \int_{D_0}^{D} \theta_q dD$$

comp. (I.1.6). Substitution of the formula (II.2.4), integration by parts and application of the expressions (I.2.1) and (II.1.18) yield the following result

(II.2.7) $$W_{\theta q} = \int_0^{\varepsilon} \sigma d\varepsilon - \frac{1}{2}E_d\varepsilon^2$$

Thus, the energy converted into heat at damage development corresponds to the area included between the $\sigma(\varepsilon)$ - curve and the secant line (0r), s. Fig. I.2.2.

The *work of structural changes*, i.e. internal damage, is due to the energy input

(II.2.8) $$W_{\theta s} = \Delta\Phi(D)$$

comp. (II.2.5) and (I.2.6), and are related to the entropy increment Δs, comp. (II.1.16) and (I.1.17).

Let us investigate the consequences of the condition (I.2.11) in the case of non-vanishing damage development energy, i.e. $\Delta\Phi > 0$. At linear unloading, the formula (I.2.13) for residual strain ε_r can be presented now in the following form

(II.2.9) $\varepsilon_r = 2\,\Delta\Phi(D) \,/\, E(1 - D)\varepsilon$

The natural condition that the residual strain ε_r must not exceed the strain ε at the point (r), corresponds to the inequality

(II.2.10) $W_{perf} \geqslant \Delta\Phi(D) \,/\, (1 - D)$

derived upon substitution of (II.1.1) into the formula (II.2.9).

Observe, that in view of the formula (II.1.17) for the specific free energy W, condition (II.2.10) is equivalent to the demand $W \geqslant 0$.

Consider now a process $\varepsilon(\tau)$, $D(\tau)$ in the (D, W_{perf}) - plane. Condition (II.2.10) obviously represents a constraint on the process, Fig. II.2.1, at any non - vanishing values of $\Delta\Phi$. We can distinguish now a class of *admissible processes*, obeying the condition $W(W_{perf}, D) \geqslant 0$. The region of inadmissibility, $W < 0$, vanishes only at $\Delta\Phi = 0$, when all processes fulfill (II.2.10).

The expression (I.2.14) for the unloading modulus E_u can be written now as

(II.2.11) $E_u / E = (1 - D)[1 - \Delta\Phi \,/\, W_{perf}(1 - D)]^{-1}$

Obviously, the unloading modulus E_u exceeds the secant modulus E_d, corresponding to the damage D at the reversal point (r), s. Fig. I.2.2, as long as $\Delta\Phi$ is greater than zero. For a material with initial damage D_o it may exceed also the initial loading modulus $E_o = E\,(1 - D_o)$, s. (II.1.18).

Even the intrinsic modulus E may not represent a limit to E_u, should the expression included in the square bracket of the formula (II.2.11) be allowed to vanish, as it is the case, when the condition (II.2.10) turns into equation, i.e. at $W = 0$. Should, however, the virgin modulus E be treated as an upper bound for the unloading modulus E_u, it would lead immediately to the following constraint in the (D, W_{perf}) - plane

(II.2.12) $W_{perf} > \Delta\Phi(D) \,/\, D(1 - D)$

Another constraint in the (D, W_{perf}) – plane is introduced for loading processes by their definition itself, s. Ch. I.1, not allowing damage D to decrease below its initial value D_o as long as the tensile strain grows. As for unloading, we shall consider here materials, which allow neither damage recovery nor damage development at decreasing tensile strain. With exception of polymers, possibly exhibiting damage recovery, at least at slow processes, this assumption seems to be fairly well applicable to a wide class of materials, [1].

The constraints (II.2.10) and (II.2.12), together with the condition of the loading, allow to narrow down the class of admissible processes $\varepsilon(\tau)$, $D(\tau)$ and to determine some of their characteristic features.

First of all, if there exists a finite strain limit ε_f, then also the extent of the damage cannot exceed a certain upper bound, D_f, which can be determined from the inequality (II.2.10). This upper bound, called here the *damage at failure* , can be calculated from the equation

$$W_f = \Delta\Phi(D_f) / (1 - D_f) \qquad \text{(II.2.13)}$$

where W_f denotes the limiting value of the strain energy W_{perf}.

It is obvious, that for any finite $W_f > 0$ one obtains $D_f < 1$, provided the condition $\Delta\Phi > 0$ holds. It follows also from (II.2.13), that the damage D_f at failure depends on D_o at a given value W_f, due to the definition (I.2.6).

With the damage D obeying the inequalities

$$0 < D_o \leqslant D \leqslant D_f < 1 \qquad \text{(II.2.14)}$$

we observe, that the condition (II.2.12) demands larger values of W_{perf} at given D, than the condition (II.2.10), s. Fig. II.2.1. This should lead to another upper bound of the damage D, denoted as D_E, comp. Fig. II.2.1, at which the unloading stiffness E_u would approach the Young's modulus E of the virgin (perfect) material. The damage D_E can be determined from the following equation

$$W_f = \Delta\Phi(D_E) / D_E(1 - D_E) \qquad \text{(II.2.15)}$$

resulting from (II.2.12). Comparison with (II.2.13) shows that $D_E < D_f$.

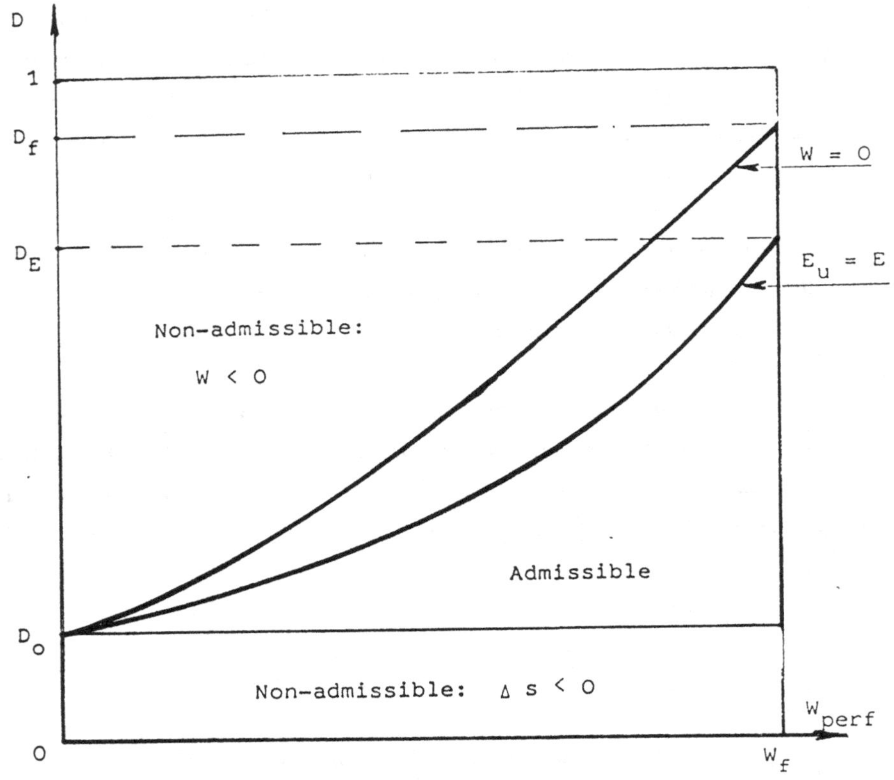

Fig. II.2.1 Constraints on loading processes.

A range of damage limits has been proposed, starting from various approaches to the continuous damage modelling, s. [2], [7], [8]. From the point of view of the stress - controlled experimental verification it seems to be the the quantity D_m of damage at the stress maximum σ_m, that could constitute the practical limit of the model, [2]. However, neither from the theoretical point of view nor for applications to kinematically - controlled experiments, [9], could the damage D_m be accepted as an indication of the limiting damage.

As the considerations of the Ch. II.1 clearly show, the damage D could, formally at least, grow up to the value D = 1. This is the case in Kachanov's analysis, [7], which has laid the foundations to the continuous damage theory.

This is also a rather common assumption in the recent literature, s. e.g. [8]. Despite its formal advantages, s. Ch. II.3, it leads to conceptual difficulties related to the notion of "complete" damage at D = 1, particularly when the damage D is defined as a ratio of the volume of flaws to the volume of the material, i.e. as a specific volume of flaws, [11]. It leads namely to a physically inconsistent notion of a "disappearing" material of a given volume or cross-section, independently of any mass transport mechanism.

The present definition of damage, s. (II.1.22), which is based on the notion of the strain energy deviation ΔW_{ε} from the perfect undamaged state, does not experience any difficulties of this kind. Indeed, in the case of cylindrical circular holes, comp. (II.1.13), one would obtain for D = 1, if it were attainable, a specific volume of the flaws equal to 1/3, , while the densest packing (with contact) would correspond to a specific volume of the flaws equal to ca. 0.91. For spherical flaws the highest packing density is equally remote from the specific volume of the flaws at D = 1, while for cracks the volumetric definition of damage is inapplicable at all. The thermodynamic constraints discussed in this Chapter put a natural limit on the damage development at the damage at failure D_f, comp. (II.2.13). It corresponds namely to the point of the damage development at which *no energy can be recovered* from the material, since W = 0, comp. (II.2.10).

Another, even lower, limit of damage development could be assumed at $D = D_E$, s. (II.2.15). It would correspond to the condition, that an unloading wave in the damaged material must not propagate faster than the elastic waves in the matrix material. Although, to our knowledge, there is no basic law that would imply this restriction, its occurence would have been obviously limited to very local events anyway. We shall therefore decline the limit D_E as unnecessarily restrictive and consider in the following the damage at failure D_f as the damage limit within our model.

Consequently, failure occurs at non-vanishing values of the secant modulus E_d, namely at

(II.2.16) $$E_{df} = E\,(1 - D_f) > 0$$

comp. (II.1.18). This corresponds to failure at some non-vanishing *failure stress*

(II.2.17) $\sigma_f = E(1 - D_f)\,\varepsilon_f$

comp. (I.2.1). Experimental observations on construction materials such as concrete or rocks as well as some modelling assumptions, [13,14], agree with the present result.

Concluding these general remarks on the elastic damage note that the extent of the damage can be measured directly in the loading experiment either from the value of the secant modulus E_d, comp. (II.1.18), or from the energy ratio given by the formula (II.1.22), provided the virgin elastic modulus E of the material is available. It can be determined also in the unloading, as proposed in [17], on the basis of the unloading stiffness E_u, comp. (II.2.11), if the energy $\Delta\Phi$ can be calculated. We shall find later, s. Ch. IV.3, that also the *increment* ΔD of damage within the process can be deduced from the stress – strain curve, without the knowledge of the particular micro – mechanisms leading to it.

II.3. Brittle damage: basic assumptions.

The history $D(\tau)$ of the damage parameter D defined in (II.1.13 – 15) depends on the development of the size of the flaws $a(\tau)$ and their number $n_v(\tau)$, which, in turn, is determined by micro – physical processes specific to the material under consideration.

We shall confine our analysis to the case of *brittle damage*, with possible applications to elastic – brittle solids under *low dynamic* tensile loading. It should apply to processes with the *duration* τ_f of the *strain pulse* $(0, \varepsilon_f)$ and the corresponding duration of the damage increment ΔD *shorter* by orders of magnitude than the *relaxation time* τ_r of rheologic phenomena in the material.

Accordingly, it shall be assumed that brittle damage is governed by an *instantaneous response* to tensile strain. The *influence of the strain rate* $\varepsilon_{,\tilde{\iota}}$ is allowed to be implicit only, which means that the instantaneous response may be solely *quantitatively* dependent on the strain rate, e.g. through the *rate – dependence of the parameters* of the model. Excluding on one side non – isothermal processes, comp. Ch. I.1, and on the other side relaxation phenomena, we confine thus our attention to *moderate strain rates*, corresponding to low dynamic effects.

As a distinctive feature of brittle damage, differing it from other kinds of dissipative processes, e.g. viscous or perfect plastic flow, we consider its property to be *fully accumulated* at cyclic loading. Full memory of the history of the dissipation state variable certainly does not characterize neither viscous behaviour, usually governed by the laws of the fading memory response, [16], nor perfect plasticity, which reveals no memory of past loading at all, [18].

In accordance with the definition of unloading, comp. Ch. II.2, we can take now the damage D at the *completion* of a loading cycle as the *initial damage* D_o for the following *reloading*. Since there is no distinction between loading and reloading for any real material and in view of the accumulative property of damage, we observe that an arbitrary process begins at some *positive* initial damage D_o:

$$\text{(II.3.1)} \qquad D = D_o > 0 \qquad \text{for} \qquad \varepsilon = 0$$

A loading – unloading – reloading diagram in the (ε, D) – plane is represented thus by a saw – tooth curve, corresponding to the damage accumulation between the subsequent points of strain reversal.

Conversely, we should state that *brittle damage can develop only in elastic materials with imperfections*, i.e. at some initial damage $D_o > 0$. A perfect flawless elastic material, if it could exist, should not be able to develop any damage:

(II.3.2) $$D = 0 \quad \text{for any } \varepsilon > 0 \quad \text{at } D_o = 0$$

Substantial specification of brittle damage properties should be expected from the application of one of the general extremal principles of thermodynamics, s. e.g. [4], [5]. We shall apply here explicitly only Onsager's *principle of minimum dissipation energy*, as formulated in [6]. It demands, that within the set of admissible processes the *real* one should follow the path in the space of state variables, which minimizes the dissipation energy over the process, provided the *physical constraints* on the material have been taken into account.

It shall be shown later, comp. Ch. III, that the results of the minimization procedure for the dissipation energy coincide, at least within the class of damage laws considered here, with the demands of another extremal principle, formulated by Ziegler in [15] for non – linear processes as the *principle of maximal entropy production.*

Let us formulate now the principle of minimum dissipation energy for our purposes. Consider all processes $\varepsilon(\tau)$, $D(\tau)$ beginning at the same initial damage D_o corresponding to $\varepsilon = 0$ and ending at the same failure damage $D = D_f$ determined by the failure strain ε_f, comp. (II.2.13). The dissipation energy W_θ consists of two components, given by expressions (II.2.6) and (II.2.8) with the upper integration limit $D = D_f$. Substituting the formulae (II.2.4) and (II.1.1), as well as the expression (I.2.6), we obtain the following expression for the energy of dissipation $W_{\theta f}$ over the whole process up to failure

(II.3.3) $$W_{\theta f} = \int_{D_o}^{D_f} W_{perf}\, dD + \Phi(D_f) - \Phi(D_o)$$

The real process corresponds to a minimum of $W_{\theta f}$ with respect to all admissible processes at given D_o and D_f.

The particular form of the expression (II.3.3) allows a qualitative analysis of the minimization procedure, when considering the admissible processes in the (D,W_{perf}) - plane, s. Fig. II.2.1. Let an arbitrary process $D(\tau),W_{perf}(\tau)$ obeying the constraints, which have been derived in Ch. II.2, starts at $(D_o,0)$ and develops into full failure at (D_f,W_f), s. (II.2.13).

In order to minimize the first component of the expression on the right - hand side of the eqn. (II.3.3), the process $D(\tau),W_{perf}(\tau)$ should correspond to the lowest possible increment $\Delta D_f = D_f - D_o$ of the damage D and to the lowest possible values of the strain energy W_{perf}, which leads consequently to the lowest possible values of the strain energy at failure W_f.

The other two components on the right - hand side of (II.3.3) correspond to the increment $\Delta\Phi$ of the monotonically increasing function $\Phi(D)$, therefore the lowest possible increment of damage leads also to the minimization of this part of the dissipation energy.

In geometrical representation, the principle of the minimum dissipation energy chooses of all virtual paths in the admissible region of the (D,W_{perf}) - plane that one, which represents the *left - hand side envelope* obeying the physical constraints on the material, Fig. II.3.1.

Given the material constraints, the above result yields a *unique dependence* of the damage D on the strain energy W_{perf}, with parameters D_o and W_f

(II.3.4) $$D = D(W_{perf};D_o,W_f)$$

To be consistent with the results of Ch. II.2, the dependence of D on W_{perf} should be represented by a monotonically increasing and continuously differentiable function.

There are three conditions, that this dependence should fulfill in the light of the above considerations. The first is equivalent to the definition of the initial damage D_o

(II.3.5) $$D(0;D_o,W_f) = D_o$$

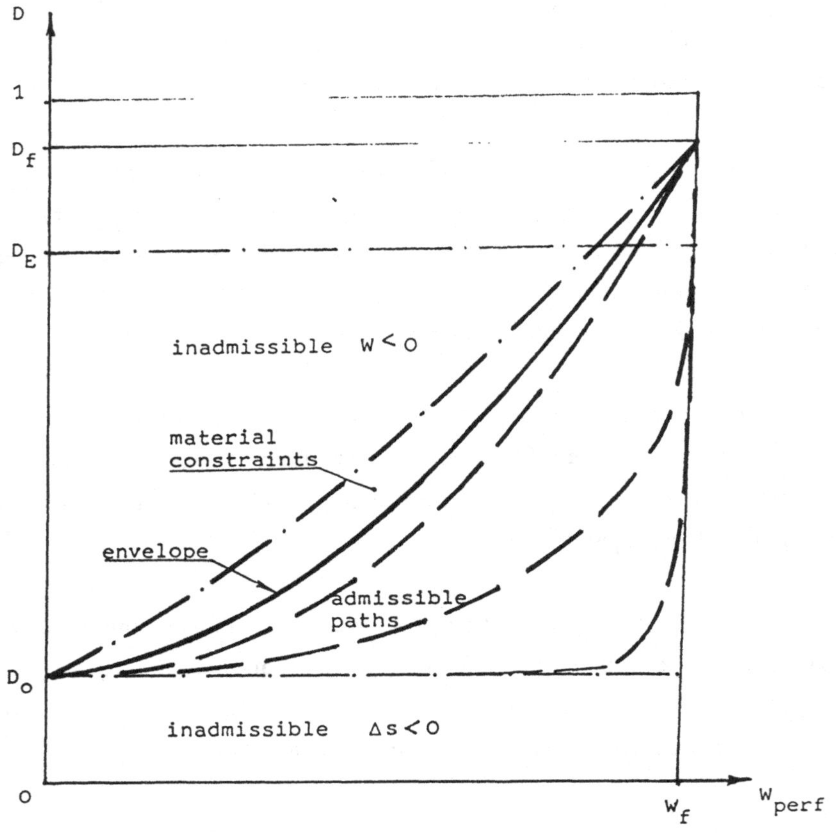

Fig. II.3.1 Principle of minimum dissipation energy applied to admissible paths.

comp. (II.3.1). The second reflects the demand (II.3.2) for perfect (flawless) elastic materials

(II.3.6) $$D(W_{perf};0,W_f) = 0$$

while the third repeats the definition (II.2.13) of the damage D_f at failure

(II.3.7) $$D(W_f;D_o,W_f) = D_f$$

As a result of the demands (II.3.5) and (II.3.7) we obtain the following relation

(II.3.8) $$W_f = 0 \quad \text{for} \quad D_o = D_f$$

It is consistent with the definition (II.2.13) of damage at failure D_f.

Relation (II.3.8) reflects the physically obvious fact, that a severely damaged material needs almost no loading to be brought to complete failure. It indicates at the same time, that the parameters W_f and D_o cannot be treated as *independent* quantities in the eqn. (II.3.4).

This finding can be strengthened by the following observation regarding condition (II.3.6). Since we assume, that no damage can be developed in a flawless (perfect) elastic material, comp. (II.3.2), it would be contradictory to assume for the perfect material any finite strain limit ε_f, i.e. a finite value of W_f. Thus, W_f should necessarily tend to infinity, as D_o tends to zero, emphasizing once more the interdependence of these parameters.

It should be postulated therefore

(II.3.9) $$W_f = W_f(D_o)$$

with corresponding changes in the relationships (II.3.4 – 7). We obtain thus, instead of (II.3.4), the dependence

(II.3.10) $$D = D(W_{perf},D_o)$$

fulfilling the conditions

(II.3.11) $D(0,D_o) = D_o$

(II.3.12) $D(W_{perf},0) = 0$

(II.3.13) $D(W_f,D_o) = D_f$

Note, that the dependence (II.3.9) substituted into the definition (II.2.13) yields the relation

(II.3.14) $D_f = D_f(D_o)$

It is easy to see that the quantity W_f decreases from infinity to zero with initial damage D_o increasing from 0 to D_f, comp. (II.3.8) and the above remarks on the limit transition to a flawless state. One can deduct therefore from eqn. (II.2.13), that damage at failure D_f decreases with the increase of initial damage. There exists therefore an *upper bound* D_{of} of initial damage D_o, equal to the *lower bound* of the damage at failure D_f.

It can be shown, that the demands (II.3.11 - 13) lead to the multiplicative form of the eqn. (II.3.10), i.e.

(II.3.15) $D = f_1(W_{perf})\, f_2(D_o)$

which, in turn, leads immediately to the following *damage law*

(II.3.16) $D = D_o\, F(W_{perf})$

Here the *damage function* F has the property

(II.3.17) $F(0) = 1$

comp. (II.3.11), and should be monotonically increasing and continuously differentiable in order to comply with the conditions derived in Ch. II.2. A qualitative characterization of the damage function is illustrated in the Fig. II.3.2.

Knowing the damage function F one can determine the damage at failure D_f from the following relation

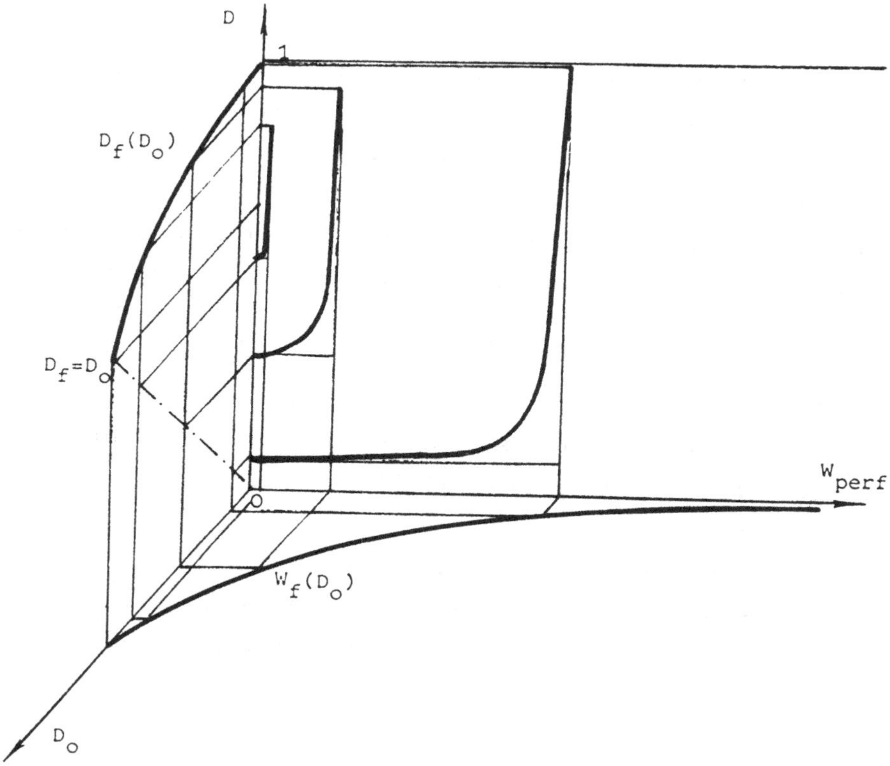

Fig. II.3.2 Damage in dependence on perfect strain energy and initial damage.

(II.3.18) $F^{-1}(D_f/D_o) = \Delta\Phi(D_f) / (1 - D_f)$

comp. (II.2.13), provided the damage development energy $\Delta\Phi(D)$ is known. Here F^{-1} denotes the inverse function of F.

Consider now loading - unloading - reloading cycles at a constant amplitude W_N of the strain energy W_{perf}. Let us determine the number N of cycles needed to bring the sample to complete failure. Obviously, after the first cycle the damage is $D = D_oF(W_N) = D_1$, after the i-th cycle we have $D = D_oF^i(W_N) = D_i$ and at the end of the process we should have $D_N = D_f$, where D_f needs to be determined from $D_o = D_{N-1}$ by a recurrence procedure based on the equation (II.3.18), Fig. II.3.3.

An exact evaluation of the number N of cycles needed for complete failure of the specimen in cyclic tensile loading shall be given in Ch.III.4 after specification of the damage energy function $\Phi(D)$.

Still, we can obtain an upper bound on the number N at this stage already, by assuming $D_f = 1$, i.e. by allowing the damage to exceed the thermodynamic constraints derived in Ch.II.2. From the above relationships we derive immediately

(II.3.19) $N < -\ln D_o / \ln F(W_N)$

The number of cycles N leading to complete failure depends thus on the amplitude W_N, on the initial damage D_o and on the damage function F.

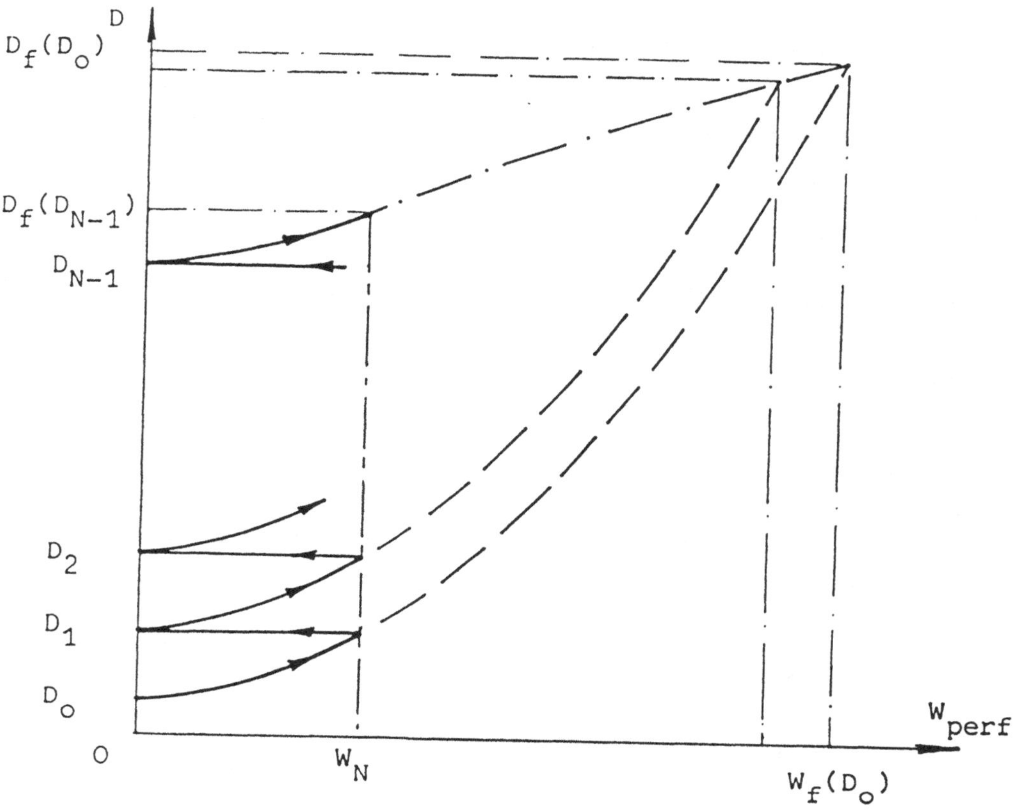

Fig. II.3.3 Cyclic loading at constant strain amplitude.

II.4. Damage energy function.

Let us examine, whether the principle of minimum dissipation energy could be applied to each of the components of $W_{\theta f}$ in (II.3.3) *separately* in order to establish the representation of the damage energy function $\Phi(D)$, comp. Ch. II.2, without specifying the damage law.

We shall limit our attention to the class of simplest power functions

(II.4.1) $$\Phi(D) = W^* D^{1-k}$$

obeying the conditions (II.2.1 – 2) if the coefficient W^* is positive

(II.4.2) $$W^* > 0$$

and the power k is not greater than 1. It is easy to see that since damage D fulfills inequality (II.2.14), the quantity W^* represents the upper bound for the damage energy at nominal full damage $D = 1$. We shall call it therefore the *nominal damage energy.*

The energy of damage development $\Delta\Phi$ at loading is equal to

(II.4.3) $$\Delta\Phi(D) = W^*(D^{1-k} - D_o^{1-k})$$

comp. (I.2.6). The nominal damage energy W^* represents the upper limit of the damage development energy $\Delta\Phi$, corresponding thus to the nominal *capacity* of the material to store energy in the form of internal structural damage considered here.

Condition (II.2.13) for the damage at failure is now

(II.4.4) $$W_f(1 - D_f) = W^*(D_f^{1-k} - D_o^{1-k})$$

In order to determine failure damage D_f from (II.4.4) we need to know the nominal damage energy W^* and the perfect strain energy W_f at failure. Given these parameters, damage at failure D_f and with it the amount of energy $\Delta\Phi(D_f)$ needed to reach failure, depends now on the choice of the power k, Fig. II.4.1 – 2.

With the power k growing from 0 to 1, damage at failure D_f increases from $(\omega_f + D_o)/(\omega_f + 1)$ to 1, where

$$(II.4.5) \qquad \omega_f = W_f / W^*$$

denotes the ratio of the perfect strain energy at failure to the nominal damage energy. It follows from relation (II.3.8) and the following considerations, comp. Ch. II.3, that the parameter ω_f varies between zero and infinity, as initial damage D_o varies between its upper bound D_{of} and 0.

At higher values of initial damage D_o damage at failure D_f grows monotonically with k, which corresponds to a monotonic drop of the damage development energy at failure $\Delta\Phi(D_f)$ from the value $W^*(1 - D_o)\,\omega_f/(1 + \omega_f)$ to zero, s. Fig. II.4.1 - 2.

At lower values of initial damage D_o, however, the changes are not monotonic: damage at failure D_f reaches a minimum at some k in the range (0,1). Correspondingly, damage development energy at failure $\Delta\Phi(D_f)$ has a maximum within the range.

In the range $k < 0$ damage at failure D_f tends to 1 as power k tends to minus infinity. Simultaneously, damage development energy at failure $\Delta\Phi(D_f)$ vanishes, after having reached its maximum either in the (0,1) range for low values of D_o or at values of k between 0 and -0.4 as initial damage D_o increases to its upper bound D_{of}.

We observe here, that there exists no minimum of the damage development energy $\Delta\Phi(D_f)$ within the range of finite $k<1$. Although in the limit transition with k to 1 or to minus infinity the quantity $\Delta\Phi(D_f)$ tends to its absolute minimum at zero, we cannot accept these limiting values of k for obvious physical reasons. Vanishing $\Delta\Phi(D_f)$ would correspond to the conversion of the *whole* dissipation energy into heat leaving no energy available for structural changes related to the damage itself.

Thus, the principle of minimum dissipation energy cannot be applied *separately* to the components of $W_{\theta f}$ in the expression (II.3.3).

On the other hand, alone the choice of the damage energy function

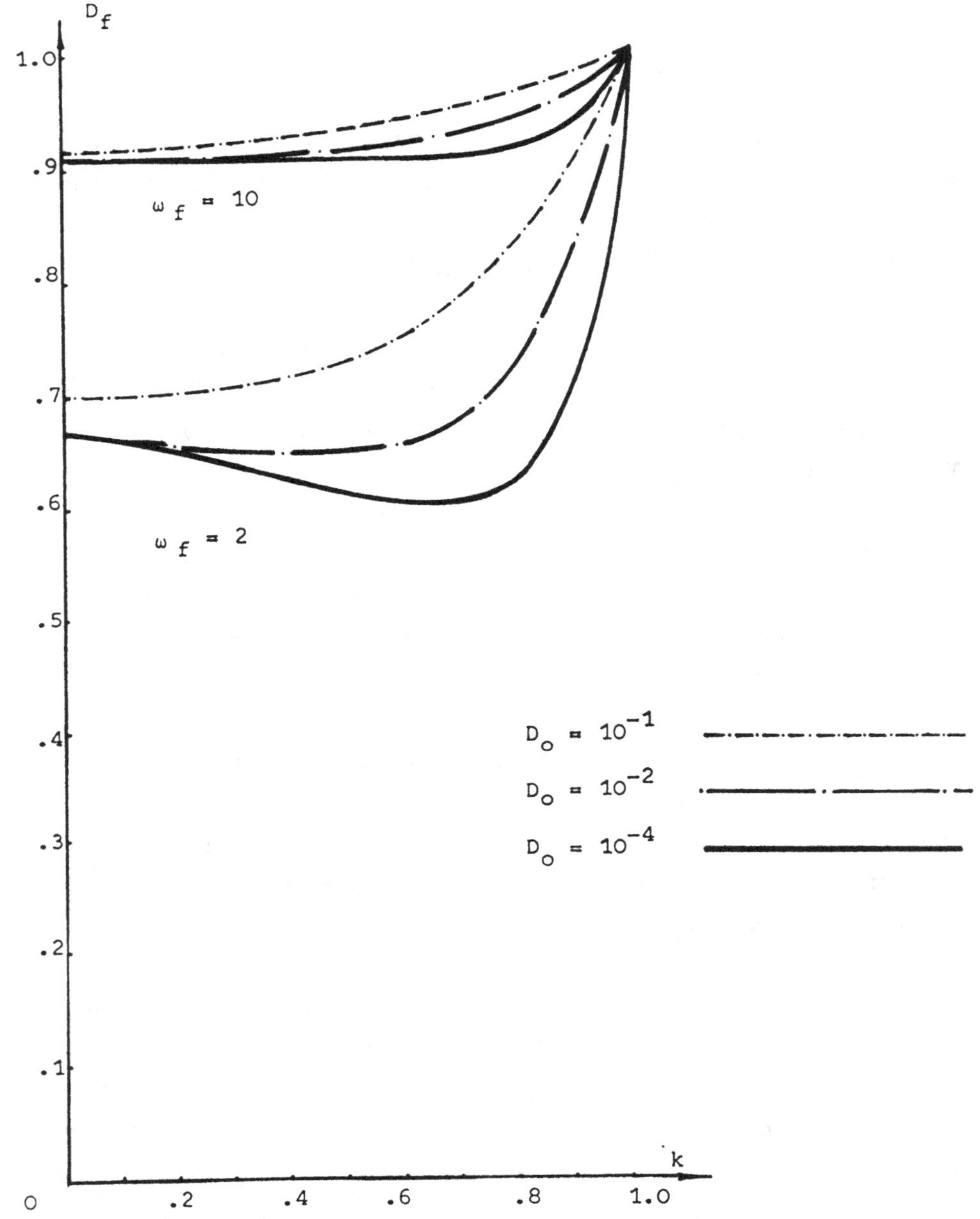

Fig. II.4.1 Failure damage at given failure strain.

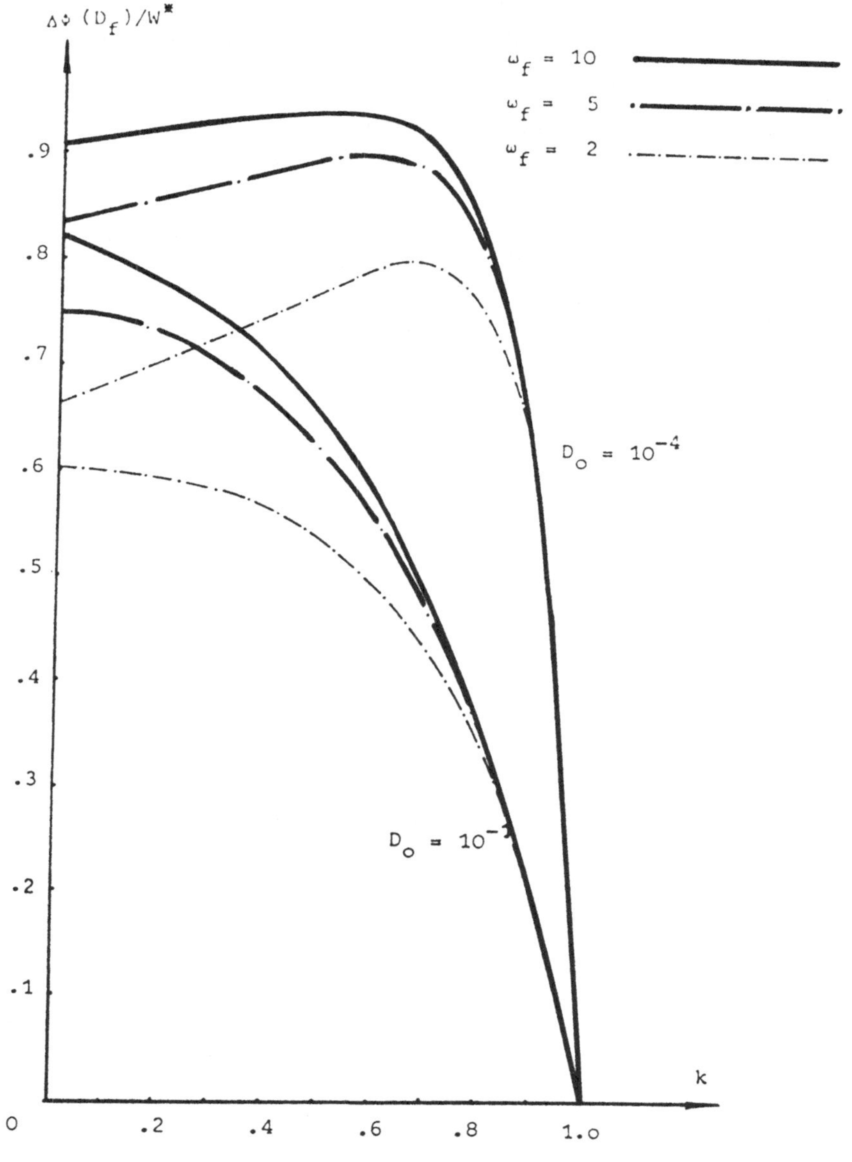

Fig. II.4.2 Damage development energy at given failure strain.

$\Phi(D)$ in the form (II.4.1) is not sufficient for unique determination of the quantity $W_{\theta f}$ from the expression (II.3.3). Indeed, the upper limit of the integration, i.e. damage at failure D_f, depends according to equation (II.4.4) on the perfect strain energy W_f at failure, which has not been determined yet. According to the relation (II.3.18), only the knowledge of the damage function F(D) in the damage law (II.3.16) may provide these quantities and allow thus the computation of $W_{\theta f}$.

Therefore, the minimization principle (II.3.3) cannot be applied at this stage of the analysis. Obviously, the choice of the power k must be performed together with the specification of the damage law.

III. BRITTLE DAMAGE DEVELOPMENT.

III.1. Damage evolution rate law.

In the analysis of the general properties of brittle damage, comp. Ch. II.3, we have come to the conclusion, that the development of damage along the path $D(\tau)$ should be related to the strain development along the path $\varepsilon(\tau)$ through the expression of the strain energy W_{perf} of the undamaged (perfect) material undergoing the same loading history $\varepsilon(\tau)$.

It is readily seen, however, that the damage law in the form (II.3.16), derived from the principle of minimum dissipation energy, needs a correction from the point of view of the *dimensional analysis*. Indeed, since the damage D and the initial damage D_o are dimensionless, also the energy - valued argument W_{perf} of the damage function F needs to be expressed in non - dimensional form. This implies the existence of an additional energy - valued parameter in the damage law (II.3.16). Having introduced already a similar quantity into the damage energy function $\Phi(D)$, namely the nominal damage energy W^*, comp. (II.4.2), we may employ it here as well.

Let us introduce therefore the following dimensionless variable

(III.1.1) $$\omega = W_{perf} / W^*$$

representing the perfect strain energy. It varies in the range $(0,\omega_f)$, comp. (II.4.5), simultaneously with the variation of the energy W_{perf} between 0 and W_f, comp. (II.2.13), and the variation of the strain ε in the range from zero to the strain at failure ε_f.

The damage law (II.3.16) can be expressed now as

(III.1.2) $$D = D_o\, f(\omega)$$

where f is a monotonically growing and continuously differentiable function of its argument, obeying the condition $f(0) = 1$, comp. (II.3.17), and complying with the demands of the principle of minimum dissipation energy, in the sense presented in Ch. II.3.

The inversion f^{-1} of the damage function f is therefore uniquely determined and yields the inverse damage law

(III.1.3) $\omega = f^{-1}(D/D_o)$

It indicates that a given value of the dimensionless strain energy ω is related to the *ratio* of the damage D to its initial value D_o, but not to the damage D itself.

In a general approach to thermomechanical processes in continua, s. e.g. [2], [11], internal state variables are usually assumed to be governed by evolution equations of the rate-type rather than by functional dependences of the kind presented by the damage law (III.1.2). It is obvious, however, that the results of our analysis of brittle damage could be equivalently expressed in the form of the following *rate-type evolution equation*

(III.1.4) $dD/d\omega = D_o\, g(D/D_o)$

where the *damage rate function* $g = f'f^{-1}$ is a superposition of the derivative f' of the damage function f from (III.1.2) and its inverse function f^{-1}. It follows from the principle of minimum dissipation energy, comp. Fig. II.3.1, that both f' and f^{-1} are monotonically increasing and continuous functions, assigning thus the same properties to the damage rate function g. The role of the independent variable, played usually by time τ, is taken here by the strain energy variable ω, i.e. by the strain ε, in accordance to the non-viscous nature of the damage, comp. Ch. II.3.

Let us consider the class of simple power damage rate functions, analogous to those in the damage energy expression (II.4.1). The evolution equation (III.1.4) takes now the form

(III.1.5) $dD/d\omega = \gamma\, D_o\, (D/D_o)^{1-\varkappa}$

where γ is a positive coefficient, specifying the damage rate $dD/d\omega$ at the onset of the process, i.e. at $D = D_o$. The power $\varkappa$ shall be considered as a parameter of the process, which needs to be determined from the minimization principle, together with the power k of the damage energy function.

It is obvious, however, that the choice of γ corresponds to a change of scale of the variable ω. It is equivalent therefore to the choice of the nominal damage energy W^* as a reference quantity in the definition (III.1.1). Having as yet no conditions on W^* we can assume further $\gamma = 1$ without loss of generality.

Substitution of (III.1.5) into the evolution equation (III.1.4) and integration over the process with the initial condition, requiring that damage D is equal to D_0 at $\omega = 0$, yields the following array of damage laws

$$\text{(III.1.6)} \qquad D = \begin{cases} D_0 \exp(\omega) & \text{at } \varkappa = 0 \\ D_0 (1 + \varkappa\omega)^{1/\varkappa} & \text{at } \varkappa \neq 0 \end{cases}$$

The corresponding inverse laws, comp. (III.1.3), are

$$\text{(III.1.7)} \qquad \omega = \begin{cases} \ln(D/D_0) & \text{at } \varkappa = 0 \\ \frac{1}{\varkappa} [(D/D_0)^{\varkappa} - 1] & \text{at } \varkappa \neq 0 \end{cases}$$

Observe, that at $\varkappa < 0$ damage D should formally tend to infinity at ω tending to $-1/\varkappa$. Therefore, the strain energy at failure ω_f, comp. (II.4.5), corresponding to any *finite* damage at failure D_f, comp. (II.2.13), must be lower than $-1/\varkappa$, independently of the initial damage D_0. In *contradiction* to the results of our analysis, comp. Ch. II.3, we would obtain at vanishing initial damage D_0 that a perfect flawless material could fail at finite ω_f, i.e. at finite strain ε_f. The latter can be estimated by the energy W^*, virgin Young's modulus E and the power $\varkappa$ as $\varepsilon_f < \sqrt{2W^*/KE}$, where $K = -\varkappa$, comp. (II.1.1) and (III.1.1).

This type of theory would imply a strain limit based on the *tensile properties of atomic lattices* rather than one taking into account the damage characteristics of flawed brittle materials in the continuum mechanics approach, as considered in Ch. II.1. The experience shows, s. e.g. [12], that the corresponding calculations of the strength of engineering materials tend to strongly overestimate the real performance of the material, see also remarks in [16].

A reasonable theory of brittle damage with engineering applications should rely rather on the flaw development as a cause of damage failure. We shall therefore limit our attention to the range

(III.1.8) $\varkappa \geqslant 0$

The corresponding damage laws are illustrated in Fig. III.1.1.

With increasing values of the power $\varkappa$ the *nominal limit* ω_1 of the strain energy variable ω, corresponding to the upper bound of damage $D = 1$, increases rapidly at given initial damage D_o. Should this tendency be reflected in the behaviour of the strain energy at failure ω_f, we could immediately predict an increase of the damage at failure D_f, comp. eqn. (II.4.4) and Fig. II.4.1, at growing $\varkappa$. This would have as a consequence increasing dissipation energy at failure $W_{\theta f}$, comp. (II.3.3), with corresponding implications from the point of view of the minimization principle.

Before investigating these possibilities, however, let us look closer at the properties of the derived damage evolution equation. For a given value of the initial damage D_o the damage D develops with growing strain energy variable ω according to the exponent $\varkappa$: the larger $\varkappa$, the slower the increase of damage, Fig. III.1.1. This finds its reflection in the expressions for the damage rate in terms of the strain energy variable

(III.1.9)
$$dD/d\omega = \begin{cases} D_o \exp(\omega) & \text{at} \quad \varkappa = 0 \\ D_o\,(1 + \varkappa\omega)^{(1-\varkappa)/\varkappa} & \text{at} \quad \varkappa > 0 \end{cases}$$

complementary to the evolution equation (III.1.5).

It follows directly from both forms of the evolution equation that the damage rate at the onset of the process, i.e. at $D = D_o$ and $\omega = 0$, depends only on the initial damage D_o

(III.1.10) $$dD/d\omega\big|_{\omega=0} = D_o$$

and does not depend on the power $\varkappa$, comp. Fig. III.1.1. The larger the initial damage of the material, the higher the damage rate already at the onset of loading.

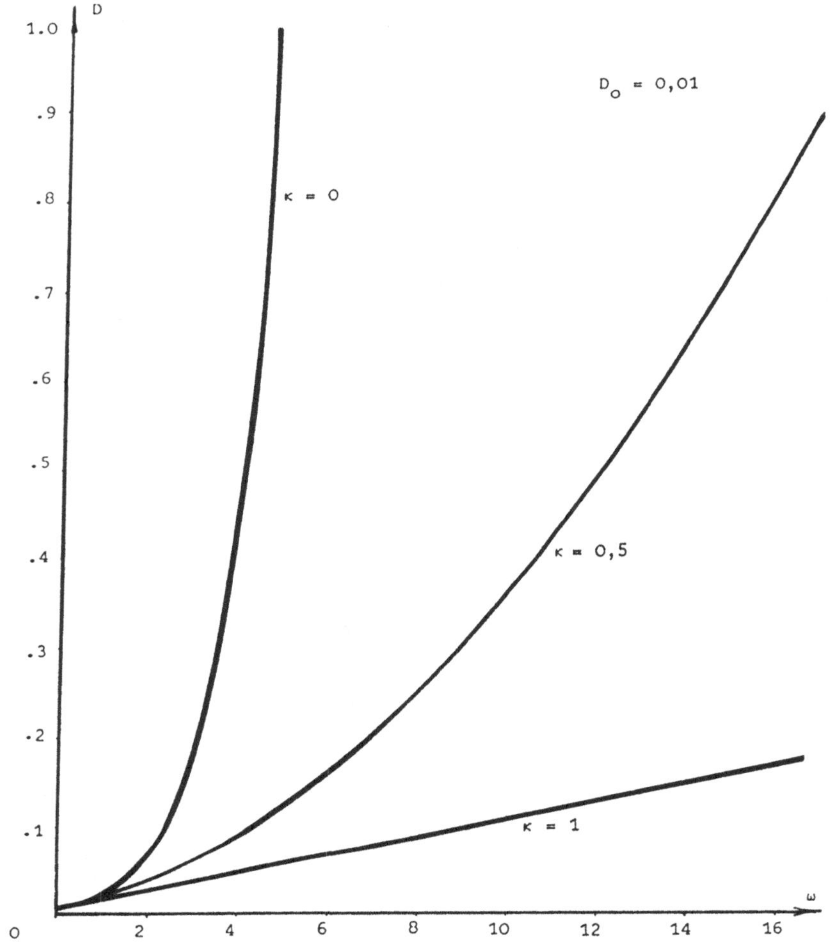

Fig. III.1.1 Damage development in dependence on dimensionless perfect strain energy.

Having in mind the definitions (II.1.1) and (III.1.1), one can express now the damage development in terms of the strain ε, for given initial damage D_0 and selected values of the elasticity modulus E and nominal damage energy W^*. Fig. III.1.2 illustrates this development for parameters corresponding to hard coal.

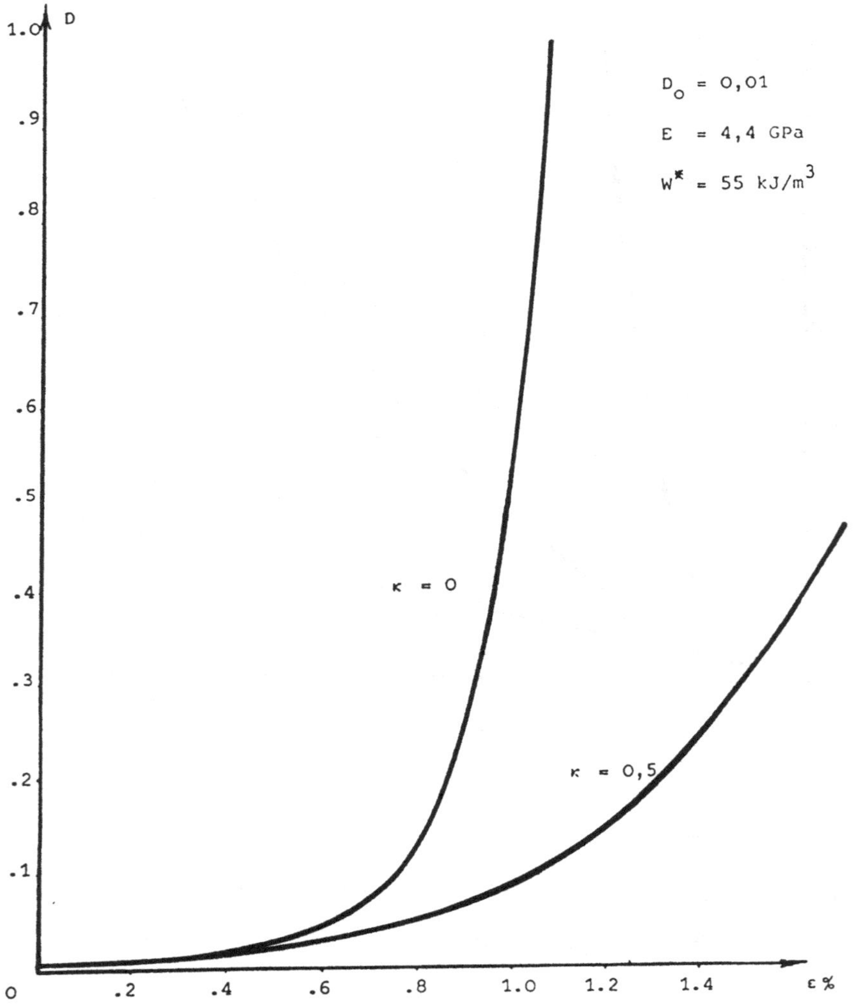

Fig. III.1.2 Damage development as a function of strain.

III.2. Damage and strain energy at failure.

We have shown in Ch. II.4 and Ch. III.1 that the simple forms of the damage energy function $\Phi(D)$ and of the damage function $D(\omega)$ given by (II.4.1) and (III.1.6) satisfy the general demands of the theory of brittle damage, covering at the same time a rather wide spectrum of material properties and process characteristics.

The choice of these functions allows to determine more precisely the features of the model derived here. The most interesting of them seem to be the phenomenon of failure at the damage D_f well *below* the nominal ultimate damage at $D = 1$, comp. Ch. II.2 and its connection with the strain at failure ε_f. In this Chapter we should like to investigate the dependence of the quantities D_f and ε_f on the parameters of the model.

Let us begin with the dependence of the failure damage D_f on the initial damage D_o. Substituting the definition (II.4.5) of the strain energy at failure ω_f and the inverse damage function from (III.1.7) into the equation (II.4.4) for D_f we obtain the following equation

$$\text{(III.2.1)} \qquad (D_f^{1-k} - D_o^{1-k})/(1 - D_f) = \begin{cases} \ln(D_f/D_o) & \text{at } \varkappa = 0 \\ \frac{1}{\varkappa}[(D_f/D_o)^{\varkappa} - 1] & \text{at } \varkappa > 0 \end{cases}$$

which determines D_f as a function of three parameters: the initial damage D_o and the powers k and $\varkappa$, comp. (II.3.14). The dependence on the first of these variables is shown in Fig. III.2.1

One finds that as the initial damage D_o tends to zero, the damage at failure D_f tends to the ultimate value of damage $D = 1$. With increasing D_o there is a monotonic decrease in the value of D_f for all admissible values of k and $\varkappa$, until the two damage variables meet at the upper bound D_{of} of the initial damage, comp. Ch. II.3, equal to the lower bound for damage at failure D_f.

By the limit transition with D_o to D_f in (III.2.1) it can be easily shown that the upper bound D_{of} of the initial damage D_o satisfies the following equation

$$\text{(III.2.2)} \qquad 1 - D_{of} = (1 - k)\, D_{of}^{1-k}$$

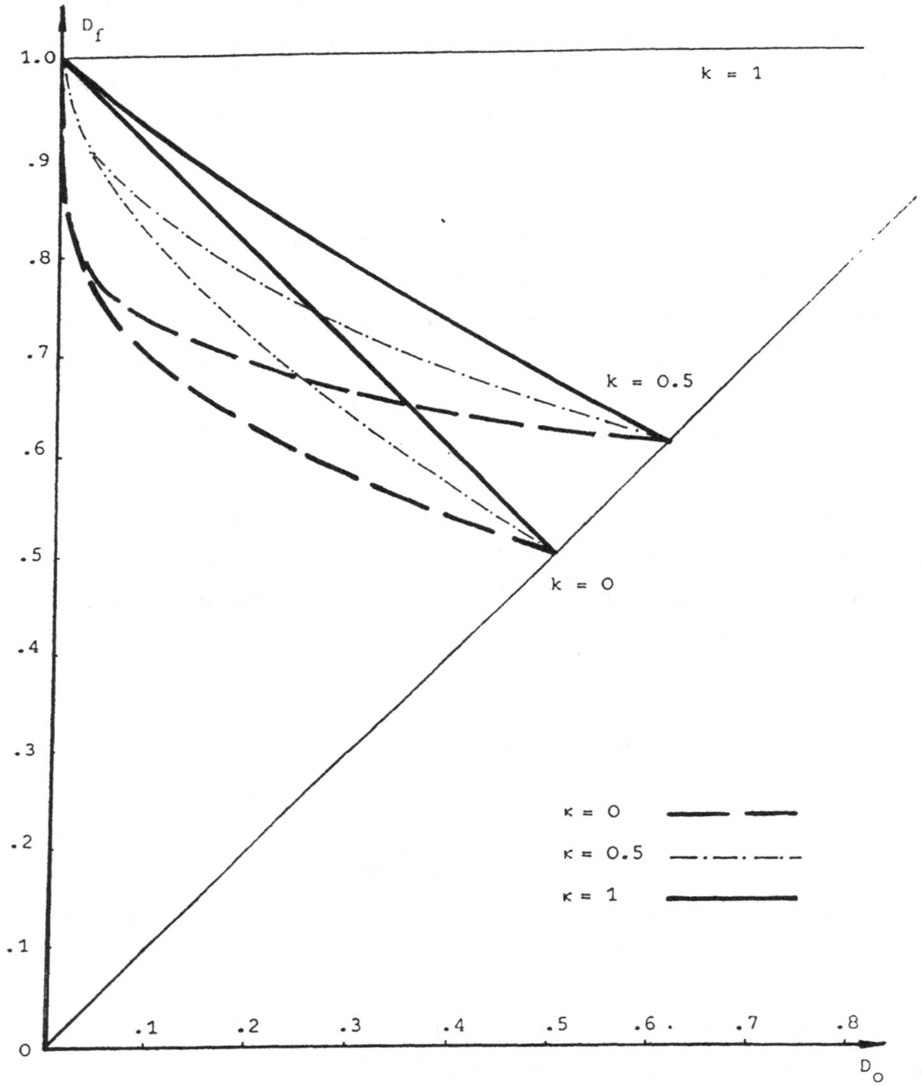

Fig. III.2.1 Damage at failure in dependence on initial damage.

Characteristically, the upper bound D_{of} of the initial damage depends only on the power k of the damage energy function $\Phi(D)$, but not on the power $\varkappa$ of the damage function $D(\omega)$. It can be shown, that this effect is not connected with the particular power form of the evolution equation (III.1.5), reflecting a rather general property of brittle damage: *any monotonically increasing function* $\Phi(D)$, independently of the form of the damage law, will *yield damage at failure* D_f *decreasing* with increasing values of the initial damage D_o. Consequently, there always exists an upper bound D_{of} of initial damage, equal to the lower bound of the failure damage

(III.2.3) $$D_o \leqslant D_{of} \leqslant D_f$$

comp. (II.2.14). The dependence of the upper bound of initial damage on the power k in the damage energy function (II.4.1) is shown on the Fig. III.2.2. As already demonstrated in Ch. II.4, damage D_{of} tends to 1 when the power k tends either to 1 or to minus infinity. Its *absolute minimum* value of ca. 0.4877 corresponds to approximately $k = -0.4$.

Having determined the dependence $D_f(D_o)$, we can calculate also the dependence of the dimensionless strain energy at failure ω_f on the initial damage D_o. Comparing the definition (II.4.5) with equation (II.4.4) we see immediately, that the quantity ω_f can be computed from the following relation

(III.2.4) $$\omega_f = (D_f^{1-k} - D_o^{1-k}) / (1 - D_f)$$

This represents a particular form of the relation (II.3.9) postulated in the general theory. The dependence of the strain energy at failure on the initial damage $\omega_f(D_o)$ is shown in Figs. III.2.3 - 4.

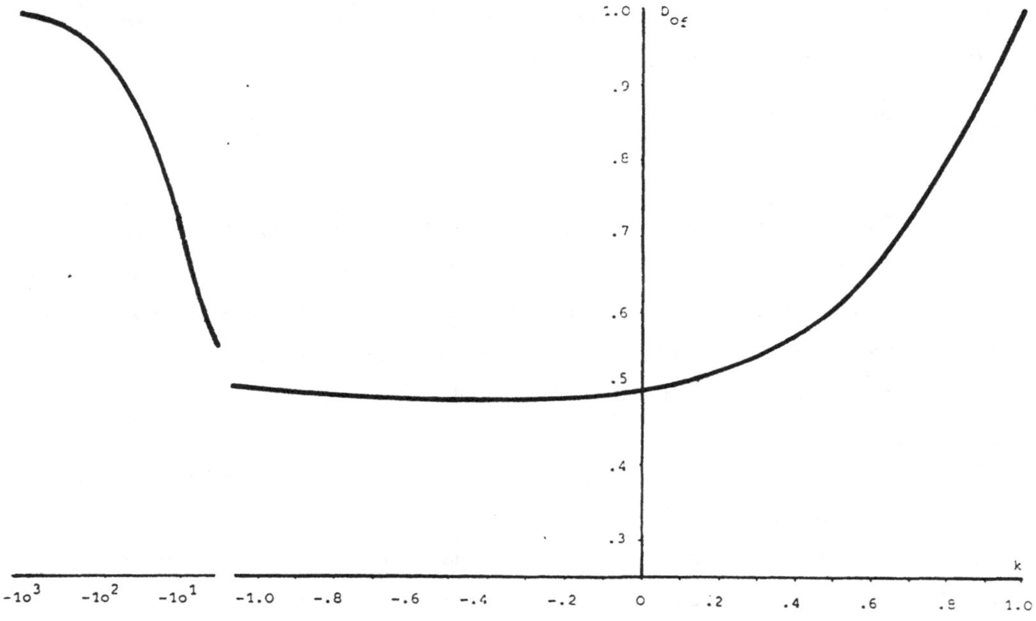

Fig. III.2.2 Upper bound of initial damage in dependence on the damage energy power.

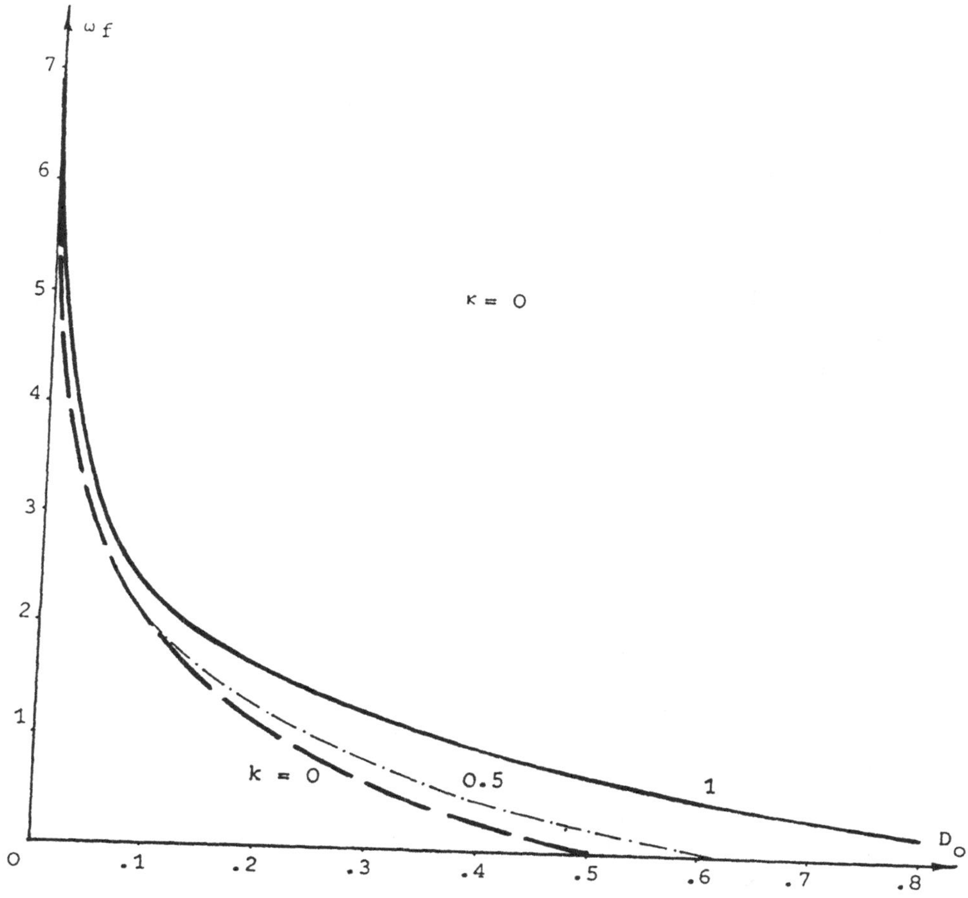

Fig. III.2.3 Dimensionless perfect strain energy at failure in dependence on initial damage at $\varkappa = 0$.

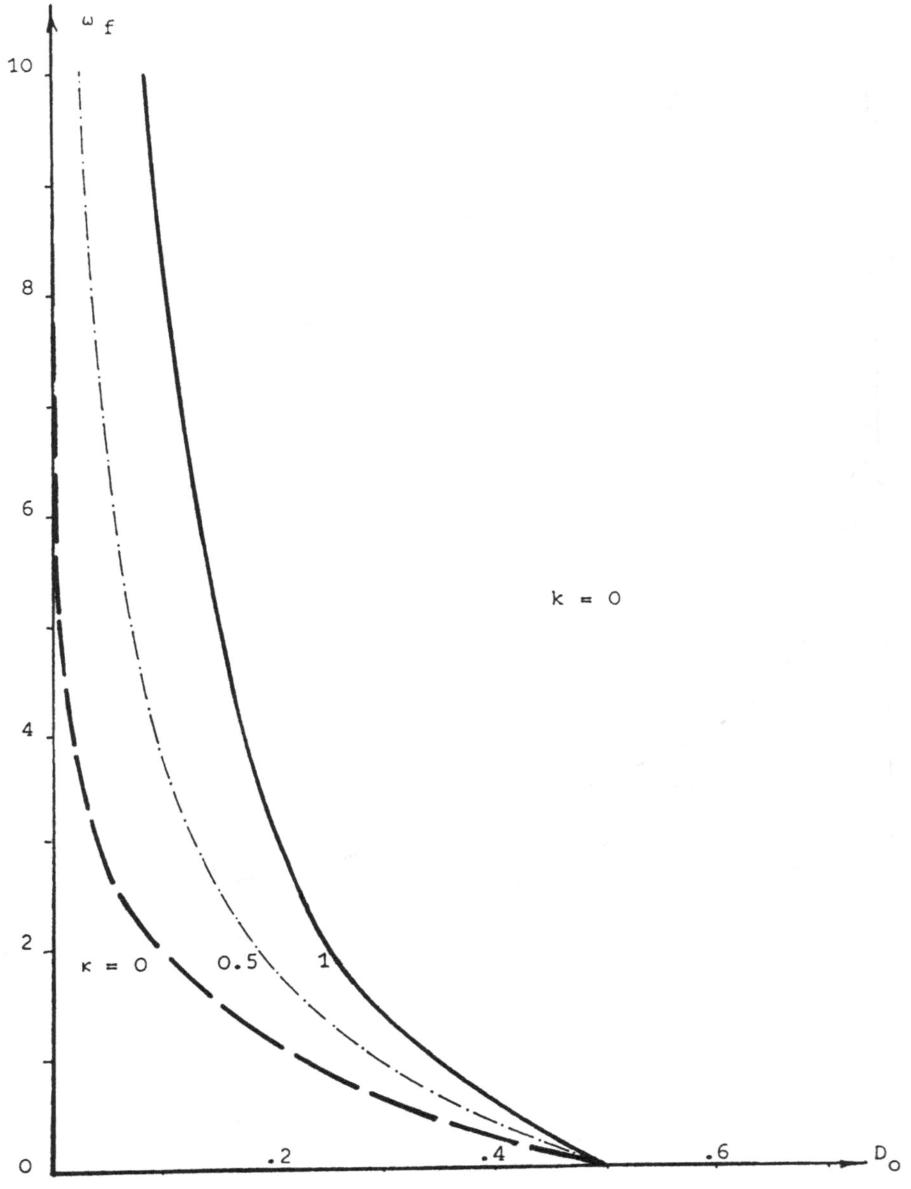

Fig. III.2.4 Dimensionless perfect strain energy at failure in dependence on initial damage at k = 0.

Observe, that the perfect strain energy at failure ω_f, and with it the strain at failure ε_f, comp. (II.1.1) and (III.1.1), depends not only on initial damage D_o and power k, which enter into the relation (III.2.4) explicitly. Also the power $\varkappa$, through its influence on the failure damage D_f, comp. (III.2.1), determines these quantities.

It can be seen, that ω_f is *least* at $\varkappa = 0$ for given initial damage D_o and a fixed parameter k, growing steeply as the power $\varkappa$ of the damage law increases, Fig. III.2.4. We have indicated at this effect and its consequences in Ch. III.1.

On the other hand, for a fixed power $\varkappa$, the strain energy at failure decreases according to the decrease of the power k in the range between 1 and 0. Depending on the value of $\varkappa$, it arrives at a minimum positioned in the range of k between -0.4 and 1.

Plotting the dependence of the damage at failure D_f on the power k of the damage energy function $\Phi(D)$ for fixed values of initial damage D_o and $\varkappa$, Fig. III.2.5, one observes a systematic shift of the position of the *minimum* of the $D_f(k)$-curve: with growing initial damage the minimum moves towards lower values of k, approaching -0.4 as D_o tends to D_{of}, comp. Fig. III.2.2.

Finally, the dependence of the damage at failure D_f on the power $\varkappa$ of the damage function corresponds to a monotonic increase of D_f with $\varkappa$, at fixed D_o and k, comp. Fig. III.2.1.

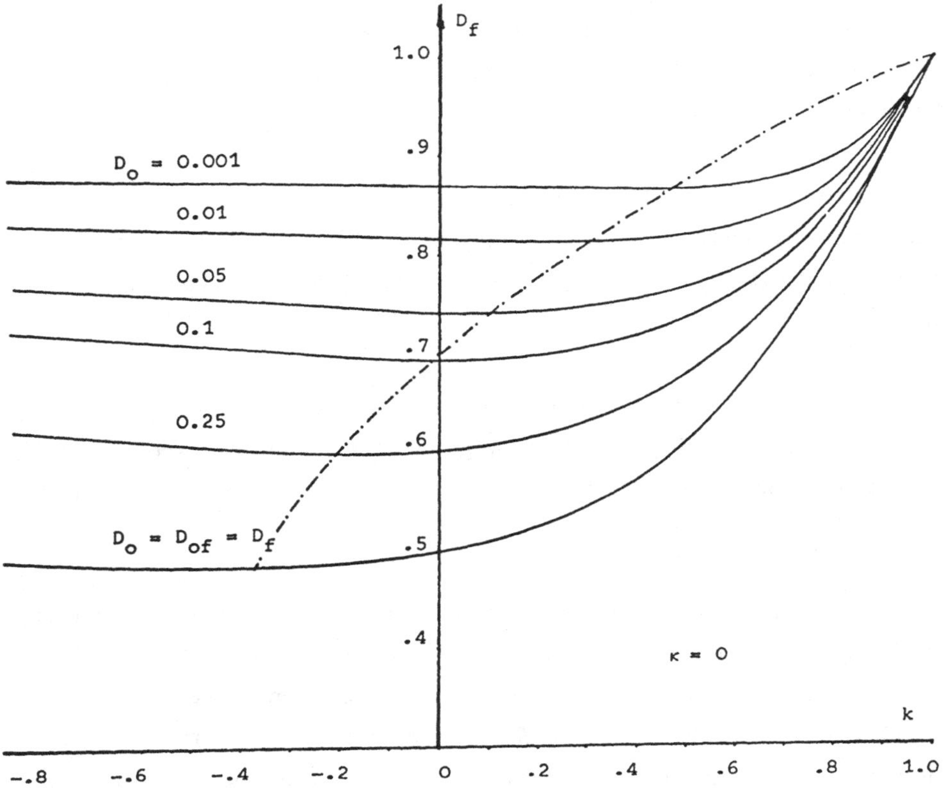

Fig. III.2.5 Damage at failure in dependence on the damage energy power.

III.3. Application of extremal principles.

We should like to specify now the value of the power k of the damage energy function $\Phi(D)$ and the power $\varkappa$ of the damage function $D(\omega)$. This can be done on the basis of the principle of minimum dissipation energy in its form (II.3.3), as applied to a single loading process up to failure.

It is more convenient, hovewer, to express first the formula (II.3.3) in a dimensionless form. Let us introduce therefore the following quantity

(III.3.1) $$\omega_{\theta f} = W_{\theta f}/W^{*}$$

defined as the dimensionless *dissipation energy at failure*. Substituting (III.1.1) and (II.4.3) into (II.3.3) we obtain now the following formula for the energy $\omega_{\theta f}$

(III.3.2) $$\omega_{\theta f} = \int_{D_o}^{D_f} \omega(D)\, dD + D_f^{1-k} - D_o^{1-k}$$

Here $\omega(D)$ corresponds to the inverse damage function from (III.1.7).

The dimensionless dissipation energy at failure $\omega_{\theta f}$ consists of two components, denoted further $I_{\theta f}$ and $II_{\theta f}$, which correspond to the heat-convertible and to the damage-related components of the dissipation energy W_{θ}, respectively, comp. Ch. II.2.

The first component $I_{\theta f}$ of the energy $\omega_{\theta f}$ can be calculated by integration within the indicated limits and upon substitution of (III.1.7) into (III.3.2), which yields

(III.3.3) $$I_{\theta f} = \begin{cases} D_f \ln(D_f/D_o) - D_f + D_o & \text{at } \varkappa = 0 \\ \{D_f\,[(D_f/D_o)^{\varkappa} - \varkappa - 1] + \varkappa D_o\}/\varkappa(\varkappa+1) & \text{at } \varkappa > 0 \end{cases}$$

It is easy to observe, that $I_{\theta f}$ depends not only on the initial damage D_o and on the power $\varkappa$ of the damage function, but also on the power k of the damage energy function, through the influence of this parameter on the damage at failure D_f, comp. (III.2.1).

The second component of the dimensionless energy $\omega_{\theta f}$

$$\text{(III.3.4)} \qquad II_{\theta f} = D_f^{1-k} - D_o^{1-k}$$

similarly does not depend on the two parameters D_o and k alone. The power $\varkappa$ of the damage function intervenes here through the relation (III.2.1), participating in the determination of the damage at failure D_f. Observe, that $II_{\theta f}$ represents the dimensionless damage development energy at failure $\Delta\Phi(D_f)$, comp. (II.4.3). It is uniquely determined at given k and $\varkappa$, comp. the remarks at the end of Ch. II.4, due to the introduction of the damage law (III.1.6).

Applying the principle of minimum dissipation energy (II.3.3) to the equations (III.2.1) and (III.3.2 – 4) we obtain the following condition for any loading process up to failure:
among all admissible pairs $(k,\varkappa)$ the *real process* corresponds to the pair, which *yields an absolute minimum* of the sum $\omega_{\theta f} = I_{\theta f} + II_{\theta f}$.

The minimum of $\omega_{\theta f}$ is, in general, dependent only on the initial damage D_o

$$\text{(III.3.5)} \qquad \omega_{\theta f}(D_o) = \min_{k,\varkappa} \left[I_{\theta f}(D_o,k,\varkappa) + II_{\theta f}(D_o,k,\varkappa)\right] \qquad \text{at } 0 < D_o < D_{of}(k)$$

The parameters k and $\varkappa$ minimizing the expression (III.3.5) are, in general, also dependent on the initial damage D_o.

Numerical evaluation of equations (III.2.1) and (III.3.3 – 5) allows to specify these results.

Consider first the case $\varkappa = 0$ at an arbitrary admissible initial damage D_o. It follows from the numerical evaluations that within the accuracy of the computation
the value of the parameter k, which yields the *lowest* failure damage D_f, s. Fig. III.2.5, corresponds simultaneously to a *minimum* of the heat – convertible component $I_{\theta f}$ of the dissipation energy $\omega_{\theta f}$, Fig. III.3.1, and at the same time to a *maximum* of the other, damage – related component $II_{\theta f}$, Fig. III.3.2.

These two opposite tendencies in the dependence of the components $I_{\theta f}$

and $II_{\theta f}$ on the power k result in a minimum of their sum $I_{\theta f} + II_{\theta f}$ with respect to k, i.e.
result in a *minimum of the dissipation energy at failure* $\omega_{\theta f}$, at a value $k = k_{min}$

The quantity k_{min} lies very close to the value of the power k, which yields the minimum of the failure damage D_f, Fig. III.3.3. It depends on the initial damage D_o and decreases monotonically from 0.4 to -0.4, s. Fig. III.3.4, as initial damage grows from 10^{-3} to its upper bound $D_{of} = 0.4877$, comp. Ch. III.2.

Increasing now the power in the damage function $\varkappa > 0$ we observe two parallel effects, leading together to a rapid growth of the dissipation energy at failure $\omega_{\theta f}$. On one hand, the damage at failure D_f grows at fixed values of D_o and k, comp. Fig. III.2.1, resulting in the extension of the limits of integration in the expression (III.3.2), as well as in the growth of the component $II_{\theta f}$ of the energy of dissipation, comp. (III.3.4). On the other hand, the same increase of $\varkappa$, at given D_o and k, gives rise to a rapid growth of the strain energy at failure ω_f, comp. Fig. III.2.4. The latter represents the upper limit of the integrand $'\omega(D)$ in the formula (III.3.2). The consequence of these developments is a fast growth of the component $I_{\theta f}$ of the dissipation energy at failure at increasing $\varkappa$.

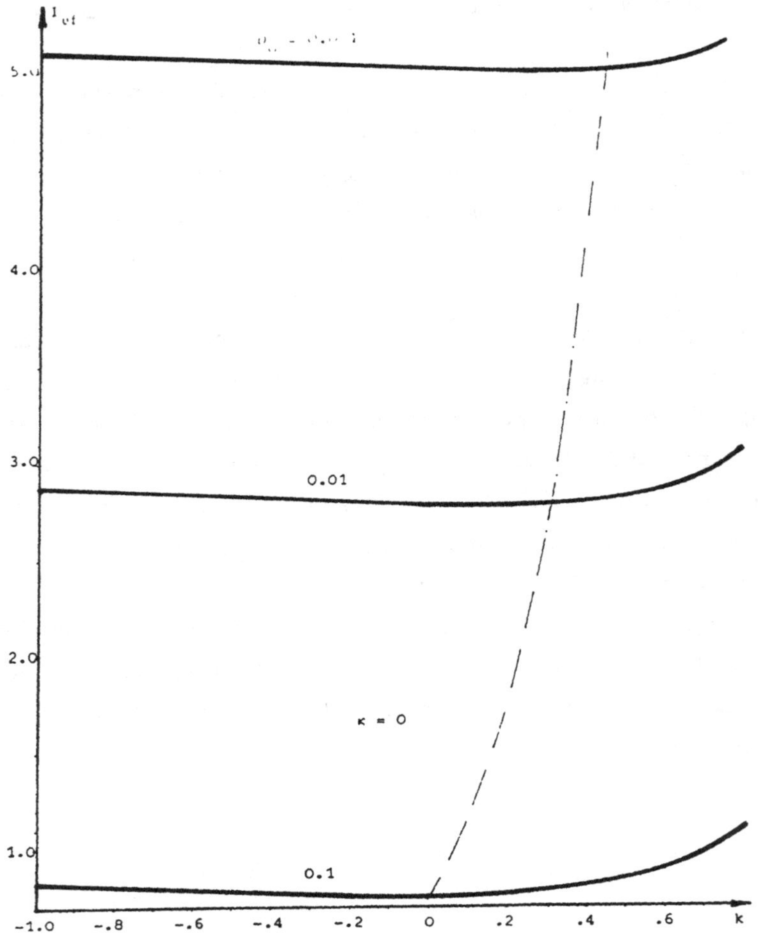

Fig. III.3.1 Heat – convertible part of dissipation at failure (dimensionless).

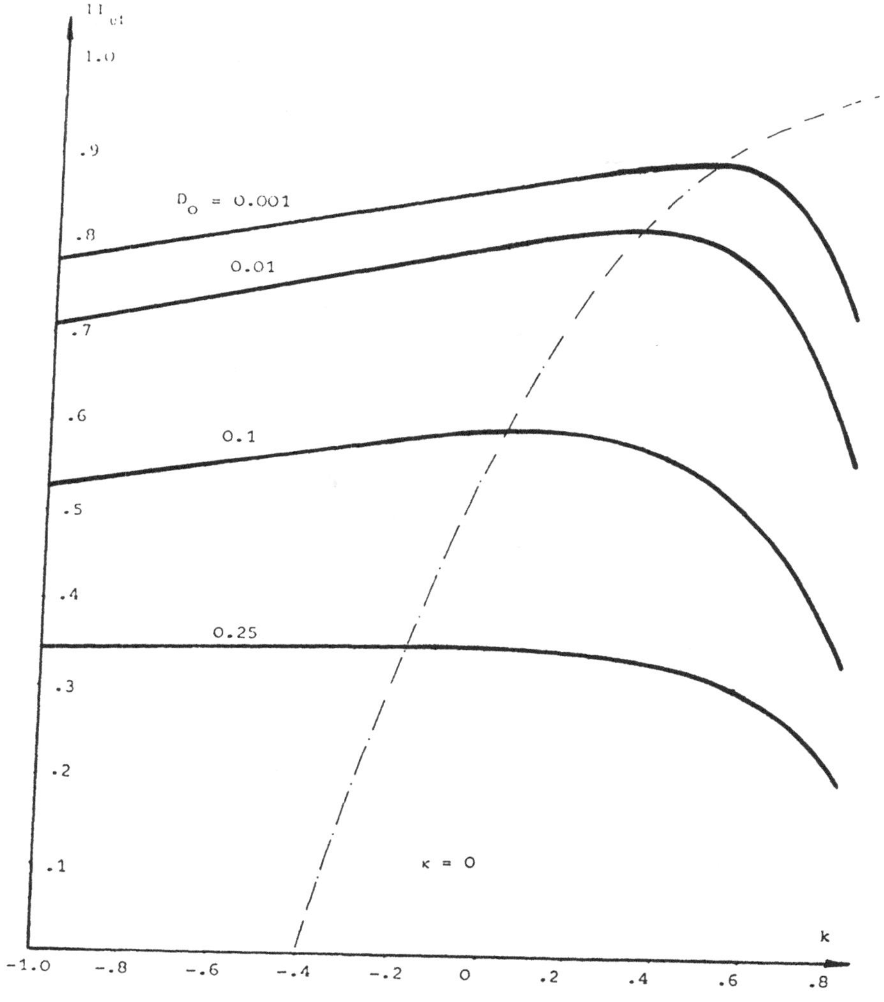

Fig. III.3.2 Damage – related part of dissipation at failure (dimensionless).

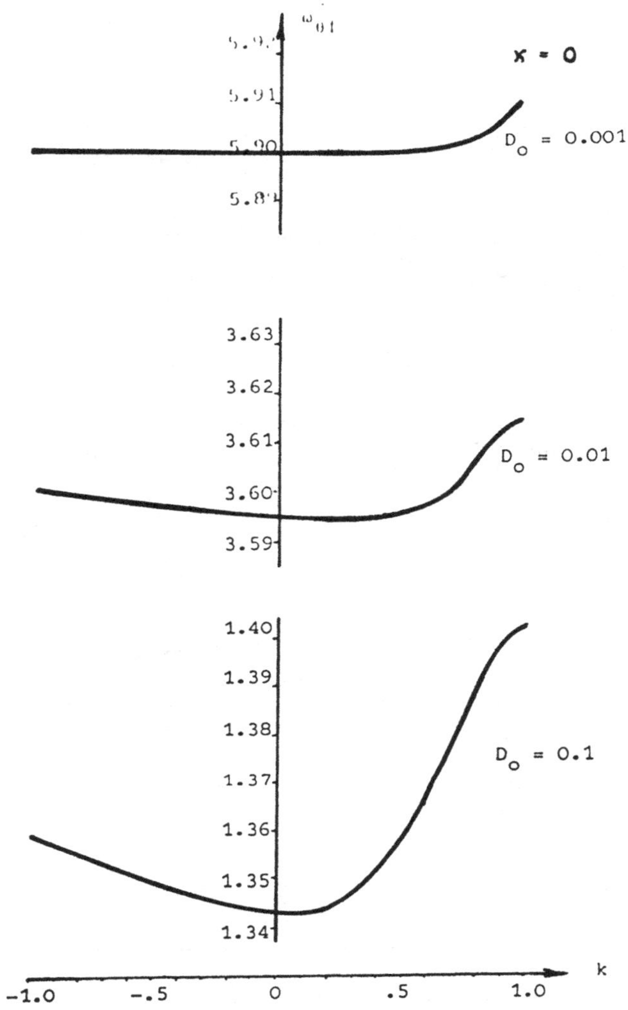

Fig. III.3.3 Dimensionless dissipation energy at failure in dependence on the damage energy power.

The above qualitative description is supported by numerical calculations. They show directly, that *the minimum of the dissipation energy at failure* with respect to the power $\varkappa$ of the damage function *can be assumed with sufficient accuracy at* $\varkappa = 0$, *independently of the initial damage* D_o.

In consequence of the minimization procedure presented above one should thus apply exclusively the power

(III.3.6) $\quad \varkappa = 0$

in the damage law (III.1.6), as well as the values

(III.3.7) $\quad k = k_{min}(D_o)$

in the damage energy function, s. Fig. III.3.4.

There is a striking coincidence in the fact that within the numerical accuracy the minimum of the dissipation energy at failure corresponds to the maximum of the damage development energy at failure $\Delta\Phi(D_f)$, as reflected in the maximum of the quantity $II_{\theta f}$, comp. Fig. III.3.2. We may recall, however, that $\Delta\Phi$ is equal to damage-related dissipation losses $W_{\theta s}$ and corresponds thus to the entropy increment Δs, comp. (II.1.16). Thus, *the minimum of the dissipation energy at failure coincides with the maximum of the entropy increment within the loading process until failure.*

It remains an open question, whether this coincidence might be associated with Ziegler's *principle of maximum entropy production* in stable systems, [15]. The question is not only with the applicability of the notion of a stable system to the material under consideration. It is also not clear, whether in this *non-linear* model the maximization of the entropy increment over the *whole process* until failure coincides with the maximization of the entropy production, representing a *local characteristic* of the process.

Observe, that in the range of initial damage $D_o > 10^{-3}$ the parameter k_{min} varies in the vicinity of $k = 0$. Its variation corresponds to an almost negligible deviation of the quantities $D_{of}(k_{min})$ and $D_f(k_{min})$ from their respective values at $k = 0$ for any given D_o, comp. Fig. III.3.4 with Fig. III.2.2 and Fig. III.2.5. It may be concluded that the damage energy power equal to

(III.3.8) $\qquad k = 0$

could provide a sufficient approximation in some applications at $D_0 > 10^{-3}$.

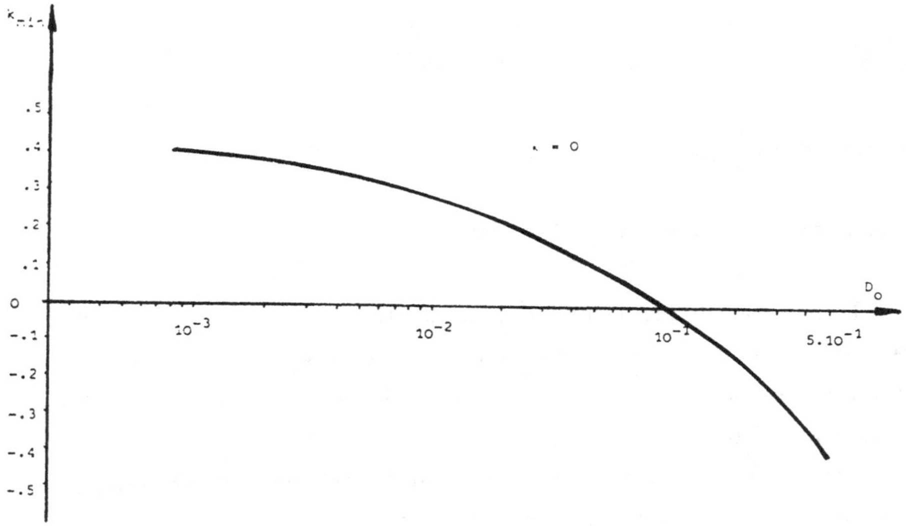

Fig. III.3.4 Damage energy exponent for minimum dissipation at failure.

BIBLIOGRAPHY.

[1] Krajcinovic, D., Continuum damage mechanics, Appl.Mech.Rev.,v.37, no.1, 1 - 6, 1984

[2] Krajcinovic, D., Continuum damage mechanics, Mechanics Today, to be published

[3] Herrmann, G., On energetics of interacting defects, Proc. 3rd Symposium on Energy Engineering Sciences,Conf. - 8510176, held at the Pennsylvania State Univ., University Park, Pa., Oct. 8 - 10, 1985

[4] Gyarmaty, I., Non - equilibrium thermodynamics. Field theory and variational principles, Springer - V., Berlin - Heidelberg - New York, 1970

[5] Prigogine, I. Introduction to thermodynamics of irreversible processes, Springfield, 1955

[6] Meixner, J., Reik, H.G., Thermodynamik der irreversiblen Prozesse, in: Handbook der Physik, Bd. 3, Tl. 2, Springer - V., Berlin - Göttingen - Heidelberg, 1959

[7] Kachanov, L.M., Izv. Akad. Nauk SSSR, Otd. Tekhn. Nauk, v.8, (1958), 26 - 31

[8] Grady, D.E., Kipp, M.E., Continuum modelling of explosive fracture in oil shale, Rock Mech. Min. Sci. Geomech. Abstr., v. 17 (1980), 147 - 157

[9] Denkhaus, H.G., The load - deformation behaviour of rock in uni - axial compression, Rock Mechanics, Suppl. v.2 (1973), 33 - 51

[10] Walsh, J.B., The effects of cracks in the compressibility of rock, J. Geophys. Res., v. 70 (1965), 381 - 389

[11] Davison, L., Stevens, A.L., Thermomechanical constitution for spalling elastic bodies, J. Appl. Phys., v. 44 (1976), 93 - 106

[12] Cherepanov, G.P., Mechanics of brittle fracture, Nauka Publ., Moscow, 1974 (in Russian)

[13] Lippmann, H., Ductility caused by progressive formation of shear cracks, in: Three - dimensional constitutive relations and ductile fracture, ed. by S. Nemat - Nasser, N. - Holland Publ. Co., Amsterdam, 1981, pp. 389 - 404

[14] Mroz, Z., Mathematical models of inelastic concrete behaviour, in: Inelasticity and non - linearity in structural concrete, ed. by Cohn, M.Z., Univ. Waterloo Press, 1972, pp. 47 - 72

[15] Ziegler, H., Some extremum principles in irreversible thermodynamics with applications to continuum mechanics, in: Progress in solid mechanics, v.IV, Ch.II, ed. by Sneddon, I.N., Hill, R., N. - Holland Publ. Co., Amsterdam, 1963

[16] Chaboche, J. - L., Lemaitre, J., Mecanique des materiaux solides, Dunod, Paris, 1985

[17] Krajcinovic, D., Ilankamban, R., Mechanics of solids with defective microstructure, J. Struct. Mech., v. 13 (1985), 267 - 282

[18] Lippmann, H., Mechanik des plastischen Fließens, Springer - V., Berlin - Heidelberg - N.York, 1981

[19] Budiansky, B., O'Connell, R., Elastic moduli of a cracked solid, Int. J. Solids Structures, v. 12 (1976), 81 - 97

[20] Kachanov, L.M., Introduction to continuum damage mechanics, M. Nijhoff Publ., Dordrecht, 1986

[21] Dems, K., Mroz, Z., Stability conditions for brittle - plastic structures with propagating damage surfaces, J. Struct. Mech., v.13 (1985), 95 - 122

[22] Hogland, R.G., Hahn, G.T., Rosenfield, A.R., Influence of micro - structure on fracture propagation in rock, Rock Mechanics, v.5 (1973), 77 - 106

ACKNOWLEDGEMENTS

Prof. George Herrmann of Stanford University co - authored Ch. II.1 and actively stimulated the Author in numerous discussions, resulting in the present shape of the report. Prof. Horst Lippmann of the Technical University, Munich, was a keen and attentive advisor in the course of the investigation, drawing the Author's attention to both principal theoretical and practical aspects of the model. It is the Author's pleasure to express hereby his appreciation for their valuable assistance.

Printed in the United States
By Bookmasters